Principles of Animal Locomotion

Principles of Animal Locomotion

....

R. McNeill Alexander

PRINCETON UNIVERSITY PRESS

PRINCETON AND OXFORD

Library of Congress Cataloging-in-Publication Data

Alexander, R. McNeill.
Principles of animal locomotion / R. McNeill Alexander.
p.   cm.
Includes bibliographical references (p.).
ISBN 0-691-08678-8 (alk. paper)
1. Animal locomotion. I. Title.
QP301.A2963   2002
591.47′9—dc21   2002016904

British Library Cataloging-in-Publication Data is available

This book has been composed in Galliard and Bulmer

Printed on acid-free paper.∞

www.pupress.princeton.edu

Printed in the United States of America

10   9   8   7   6   5   4   3   2   1

Contents

· ·

Preface

....

*T*HIS BOOK is about the mechanics and energetics of animal movement on land, in water, and through the air. Its emphasis is on understanding rather than comprehensive description, on principles rather than details. Most of it is about vertebrates, arthropods, and molluscs, because these are the groups that have been most intensely studied. If a style of locomotion is peculiar to a few species or to an obscure group, I have felt no need to include it, unless there is something particularly interesting about it. I have included only muscle-powered locomotion, excluding the movements of small animals such as planktonic larvae of invertebrates that depend on cilia for propulsion.

My aim has been to explain the mechanical principles on which locomotion depends; to account for its metabolic energy cost; and to explore the merits of different styles of locomotion, in different circumstances. I have used rough calculations and simple mathematical arguments frequently to check and clarify the explanations.

I have designed this book principally for advanced undergraduates, graduate students, researchers, and university teachers of biology. I have assumed that readers will be familiar with the major groups of animals, and that they will know a little anatomy, physiology, and mechanics.

I am grateful to two anonymous reviewers, whose percipient suggestions have improved this book.

Principles of Animal Locomotion

Chapter One

....

The Best Way to Travel

*T*HIS BOOK describes the movements of animals and of the structures such as legs, fins, or wings that they use for movement. It tries to explain the physical principles on which their movements depend. And it asks whether the particular structures and patterns of movement that we find in animals are better suited to their ways of life than possible alternatives. This chapter will, I hope, help us when we come to ask these questions about the merits of particular structures and movements.

The structures of animals and some of their patterns of movement (the ones that are inherited) have evolved. Other patterns of movement may be learned afresh by successive generations of animals, by trial and error. Evolution by natural selection, and learning by trial and error, both tend to make the animals and their behavior in some sense better. What, in this context, does "better" mean?

1.1. FITNESS

The most fundamental answer is that evolution favors structures and patterns of movement that increase fitness, and that the capacity for learning has evolved so that learning also can be expected to increase fitness. The fitness of an animal's complement of genes (its genotype) is the probability of the same group of genes being transmitted to subsequent generations. Unfortunately for the purposes of this book, it is not generally easy to measure or calculate the effect on fitness of, for example, a change in the length of an animal's legs or a modification of its gait. We can make more progress by looking at the effects of evolution in a less fundamental way.

Fitness depends largely on the number of offspring that animals produce, and on the proportion of those offspring that survive to breed. Thus, natural selection favors genotypes that increase fecundity or reduce mortality. This insight still seems rather remote from our discussions of locomotion. It seems helpful to ask at this stage, what aspects of an animal's performance in locomotion are most likely to affect fecundity and mortality, and so fitness? What qualities, in the context of locomotion, can natural selection be expected to favor? Some suggestions follow.

1.2. SPEED

For many animals, natural selection may tend to favor structures and patterns of movement that increase maximum speed. A faster-moving predator may be able to catch more prey, which may enable it to rear and feed more offspring. A faster moving prey animal may be better able to escape predators, and so may live longer. However, we should not assume that speed is important for all animals. For example, tortoises are herbivores, with no need for speed to catch prey. Their shells are sufficient protection against most predators, so they do not need speed to escape. It seems clear that maximum speed has had little importance in the evolution of tortoises, so we need not be surprised that tortoises are remarkably slow.

It is probably generally true that most animals spend very little of their time traveling at maximum speed. Lions (*Panthera leo*) are idle for most of the day, but their ability to run fast occasionally is vital to their hunting success. The antelopes and zebra on which they feed spend nearly all their time quietly grazing or traveling slowly, but depend on their ability to run fast in emergencies, to escape from lions and other predators. Ability to travel fast may be highly important to animals, although it may seldom be used.

1.3. ACCELERATION AND MANEUVERABILITY

Acceleration may be even more important than speed for predators such as lions, which stalk antelopes and then make a sudden dash from a short distance; and pike (*Esox*), which hide among vegetation and dash out to catch small fish that swim past. Acceleration must be correspondingly important for the prey. Suppose a predator dashes with constant acceleration a_{pred}, starting from rest at zero time, at a distance d from its prey. At time t its speed is $a_{\mathrm{pred}}t$, and it has traveled a distance $0.5 a_{\mathrm{pred}}t^2$. If the prey starts running at the same instant as the predator, with acceleration a_{prey}, it has traveled a distance $0.5 a_{\mathrm{prey}}t^2$ at time t. If the predator's acceleration is greater than the prey's, and if the chase is short enough for neither animal to reach top speed, the predator catches the prey when

$$0.5 \ t^2 \left(a_{\mathrm{pred}} - a_{\mathrm{prey}} \right) = d$$

$$t = \left(\frac{2d}{a_{\mathrm{pred}} - a_{\mathrm{prey}}} \right)^{0.5} \tag{1.1}$$

by which time the predator has covered a distance $a_{\mathrm{pred}} d / (a_{\mathrm{pred}} - a_{\mathrm{prey}})$. If the predator has twice the acceleration of the prey, it catches it after covering a distance $2d$; but if its acceleration is only 1.1 times that of the prey it has to run a distance $11d$.

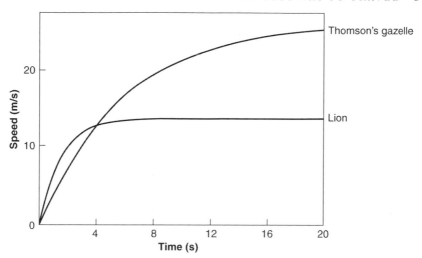

Fig. 1.1. Graphs of speed against time for lions and Thomson's gazelle, calculated from films of lions attacking prey. The curves were obtained by fitting to the data equations of the form $v = v_{max} [1 - \exp(-kt)]$, where v is the speed at time t, v_{max} is the speed that is approached asymptotically, and k is a constant. Redrawn from Elliott et al. (1977).

That analysis is grossly simplified. It assumes that both animals start moving simultaneously, and that both animals have constant acceleration throughout the chase. Elliott et al. (1977) filmed lions hunting gazelles (*Gazella thomsoni*), and used his films to calculate graphs of speed against time. These graphs curve and level off, showing that both predator and prey accelerated at decreasing rates, as they gained speed (Fig. 1.1). However, the analysis is sufficient to show that the ability of a predator to catch prey may depend more on its acceleration than on its maximum speed. Indeed, a predator with superior acceleration may be able to catch prey, even if its top speed is lower than that of the prey. Elliott found that the initial accelerations of the lions averaged 9.5 m/s², and those of the gazelles only 4.5 m/s². He estimated that the speeds they would eventually have reached were 14 m/s for the lions, and a much faster 27 m/s for the gazelles. However, these estimates of top speed depended on extrapolation of his data, and may not be accurate.

The analysis also ignored the possibility that the prey might attempt to escape by swerving. Films of gazelles (*Gazella thomsoni* again) pursued by cheetah (*Acinonyx jubatus*) show the prey swerving when the predator is close behind. Children playing the game of tag (called tig in Britain) know that a well-timed swerve is a good escape strategy.

An animal traveling at speed v on a circular arc of radius r has an acceleration v^2/r toward the center of the circle. Thus, swerving involves sideways acceleration. Suppose that a predator running at speed v_{pred} is capable of swerving with radius r_{pred}, and a prey animal running with speed v_{prey} swerves with radius r_{prey}. The prey can escape, even if v_{prey} is lower than v_{pred}, if its sideways acceleration v_{prey}^2/r_{prey} is greater than the predator's sideways acceleration v_{pred}^2/r_{pred}. Howland (1974) pointed this out, and went on to show that, to take full advantage of its superior sideways acceleration, the prey must delay swerving until the predator is very close behind. This is illustrated in Fig. 1.2, which shows the paths of predator and prey. The predator is represented as traveling faster than the prey, but with larger radius. Time intervals are marked on the animals' paths. Each animal has the same speed and radius in both diagrams. The prey escapes if it swerves at the last possible moment (B), but if it swerves too soon the predator cuts off the corner and intercepts it (A).

1.4. ENDURANCE

Animals cannot maintain their top speeds indefinitely in a prolonged chase. Figure 1.3A shows the speeds at which human athletes have run races ranging from a 100-m sprint to a marathon, plotted against the time taken for the race. Figure 1.3B shows the maximum speeds maintained by trout (*Salmo irideus*) for different times. In each case speed falls as time and distance increase.

The graph for the fish (Fig. 1.3B) is plotted on ordinary linear coordinates. It shows, for example, that the 15-cm fish's maximum speed was 180 cm/s for one-second sprints, but fell, as time increased, toward an asymptote of about 40 cm/s. The graph for human running (Fig. 1.3A) would look very similar to the fish graph, if it had been plotted in the same way. However, it has been plotted on logarithmic coordinates, which have made it possible to display data for a much wider range of times. This graph shows not only that maximum speed declines markedly in the first 100 s of running time, but also that the decline continues over a period of several hours. The point for the 100-m race (triangle) is potentially misleading because sprinters are still accelerating over most of this distance. The remaining data, for races from 200 m to a marathon, form two straight lines meeting at an angle when plotted thus on logarithmic coordinates. This suggests that the decline in speed over short times (less than about 150 s) depends on a different phenomenon from the longer term decline in speed. We will find a likely explanation in Section 2.5.

Now suppose that a predator is chasing prey over a sufficient distance for us to ignore the acceleration period. We might, for example, be considering African hunting dogs (*Lycaon*), which chase antelopes over distances of several kilometers (van Lawick-Goodall and van Lawick-Goodall 1970). Assume that both animals are able to estimate the duration of the

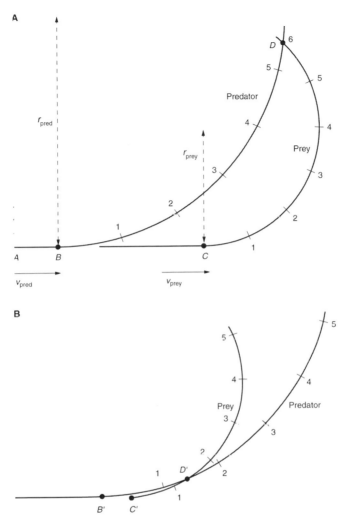

Fig. 1.2. Diagrams of a predator chasing swerving prey. The paths of the animals are seen in plan view, with the animals' positions after successive intervals of time numbered 1, 2, etc. The prey is slower than the predator ($v_{prey} = 0.75\,v_{pred}$), but can execute a tighter turn ($r_{prey} = 0.5\,r_{pred}$). In (A) the animals were initially running along the line ABC. At time zero, when both animals started swerving, the predator was at B and the prey at C. The prey reaches point D after 6 units of time. The predator would pass D after 5.4 units of time if it continued running at maximum speed, but by slowing down a little it can intercept the prey there. In (B) swerving starts when the animals are at B', C'. The prey passes point D' after 1.3 units of time, and the predator arrives there only after 1.4 units of time, so in this case the prey escapes. Modified from Alexander (1982).

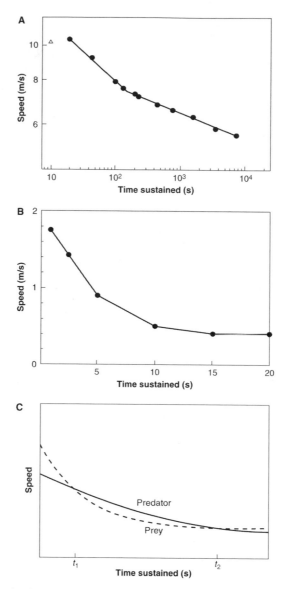

Fig. 1.3. Graphs showing how the speed at which an animal can travel falls, as the time for which it has to be sustained increases. (A) World record speeds for male human athletes in races of different lengths, plotted against the time for the race, redrawn from Savaglio and Carbone (2000). (B) The highest speeds that a trout (*Salmo irideus*) maintained for different times when swimming spontaneously in an annular tank, redrawn from Bainbridge (1960). (C) A schematic graph, which is explained in the text, showing how a predator with good endurance may be able to catch faster prey.

chase in advance, and choose the speeds that will take them furthest in that time. Though its sprinting speed may be less than that of the prey, the predator will eventually overtake the prey if its sustainable speed is greater than that of the prey. Less obviously, even though its sprinting speed and its maximum speed over long distances may be less than those of the prey, it may be able to catch the prey if it fatigues less quickly than the prey. Figure 1.3C is a schematic graph illustrating this possibility; notice how the lines cross, showing that there is a range of chase times for which the predator can travel faster than the prey.

1.5. ECONOMY OF ENERGY

Measurements of the oxygen consumption of many animals have been made, to find out how much energy they use in locomotion; the principal methods will be outlined in Section 5.3. Some striking differences have been observed. For example, Taylor et al. (1982) found that walking penguins (*Pygoscelis*) use energy about 60% faster than turkeys (*Meleagris*) of the same mass, walking at the same speed. In another comparison, this time of the energy cost of swimming at the surface of water, penguins (*Eudyptula*, in this case) performed much better; they used only 0.72 times as much energy as ducks (*Anas*) of equal mass, swimming at the same speed (Baudinette and Gill 1985). In a second comparison of swimmers, squid (*Illex*) used energy 1.75 times as fast as salmon (*Onchorhynchus*) of comparable mass, although they were swimming at only 0.6 times the speed of the fish (Webber and O'Dor 1986). Are these differences likely to be important to the animals?

Economy of energy can affect fitness in various ways, of which the most generally important is probably this: energy that is not used for locomotion is available for growth and reproduction. For example, birds rearing nestlings may have to spend all the daylight hours foraging for food, flying for much of the time. A substantial proportion of the food they collect has to be used to fuel flight, and so is not available to feed the nestlings. House martins (*Delichon urbica*) are small birds that feed on insects, which they catch on the wing. In field experiments in Scotland, Bryant and Westerterp (1980) set up nest boxes that were used by house martins. Trapdoors on the boxes enabled them to capture the birds, to make the injections and (a day or two later) collect the blood samples needed to measure their metabolic rates by the doubly labeled water technique, which is explained in Section 5.3. While they had young in the nest, the birds spent an average of 14 h per day off the nest, flying all the time, and their metabolic rates were 3.6 times the resting rate. For part of the time, the nestlings were temporarily fitted with collars that prevented them from swallowing, so that the experimenters could recover and weigh the mouthfuls of food that their parents gave them. The brood was found to be receiving food from each parent at a rate equivalent to 3.0 times the parent's resting metabolic rate, while the parents (as we have seen) were using energy at 3.6 times the resting rate for their own metabolism. A very large fraction of the energy that the parents were using, in excess of the resting rate, must have been used to power flight; and if they could have flown more economically they would have had more food to spare for the young. They might have been able to rear a larger brood, and so pass on more of their genes to the next generation.

As another example to show how economy of energy can affect fitness consider a typical fish, which, unlike the birds we have been considering, does not care for its young. The more eggs it lays (of given size and quality), the more offspring it will have and the more genes it is likely to contribute to successive generations; but the number of eggs it can produce is limited by its size. As a rough general rule, a mature female fish of mass m can be expected to produce a mass of 0.1 to $0.2m$ of eggs in the course of the season (Le Cren and Holdgate 1962). Other things being equal, the less energy it has had to use for locomotion in the course of its life, the more of its food energy intake will have been available for growth, the bigger it will have grown, and the more eggs it can lay. Alexander (1967) made a simple calculation to assess the likely effect of energy economy on fitness. I estimated that 20% of the energy content of the food eaten by a typical fish would be lost in feces and urine; 34% would be used for resting metabolism; 34% would be used to power swimming; and 12% would be available for growth and reproduction. If these estimates are realistic, three times as much energy is used for swimming as for growth and reproduction, so a 1% improvement in the efficiency of swimming can be expected to make 3% more energy available for growth and reproduction.

1.6. STABILITY

We have already noted that tortoises walk very slowly. The likely reason is that, if speed is unimportant, an animal can make do with very slow muscles. These can be very economical of energy, as will be explained in Section 2.5. Experiments with tortoise muscle have shown that it is remarkably economical (Woledge et al., 1985). We will see in Section 7.9 that stability is a problem for walking animals with very slow muscles, but that the problem can be alleviated by appropriate choice of gait. Natural selection seems to have optimized the gait of tortoises to obtain adequate stability with the slowest possible muscles.

1.7. Compromises

The discussion so far may suggest that animals should evolve to be as fast as possible, to have the best possible acceleration, maneuverability and endurance, and to be as economical as possible of energy. However, these objectives are not always compatible. The example of tortoises has already shown us that an animal designed to walk as economically as possible cannot be fast. Similarly, no human athlete is a champion both in sprinting and in distance running, and an animal adapted to sprint as fast as possible would be unlikely to have good endurance. Sprinters and distance runners differ markedly in physique, the sprinters having well-developed muscles and the distance runners being less muscular, with bigger hearts capable of pumping a greater volume of blood at each stroke (Reilly et al., 1990). Evolution can be expected to favor compromises between the requirements of speed, endurance, economy, etc.

If we were to try to express the relationship between the locomotion of animals and their fitness in mathematical terms, we would have to conclude that fitness is a function of speed, acceleration, maneuverability, endurance, energy economy, and a great many other properties. It would not be at all obvious what the function should be, and if we were to try to assess the effect on fitness of some change (for example, longer legs or bigger thigh muscles) we would find ourselves doing elaborate and highly unreliable calculations. To make our discussions manageable, we must try to identify the properties that are most important, and concentrate on the effects that adaptations have on them. We can safely assume that racehorses have been selected for speed over distances of the order of a few kilometers, but for animals designed by natural selection, as distinct from selective breeding, the criteria for selection are generally less clear-cut.

1.8. Constraints

We will have to remember in our discussions that evolution cannot bring about every imaginable change. We have already seen that squid are less economical swimmers than salmon. They are also slower; the maximum sustainable speeds of a 0.5-kg salmon and a similar-sized squid were 1.35 and 0.76 m/s, respectively (Webber and O'Dor 1986). Squid might be faster and more economical if they had evolved fishlike tails, but their evolution has been constrained by their molluscan ancestry. Evolution proceeds by relatively small steps, and there does not seem to be any conceivable evolutionary route from a squid to a fishlike animal that would not involve passing through a stage less fit than either. Again, the walking of tortoises would be more stable if they had six legs instead of four, but the evolution of tortoises has been constrained by their four-legged ancestry.

To understand how these constraints operate, think of a walker in a hilly landscape, who walks always uphill. He or she may reach the highest summit, but is much more likely to finish on some subsidiary peak. There is no route from a lower peak to a higher one that does not involve first going downhill. Similarly, an evolutionary path along which an animal species changed progressively, increasing fitness at every stage, would not necessarily lead to the fittest imaginable structure.

1.9. OPTIMIZATION THEORY

Optimization theory is the branch of mathematics that finds the best possible solutions to problems. Here is a simple example. Consider a bird gliding with fully spread wings, aiming to glide at the shallowest possible angle and so to travel as far as possible for given loss of height. It can glide faster or slower by holding its wings in slightly different positions, and this will affect its angle of descent. This angle θ is given by an equation that applies also to man-made gliders:

$$\sin \theta = Av^2 + B/v^2 \qquad (1.2)$$

where v is the speed, and A and B are constants that depend on the size and shape of the wings (Equation 10.18). Figure 1.4A is a graph of $\sin \theta$ against speed v. It shows that the angle is steep if the bird glides very slowly or very fast, and is least at an intermediate speed.

The same result can be obtained without drawing a graph. Notice that at low speeds the graph slopes downhill, and at high speeds uphill. The minimum angle of glide is obtained where the graph runs level, with zero slope. We can find the minimum by deriving an equation that gives the slope, which can be done by the mathematical process of differentiation, and then finding the value of v that makes the slope zero. Readers who do not know how to differentiate can take the process on trust, or consult a textbook of calculus or (for a very quick explanation) read Section 1.2 of my book *Optima for Animals* (Alexander 1996). The slope of a graph of $\sin \theta$ against v is represented by the mathematical expression $d(\sin \theta)/dv$. Differentiation of Equation 1.2 tells us that the slope is

$$\frac{d (\sin \theta)}{dv} = 2Av - 2B/v^3$$

which is zero when

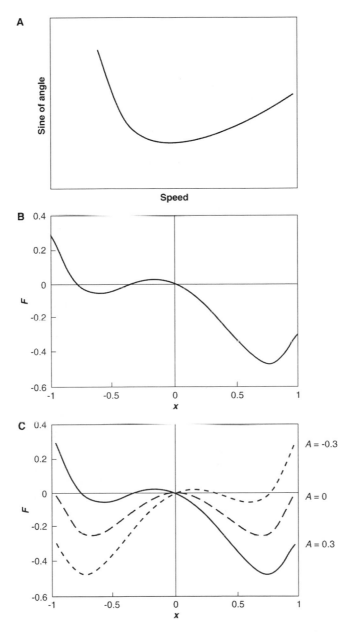

Fig. 1.4. Graphs illustrating an explanation of the basic principles of optimization theory: (A) Equation 1.2; (B) Equation 1.3 with the parameter A given the value 0.3; and (C) the same equation for several different values of A.

$$v = \left(\frac{B}{A} \right)^{0.25}$$

Thus, the speed at which the angle of glide is least is $(B/A)^{0.25}$. We will use a more complicated form of this equation in Section 10.6, when we discuss how the wings of soaring birds are adapted to their ways of life.

Instead of a graph of the sine of the angle of descent against speed, we might have drawn a graph of the distance traveled per unit loss of height. This would have shown a maximum, instead of a minimum. The slope is zero at the top of a hill, as well as at the bottom of a valley, so the speed that gives maximum glide distance could be found by differentiating the appropriate equation, in the same way as we found the (identical) speed that gives minimum glide angle.

When we discuss gaits, we will encounter more complex situations, involving graphs with more than one maximum or minimum. Figure 1.4B illustrates this possibility, by showing a graph of the function

$$F = x^4 - x^2 - Ax \tag{1.3}$$

where A is a constant that has been given the value 0.3. The previous graph had one minimum but this has two, at $x = 0.8$ and at $x = -0.6$. At both minima the graph runs level, and both would be found by differentiation. However, if the aim is to make F as small as possible, the deeper minimum should be chosen, at $x = 0.8$. This is described as the global minimum, and the other as a local minimum.

Figure 1.4C shows graphs of Equation 1.3 for several different values of A. When $A = 0.3$, the global minimum is at a positive value of x and the local minimum at a negative one, as we have already seen. As A is reduced, the minima become more equal, and when $A = 0$ they are equal. When A is negative, the global minimum is found at a negative value of x and the local minimum at a positive value. This phenomenon, in which a small change of a parameter results in an abrupt shift of the global minimum (or maximum), is called bifurcation.

1.10. GAITS

People walk to go slowly and run to go fast. Walking and running are quite different patterns of movement, which do not merge into each other; as we increase speed, we make the change from walking to running within a single stride. Similarly, horses change from walking to trotting and then to galloping as they increase speed. Walking, running, trotting, and galloping are described as gaits, and in later chapters we will see that flying birds and swimming fishes also use several distinct gaits.

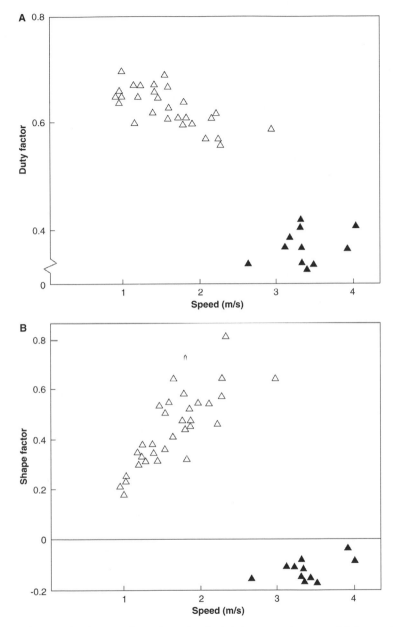

Fig. 1.5. Graphs showing how the gait of adult humans changes with increasing speed. (A) Duty factor and (B) shape factor plotted against speed. Note the abrupt changes at about 2 m/s, at the transition from walking (open symbols) to running (filled symbols). From Alexander (1989a).

The formal definition of a gait is as follows: "A gait is a pattern of loco-motion characteristic of a limited range of speeds, described by quantities of which one or more change discontinuously at transitions to other gaits" (Alexander 1989a). Figure 1.5 shows graphs of two of the quantities that change, when people change gait. Duty factor (Figure 1.5A) is the fraction of the duration of the stride, for which each foot is on the ground. In walking, each foot is on the ground for more than half the time, and in running for less than half the time. As speed increases, the duty factor falls gradually from about 0.65 in slow walks to about 0.55 in the fastest walks; but at the change to running it drops abruptly, to around 0.35. The shape factor q (Figure 1.5B) describes the pattern of force exerted on the ground; it will be explained in Section 7.3. As speed increases, it rises smoothly from about 0.2 in slow walking to about 0.8 in very fast walking; but then drops abruptly to negative values at the transition to running.

I will show in Section 7.7 that people seem to adjust their gaits so as to minimize the energy cost of traveling at their chosen speed. Thus, our gaits are solutions to optimization problems. The abrupt shift of the opti-mum from walking to running is a bifurcation.

Thus speed, acceleration, maneuverability, endurance, energy economy, and stability are aspects of locomotion that are likely to be important to animals in different circumstances. Natural selection can be expected to act on structures and patterns of movement that affect them, but the course of evolution is constrained by the animal's ancestry.

Chapter Two

Muscle, the Motor

*M*OST ANIMAL locomotion is powered by muscles. Some very small animals including rotifers and the planktonic larvae of echinoderms, annelids, and molluscs depend on cilia for swimming, and spermatozoa swim by means of the flagella that form their tails. Small flatworms crawl by means of cilia on their ventral surfaces. Protozoans, which are not regarded as animals in modern classifications, move by means of cilia, flagella, or pseudopodia. All these are ignored in this book, which is concerned only with muscle-powered locomotion.

The purpose of this chapter is to explain briefly how muscle works, what it can do, and how much energy it uses. It is concerned only with striated muscle, the kind of muscle that powers locomotion. Smooth muscle (as found in the walls of guts and blood vessels) is ignored, as is heart muscle. You will find more detailed information about muscle in Woledge et al. (1985) and Josephson (1993).

2.1. HOW MUSCLES EXERT FORCE

In this section, a brief description of the structure of muscle leads into an explanation of force production.

Striated muscle consists of cells known as muscle fibers, which are very long and slender. Some human muscle fibers are as much as 100 mm long (Richmond 1998), but few have diameters greater than about 0.1 mm. They are bundled together in fascicles that are stout enough, in large mammals, to be visible by the naked eye. These fascicles are in some cases much longer than the individual fibres, for example about 450 mm in the human sartorius muscle.

Most of the space within muscle fibers is generally occupied by myofibrils, which are composed of protein filaments a few micrometers long, lined up parallel to each other and to the long axis of the fiber. There are thick filaments of the protein myosin and thin filaments that consist largely of the protein actin, arranged in a very orderly manner with the thin filaments interdigitated between the thick ones (Fig. 2.1A). The two kinds of filaments are arranged in bands so that, when a fiber is examined under a microscope, stripes are seen running across it. The unit of this repeating, banded pattern is the sarcomere. The ends of the sarcomeres are defined by partitions known as Z disks, which have thin filaments projecting from both their faces. The bands of thick filaments lie between successive Z disks, with the ends of each thick filament connected to the disks by strands of titin, a highly extensible elastic protein (not shown in Fig. 2.1.) (Kellermayer et al., 1997).

Each thick filament consists of several hundred myosin molecules. The molecules are about 0.15 μm long, consisting of a long tail with two small heads at one end (Fig. 2.1C). They are bundled together, each with about two-thirds of its length incorporated in the main strand of the filament and one-third (the end with the heads) projecting as a cross bridge capable of attaching to adjacent thin filaments (Fig. 2.1D). The molecules are all arranged with their tails pointing toward the midpoint of the filament, so there is a bare region in the middle of the filament (about 0.2 μm long) from which no cross bridges project. For the rest of the length of the filament, rosettes of three cross bridges project at intervals of 14 nm (0.014 μm) (Fig. 2.1D). The figure shows pairs of cross bridges instead of groups of three, because of the limitations of a two-dimensional diagram.

When a muscle is shortening, each cross bridge repeatedly attaches one of its heads to an actin molecule in an adjacent thin filament, and pulls toward the midpoint of the thick filament. The cross bridges attach, pull, detach, and reattach 5 nm further along the thin filament (Kitamura et al., 1999). Their action is like people pulling in a rope hand over hand. The thick and thin filaments slide past each other without changing length (compare Fig. 2.1B with Fig. 2.1A). When a cross bridge detaches, a molecule of ATP attaches to it. The myosin functions as an enzyme, breaking the ATP down to ADP and a phosphate ion, but these are not released until the cross bridge has reattached to another actin molecule. The action of the muscle is powered by the energy released by the breakdown of ATP to ADP. Kitamura et al. (1999) have shown that a single ATP molecule may provide the energy for several 5-nm steps of a cross bridge.

There are of course other organelles in the muscle fibers, as well as the myofibrils. There are nuclei, which generally occupy only a small fraction of the cell volume. There are mitochondria containing the enzymes of oxidative phosphorylation, on which aerobic respiration depends. These may occupy a large fraction of the fiber volume in fast aerobic muscles, for example, 35% in the flight muscles of a hummingbird and 37% in the wing muscles of a beetle (Table 2.1). There is sarcoplasmic reticulum, a network of fluid-filled tubules between the myofibrils, which releases calcium ions into the cell lumen when the fiber is stimulated, to trigger contraction of the muscle; and pumps the calcium back into its tubules when stimulation ceases, to make the muscle relax. The sarcoplasmic reticulum may occupy a large fraction of the fiber volume in muscles that contract and relax at

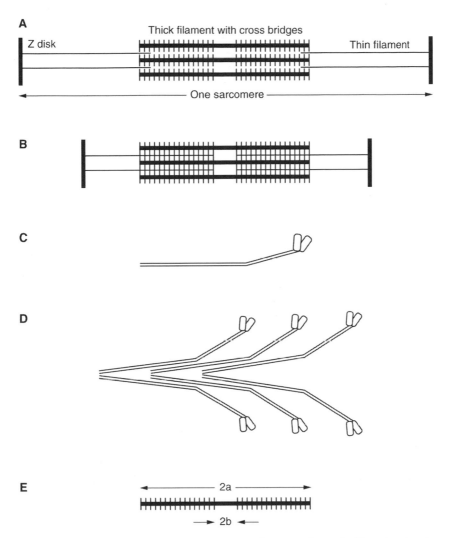

Fig. 2.1. (A, B), Diagrams of one sarcomere of a striated muscle fiber showing it (A) extended beyond the optimum length for force production and (B) at the optimum length. (C) A single myosin molecule and (D) a diagram showing how myosin molecules are arranged in a thick filament. (E) A thick filament showing lengths referred to in Equations 2.1 and 2.2.

Table 2.1.
Composition (%) by volume of various muscles

	Mitochondria	Sarcoplasmic reticulum	Myofibrils	Reference
Rattlesnake body muscle	2	11	87	Schaeffer et al. (1996)
Rattlesnake tail-shaker muscle	26	26	32	Schaeffer et al. (1996)
Hummingbird flight muscle	35	not recorded	not recorded	Suarez (1996)
Locust flight muscle	24	10	65	Josephson et al. (2000)
Beetle flight muscle	37	2	58	Josephson et al. (2000)

high frequencies, for example, 26% in the tail-shaker muscle of the rattle-snake (*Crotalus*), which vibrates at about 90 Hz (Rome et al., 1990a). There are also T tubules, invaginations of the cell surface that function like inside-out axons, transmitting electrical stimuli to the myofibrils in the interior of the fiber.

The stresses (forces per unit cross-sectional area) that muscles can exert depend on the length of the thick filaments and on the fraction of the cell volume that is occupied by the myofibrils. Consider a thick filament of length $2a$ with a bare region of length $2b$ (Fig. 2.1E). Groups of three cross bridges project from it except in the bare region, each group a distance c from the next. Thus, each half of the filament has $3(a - b)/c$ cross bridges, each of them capable of exerting a force that we will give the symbol δF. Thus, if all the cross bridges are active simultaneously, the total force F that one-half of a thick filament can exert on adjacent thin filaments is

$$F = 3 \left(\frac{a - b}{c} \right) \delta F \qquad (2.1)$$

Let the myofibrils occupy a fraction A of the cross section of each muscle fiber, and within each myofibril let there be n thick filaments per unit cross-sectional area. The stress σ exerted by the muscle is then

$$\sigma = 3An \left(\frac{a - b}{c} \right) \delta F \qquad (2.2)$$

Many of the quantities in this equation seem to be the same for most striated muscles, because they depend on the dimensions of the myosin molecules and on the way these molecules are packed together to form

Table 2.2.
Thick filament lengths and maximum isometric stresses of some muscles

	Thick filament length, μm	Maximum iso-metric stress, Nmm²	Reference
Frog leg muscles	1.6	0.15–0.36	Marsh (1994)
Rat leg muscles	1.6	0.29–0.33	Wells (1965)
Locust flight muscle	3.1	0.35	Bennet-Clark (1975)
Locust leg muscle	5.5	0.7	Bennet-Clark (1975)
Oyster yellow adductor muscle	5–8	0.5	Rüegg (1968)
Mussel byssus retractor muscle	30	1.4	Rüegg (1968)

the thick filament; $n = 560$ μm^{-2}, $b = 0.1$ μm, $c = 0.014$ μm. It seems likely that δF will also be constant. Thus, muscles whose myofibrils occupy a large fraction of their volume (large A), and which have long, thick filaments (large a), are expected to be capable of exerting large stresses.

Small muscles, and muscle fascicles, can be kept alive for substantial times if kept moist with saline solutions of appropriate osmotic strength. They can be made to contract by electrical stimulation, and the forces they exert can be measured by transducers. If the cross-sectional area is measured, the stress can be calculated. Table 2.2 shows results from experiments of this kind. In these experiments, the muscles were held at constant length (they were contracting isometrically) at the length at which they could exert most force. It shows, as expected, that muscles with long, thick filaments can exert large stresses. The thick filaments of the mussel muscle are exceptionally long and the stress it exerts is very high, but not as much higher than the stresses exerted by other muscles as Equation 2.2 might lead us to expect. The reason is that the thick filaments are exceptionally thick (due to reinforcement with the protein paramyosin), reducing the number n that can be accommodated in unit cross-sectional area of myofibril. Taylor (2000) gives further data on the relationship between sarcomere length and isometric stress, especially for Crustacea.

In other, technically remarkable, experiments myofibrils have been disaggregated to obtain separate myosin molecules, and the forces exerted by individual molecules have been measured. This has been done by means of optical tweezers, instruments that use the inertia of the photons in a beam of light to balance the tiny force exerted by the molecule. This force (δF of our equations) has been found to be at least 1.7 pN (1.7×10^{-12} N) (Molloy et al., 1995). By putting this into Equation 2.2 with $2a = 1.6$ μm (the length of the thick filaments in vertebrate striated muscle) and the values of n, b, and c given above, we can estimate that if all the cross

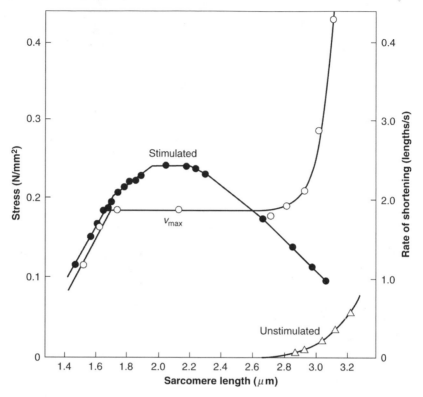

Fig. 2.2. A graph showing the stress exerted by a frog striated muscle fiber at different sarcomere lengths, both when maximally activated by electrical stimulation and when inactive. Another curve (scale at right) shows the unloaded rate of shortening v_{max} at different sarcomere lengths. From Edman (1979).

bridges were simultaneously attached in a muscle in which the myofibrils occupied the whole of the cross section ($A = 1$), the stress would be 0.14 MN/m². This is just below the range of isometric stresses that have been measured for frog leg muscles (Table 2.2). It would be lower if we had allowed for the fact that a proportion of the cross bridges would be detached, at any instant. However, it should be remembered that it is based on a minimum estimate of cross bridge force.

In the experiments on which Table 2.2 is based, the muscles were held at the lengths at which they could exert the most force. In other experiments, a muscle fiber has been held at a range of different lengths, and the force it could exert at each length has been measured. Results from an experiment with a frog muscle fiber are shown in Fig. 2.2. Frog muscle exerts maximal stresses when its sarcomeres are 2.0–2.2 μm long, and lower stresses at longer and shorter lengths. Gordon et al. (1966) showed how this could be explained. At the optimum sarcomere length (Fig. 2.1B) every cross bridge is alongside a thin filament, to which it can attach. When the sarcomere is stretched longer (Fig. 2.1A), some cross bridges are no longer within reach of a thin filament, and the stress falls in proportion to the number that can still attach. When it is shorter than the optimal length, various undesirable things happen. Thin filaments from the two ends of the sarcomere overlap and may reach cross bridges on the wrong half of the thick filaments, which push them in the wrong direction. Eventually the ends of the thick filaments collide with the Z partitions. It has also been suggested that titin is arranged in such a way as to resist extreme shortening of the sarcomere (Spierts and van Leeuwen 1999).

As well as the force exerted by the cross bridges in active muscle, Fig. 2.2 shows the passive elastic force that resists extreme stretching even of inactive muscle. This seems to be due in part to stretching of the titin and in part to stretching of the sarcolemma, the network of connective tissue that is wrapped around the fiber.

Note that Fig. 2.2 refers specifically to frog muscle. Muscles with longer thick and thin filaments have longer optimum sarcomere lengths.

We can use Fig. 2.2 to estimate the work that a muscle can do in a single contraction, if it shortens very slowly. (We will see in the next section that the faster a muscle shortens, the less force it can exert.) Work is done when the point of application of a force moves in the direction of the force; the work is the force multiplied by the distance moved. Suppose that a frog muscle fiber contracts very slowly through the range of sarcomere lengths at which it can exert at least half the maximum force, that is, from 3.0 μm (1.4 times its optimal length) to 1.5 μm (0.7 times its optimal length). As it contracts the stress rises and then falls as shown in Fig. 2.2. We can calculate the work the muscle does from the area under the graph. (Because work is force multiplied by distance, the area under a graph of force against length represents work.) The area under this particular graph gives us the work that could be done in a single contraction by a piece of muscle one sarcomere long and with a cross section of a square millimeter. From this we can calculate that frog muscle should be capable of doing about 120 J of work per kilogram of muscle, in a single contraction. This is an extreme estimate, for infinitely slow contraction in which the muscle shortens to half its initial length. Some vertebrate striated muscles shorten as much as this, in some movements, but it seems to be more usual for muscles to shorten by 25% or less (Burkholder and Lieber 2001). Figure 2.2 shows that a muscle shortening slowly by 25% of its length could do up to 70 J/kg. Muscles with longer thick filaments, capable of exerting larger stresses, could do more work.

2.2. SHORTENING AND LENGTHENING MUSCLE

Figure 2.2 showed that the force that a muscle can exert, when contracting isometrically, depends on its current length. Figure 2.3A shows results from a different kind of experiment: a bundle of muscle fibers is stimulated electrically while being allowed to shorten at a constant rate, and the force it exerts is measured as it passes through the range of lengths in which it could exert maximum force in isometric contractions. The graph shows results for negative rates of shortening (forcible stretching of the muscle) as well as for positive rates of shortening. The graph shows that when the muscle is shortening it exerts less force than in isometric contraction. The faster it shortens, the less force it exerts, until at the maximum possible rate of shortening it exerts no force at all. This maximum rate may be described as the unloaded rate of shortening, and is conventionally represented by the symbol v_{max}. When the muscle is being stretched, however, it exerts forces that are greater than the isometric force, approaching 1.8 times the isometric force at high rates of stretching. The relationship between the force F and the rate of shortening v is quite well represented by the empirical equations

$$\text{for } v \geq 0, \; F = F_{iso} \frac{v_{max} - v}{v_{max} + Gv} \qquad (2.3a)$$

$$\text{for } v < 0, \; F = F_{iso} \left[1.8 - 0.8 \left(\frac{v_{max} + v}{v_{max} - 7.6Gv} \right) \right] \qquad (2.3b)$$

where F_{iso} is the isometric force, v_{max} is the unloaded rate of shortening, and G is a constant (Alexander 1997b). G has different values for different muscles but generally lies between 2 and 6 (Woledge et al. 1985). Figure 2.3 has been drawn for $G = 4$, a good average value for skeletal muscles of vertebrates. The higher the value of G, the more concave is the graph for positive rates of shortening. Equation 2.3a is known as Hill's equation in honor of A. V. Hill, the great muscle physiologist who introduced it. It has been shown to give a reasonably good description of the force/rate of shortening curves for a wide range of muscles, but deviations from its predictions have been noted (Edman et al. 1976). Equation 2.3b is based on fewer data. Forces are less at sarcomere lengths outside the range for maximum isometric force, but v_{max} is more or less constant over a wide range of sarcomere lengths (Fig. 2.2; the very high unloaded rates of shortening at long sarcomere lengths are due to elastic recoil of titin fibers and other passive elastic structures).

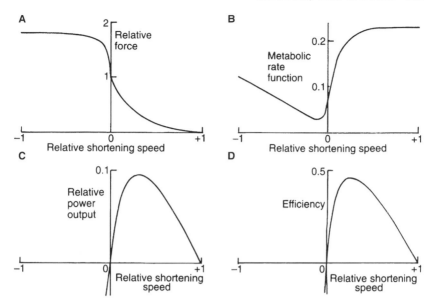

Fig. 2.3. Properties of vertebrate striated muscle fibers, maximally activated by electrical stimulation, plotted against the rate at which the fiber is being allowed to shorten. (Negative shortening rates mean that it is being forcibly stretched.) Relative shortening speed means speed/v_{max}. (A) Relative force (force/F_{iso} , Equations 2.3a and 2.3b); (B) metabolic rate function (metabolic rate/F_{iso} v_{max} , Equations 2.10a and 2.10b); (C) relative power output (force × rate of shortening/F_{iso} v_{max}); and (D), efficiency (power output/metabolic rate). The parameter G has been given the value 4, which is typical for reasonably fast muscles. Metabolic rate is expressed in terms of ATP consumption. From Alexander (1999).

The force/rate of shortening relationship is reasonably well predicted by a model of muscle contraction proposed by Sir Andrew Huxley (1957). I will describe the model in outline only, without the mathematics, which has been expounded very clearly by McMahon (1984). Huxley represented the cross bridge as being mounted on springs that tend to pull it back to its equilibrium position, if it is displaced (Fig. 2.4). Let x be the displacement of the cross bridge from its equilibrium position; when x is positive, the cross bridge pulls in the direction required to shorten the muscle, and when x is negative it pulls in the reverse direction. A detached cross bridge oscillates about its equilibrium. At any stage in its oscillation when x lies between 0 and some value h, it may attach to an adjacent actin molecule. The probability that it will attach in a small increment of time is proportional to x, but the probability that an attached cross bridge will detach in a small increment of time is also proportional to x; therefore, in a muscle

A   B

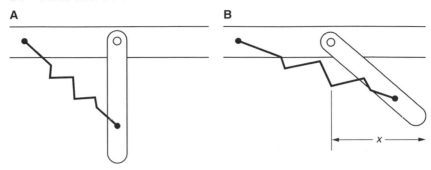

Fig. 2.4. Diagrams of a cross bridge (A) in its equilibrium position and (B) displaced to a position in which it could pull on a thin filament.

that is contracting isometrically an attached cross bridge is equally likely to have any value of x between 0 and h (Fig. 2.5A).

As the muscle shortens, an attached cross bridge moves to lower values of x at which it exerts less force. If it moves to negative values of x at which it exerts reverse forces, the probability that it will detach in a given increment of time is greatly increased. As the rate of shortening increases (Fig. 2.5B–D), the total force exerted by the cross bridges decreases, because cross bridges are moving to positions at which they exert less force and in some cases reverse force. Also, the increased rate of detachment at negative values of x results in there being fewer cross bridges attached at any instant. When $v = v_{max}$ (Fig. 2.5D), the forces exerted by cross bridges at positive values of x are balanced by reverse forces exerted by cross bridges at negative values of x, and the muscle exerts no force.

When an active muscle is stretched, cross bridges are pulled to greater values of x at which they exert increased forces, but the force does not increase indefinitely as speed of stretching increases because rates of detachment are greater at high values of x.

With appropriate constants in the equations, this model predicts empirical force-rate of shortening relationships well for positive rates of shortening; but the forces it predicts for muscle stretching are much too high.

The unloaded rate of shortening v_{max} is often expressed as fractional length change per unit time; that is, as a negative strain rate. Its value for any particular muscle fiber depends on the temperature; for example, for fast fibers in the iliofibularis muscle of the clawed toad *Xenopus* it is 1.5 lengths per second at 5°C, 2.6 s^{-1} at 10°, and 5.2 s^{-1} at 20°C (Lännergren 1978). It varies enormously between muscles, for example (considering only vertebrate striated muscles), from 0.6 muscle lengths per second for a tortoise penis retractor muscle at 16°C to 24 lengths per second for a mouse digital extensor muscle at 35°C (Woledge et al. 1985). These differences in v_{max} depend largely on differences in the activity of the myosin as

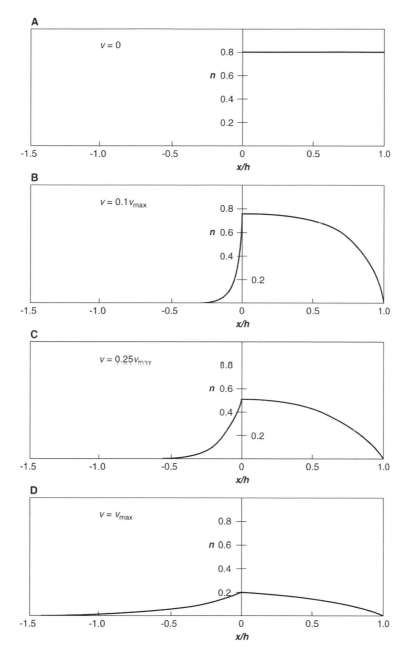

Fig. 2.5. The distributions of displacements of attached cross bridges from their equilibrium positions, when a muscle is shortening at four different rates. From Huxley (1957).

an enzyme that breaks down ATP. Such differences account for mouse muscles being about twice as fast as homologous muscles in cats, and for mouse digital flexor muscles being twice as fast as mouse soleus (Close 1972). Even among the fibers of a single muscle, large differences of v_{max} may be found. In the soleus muscle of horses, Rome et al. (1990b) found fibers with unloaded shortening speeds ranging from 0.2 to 3.8 s^{-1}.

Muscle fibers with equal ATPase activity may nevertheless have different unloaded rates of shortening, if their sarcomeres differ in length. Consider a muscle fascicle of length L with sarcomeres of length λ, so that it has L/λ sarcomeres in series along its length. Let its cross bridges be capable of making adjacent thin filaments slide past the thick filaments at a speed v_{fil}. The unloaded shortening speed will be

$$v_{max} = 2 \left(\frac{L}{\lambda} \right) v_{fil} \tag{2.4}$$

(The factor 2 arises because there are cross bridges pulling on thin filaments at both ends of the thick filament.) Muscles with short thick and thin filaments, and correspondingly short sarcomeres, can be much faster than muscles with long filaments and sarcomeres. For example, the tentacle extensor muscles of the squid *Loligo* have thick filaments 0.5–0.9 μm long with v_{max} around 45 s^{-1} (van Leeuwen and Kier 1997), while the anterior byssus retractor muscle of the mussel *Mytilus* has thick filaments around 30 μm long and v_{max} about 0.3 s^{-1} (Ruegg 1968). Equation 2.2 told us that muscles with long thick and thin filaments can exert high stresses. Now we see that the penalty for high stress is low speed.

2.3. POWER OUTPUT OF MUSCLES

A muscle exerting a force F while shortening at a rate v has a power output Fv (power is the rate of doing work). Figure 2.3C shows how the power depends on the rate of shortening; it is zero in isometric contraction (when $v = 0$), and also in contraction at v_{max} (when $F = 0$), but has a maximum value at an intermediate speed. Using Hill's equation (2.3a), the power output P is

$$P = \frac{F_{iso} v (v_{max} - v)}{v_{max} + Gv} \tag{2.5}$$

It can be shown by calculus that when $G = 2$, P has its maximum value when $v = 0.37 v_{max}$. At this rate of shortening, the force is $0.36 F_{iso}$ and the power is $0.134 F_{iso} v_{max}$. When $G = 6$, the maximum power is $0.075 F_{iso} v_{max}$ and occurs when the rate of shortening is $0.27 v_{max}$ and the force is $0.28 F_{iso}$.

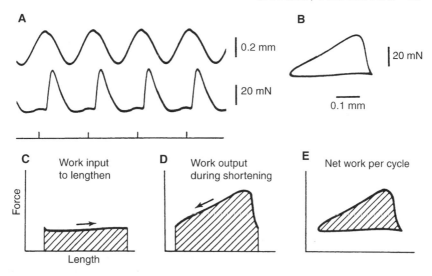

Fig. 2.6. Results of work loop experiments on a muscle fiber. (A) Length changes of the fiber (above) and the force it exerted (below), plotted against time. (B) Anticlockwise loop obtained by plotting the force against the length. (C–E) are explained in the text. From Josephson (1985).

Figure 2.3 shows results of experiments in which a bundle of muscle fibers made a single contraction at constant speed. Muscle action in locomotion is generally not like this. Whether moving a walking leg, flapping a wing, or beating a fin, muscles generally contract and are reextended repeatedly, in a regular cycle. Figure 2.6 illustrates a different type of experiment, which tests the properties of the muscle in conditions more like locomotion. A bundle of muscle fibers is fixed in an apparatus that alternately stretches it and allows it to shorten. It is stimulated electrically at an appropriate stage in each cycle, starting a little before the start of the shortening phase. A force transducer records the force exerted by the muscle, throughout the cycle. Figure 2.6A shows the length changes of a muscle (above) and the force it exerted (below), plotted against time. Figure 2.6B shows the force plotted against the length. This graph forms an anticlockwise loop. The force is low while the muscle is inactive, being stretched, and high while it is shortening actively. Each circuit of the loop represents a cycle of contraction and reextension. Figure 2.6C and D shows parts of the same loop, with hatched areas that represent work. (Remember that because work is force multiplied by displacement, areas on graphs of force against length represent work.) The area in Fig. 2.6C represents the work that the apparatus had to do on the muscle, to stretch it. The area in Fig. 2.6D represents the work that the muscle did on the apparatus, as it shortened. And the area of the loop (Fig. 2.6E) represents the net work done by the muscle. For a reason that should by now be obvious, experiments of the kind that I have been describing are referred to as work loop experiments.

The number and timing of electrical stimuli in each cycle is generally varied, to find the pattern of stimulation that maximizes work per cycle at each frequency. The amplitude of the length changes may also be varied. For given amplitude, the work that can be obtained in a cycle is highest at low frequencies, because the forces that muscles can exert are highest at low rates of shortening (Fig. 2.3A). However, the power output (work per cycle multiplied by cycle frequency) has a maximum at a moderate frequency.

Askew and Marsh (1998) performed work loop experiments on mouse soleus muscles, adjusting conditions to get as much power output as possible. They found that a sawtooth pattern of lengthening and shortening gave more power than the sinusoidal pattern that previous experimenters had used, and obtained a maximum power output of 94 W/kg at a frequency of 5 Hz. This corresponds to a work output in each cycle of 19 J/kg. This is much less than could be obtained if the muscle contracted very slowly (see above), but fairly fast contractions are needed to give maximum power . It seems likely that a work output of 20–25 J/kg in each cycle of repetitive contraction is typical of vertebrate striated muscle in strenuous locomotion (Alexander 1992b).

It would be reasonable to expect that the force exerted by a muscle, at any stage in the shortening phase of a work loop experiment, would be the same as in a single contraction at the same rate of shortening. However, this is not quite the case, because the force exerted by a muscle depends to some extent on its recent history. A shortening muscle can exert more force if it has been stretched immediately prior to shortening; this is called stretch activation. Also, shortening has a deactivating effect, reducing the force. Consequently, Hill's equation (Equation 2.3a) does not give accurate predictions of the forces exerted in cycles of shortening and stretching (Askew and Marsh 1998).

2.4. PENNATION PATTERNS AND MOMENT ARMS

In the previous section we saw that a muscle can deliver its maximum possible power output only if its rate or frequency of shortening is optimal for its physiological properties. In this section we will see how the structure and arrangement of muscles can be optimized to take the fullest advantage of the muscles' properties.

Figure 2.7A–C shows some of the ways in which the fascicles can be arranged within a muscle. All these muscles are represented with tendons

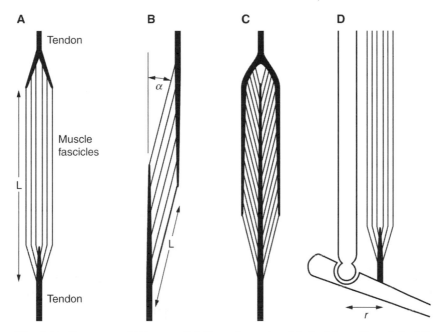

Fig. 2.7. Diagrams of (A) a parallel fibered muscle with fascicles of length L; (B) a unipennate muscle with an angle of pennation α; (C) a bipennate muscle; and (D) a muscle that has a moment arm r about a joint.

at both ends, which attach them to the skeleton, although there are some muscles that attach directly to the skeleton without tendons. Figure 2.7A represents a parallel fibered muscle, in which all the fascicles run more or less parallel to the force that the muscle exerts. Figure 2.7B represents a unipennate muscle. The tendons attaching the two ends of the muscle to the skeleton run along opposite faces of the muscle, with the fascicles running obliquely between them, attaching to the tendons at an angle α which is called the angle of pennation. Figure 2.7C represents a bipennate muscle, with fascicles converging from both sides on a central tendon. More complex, multipennate muscles are also found. It seems important to explain that these are schematic diagrams, drawn with simple geometry designed for ease of analysis. In real muscles, the fascicles and tendons are all slightly curved, for reasons explained by van Leeuwen and Spoor (1992, 1993). Though the three muscle designs in Figure 2.7 look very different from each other, the following calculations can be applied to them all.

Consider a muscle of volume V, with fascicles of length L. The muscle has angle of pennation α; if the muscle is parallel fibered, this angle is zero. Let the fascicles exert stress σ, and shorten at a strain rate $d\varepsilon/dt$ (strain ε is fractional change of length, so strain rate is the rate of change of length divided by the length). The total of the cross-sectional areas of the fascicles is V/L, so the total of the forces that they exert is $V\sigma/L$. These forces act at an angle α to the line of action of the muscle, so the force F transmitted along the tendons is

$$F = \frac{V\sigma}{L} \cos \alpha \qquad (2.6)$$

The rate of shortening of the fascicles is $L\, d\varepsilon/dt$, which makes the muscle as a whole shorten at a rate v (taking account of the angle of pennation)

$$v = \frac{L\, d\varepsilon/dt}{\cos \alpha} \qquad (2.7)$$

Compare muscles with equal volumes but different fascicle lengths L. Equations 2.6 and 2.7 tell us that (other things being equal) the muscle with the shorter fascicles will exert more force, but shorten more slowly.

The cosines of small angles are close to one; cos 0 = 1 and cos 30° = 0.87. Angles of pennation of muscles seem generally to be less than 30° (Yamaguchi et al., 1990), and it is often accurate enough to ignore the cosines in Equations 2.6 and 2.7. However, angles of pennation increase as muscles shorten, and it has been shown that when the human gastrocnemius muscle (in the calf) contracts, with the leg in certain positions, the angle of pennation may rise to as much as 50°. This has been shown by ultrasonic imaging of the muscle in the legs of intact living people (Maganaris et al. 1998).

Now consider the same muscle attached to a skeleton so that it crosses just one joint, with a moment arm r (Fig. 2.6D; note that the moment arm must be measured perpendicular to the muscle force). The moment M about the joint is

$$M = Fr = V\sigma \left(\frac{r}{L}\right) \cos \alpha \qquad (2.8)$$

The angular velocity ω with which the joint is moved is

$$\omega = \frac{v}{r} = \left(\frac{L}{r}\right)\left(\frac{d\varepsilon}{dt}\right) \cos \alpha \qquad (2.9)$$

Other things being equal, increasing the moment arm enables the muscle to exert larger moments about the joint, but reduces the angular velocity with which it can move the joint.

In Equations 2.8 and 2.9, fascicle length L and moment arm r appear only as the ratio r/L (or L/r). This tells us that, for a muscle of given volume, exactly the same moment and angular velocity can be obtained with long fascicles and a long moment arm or with short fascicles and a short moment arm. Which of these alternatives evolves may depend on the need to fit the muscle into the body. Human gastrocnemius and soleus muscles (which connect to the heel through the Achilles tendon) are pennate, with relatively short fascicles (about 20–40 mm) and a moment arm about the ankle of about 50 mm. The lower leg is long enough to accommodate a parallel fibered muscle with fascicles five times as long, but we would then need ridiculously long heel bones to give the muscle an appropriate moment arm.

Another important principle of muscle design that has apparently affected the course of evolution is that muscles can exert large forces only over a limited range of sarcomere lengths (Fig. 2.2). Therefore, the range of sarcomere lengths through which the fascicles have to shorten, to move the joint through its range of movement, should not be too great. As we have already seen, sarcomeres in vertebrate striated muscles commonly lengthen and shorten by around one-quarter of their length, in the movements of locomotion.

The ultimate tensile strength of mammalian tendon (the stress that breaks it, in a single pull) is at least 100 N/mm^2 (Bennett et al., 1986). This is many times larger than the stresses that mammalian muscle can exert (up to about 0.5 N/mm^2 when being rapidly stretched), so a thin tendon can transmit the force of a very much thicker muscle. Similarly in insects, very thin tendons (apodemes) are strong enough to transmit the forces of much thicker muscles (Bennet-Clark 1975). The tendons of vertebrates consist of collagen fibers in a mucopolysaccharide matrix, and the tendons of insects consist of chitin fibers in a protein matrix.

2.5. Power Consumption

Resting muscle uses metabolic energy at a low rate, for example, 3 W/kg for dog muscle (Martin and Fuhrman 1955). When the muscle is activated and exerts force, its metabolic rate increases greatly. The additional energy input is required principally for two processes: to provide the ATP that drives cross bridge cycling, and to pump calcium ions back into the sarcoplasmic reticulum after each stimulus. Cross bridge cycling is the major consumer of metabolic energy, but calcium pumping seems to account for around 30% of the metabolic rate of typical vertebrate striated muscles contracting isometrically, and around 10% of their metabolic rate when they are contracting at the rate that gives maximum efficiency (Lou et al. 1997).

Figure 2.3B summarizes the results of experiments in which the metabolic rates of active vertebrate striated muscles have been measured, while they were shortening or being stretched. The measurements were made in several different ways: some experiments measured the muscle's rate of oxygen consumption, others measured the rate at which it generated heat, and yet others measured the rate at which it used ATP. When the muscle contracts isometrically, it metabolizes at a rate of about $0.07 F_{iso} v_{max}$, where F_{iso} is the isometric force and v_{max} is the maximum (unloaded) rate of shortening. Thus, more metabolic power is needed to maintain a given force in a fast (high v_{max}) muscle than in a slow one. When the muscle shortens, doing work, the metabolic rate increases because (as we have already seen) the rate of cross bridge cycling increases. The rate of cross bridge cycling and the metabolic rate plateau at high shortening speeds because of the time needed for a detached cross bridge to reattach. When the active muscle is stretched at increasing speeds, the metabolic rate falls and then rises.

The graph (Fig. 2.3B) can be represented by the following equations (Alexander 1997b):

$$\text{for } v \geq 0 \qquad M = F_{iso} v_{max} \left[0.23 - 0.16 \exp\left(-\frac{8v}{v_{max}} \right) \right] \qquad (2.10a)$$

$$\text{for } v < 0 \qquad M = F_{iso} v_{max} \left[0.01 - 0.11 \frac{v}{v_{max}} + 0.06 \exp\left(\frac{23v}{v_{max}} \right) \right] \qquad (2.10b)$$

where M is the metabolic rate. The other symbols have the same meanings in these equations as in Equations 2.3a and 2.3b.

The efficiency of a muscle is its power output (the rate at which it is doing work) divided by the power input (the rate at which it uses metabolic energy). Figure 2.3D shows efficiencies calculated from the data of Figure 2.3B and C. The maximum efficiency, about 0.45, is obtained at a shortening speed of about $0.2 v_{max}$, rather lower than the speed that maximizes power output. It is important to note that power input in Fig. 2.3B and D is calculated as the rate at which ATP energy is used, ignoring the energy losses involved in synthesizing ATP. The efficiency with which the enthalpy of combustion of foodstuffs is converted to ATP is only about 0.5, so the maximum efficiency of converting food energy to work in a typical muscle is about $0.45 \times 0.5 = 0.23$. Values close to this are obtained when the efficiency of human muscle is calculated from the oxygen consumption of athletes pedaling bicycle ergometers (Dickinson 1929). However, muscles do vary in efficiency. A tortoise leg muscle has been shown to convert food energy to work with an efficiency of 0.40 (Woledge et al. 1985).

Some muscle fibers depend on aerobic metabolism, releasing the energy to power their contractions by oxidizing foodstuffs to carbon dioxide and water. Others in vertebrates (Bone 1966), squids (Bone et al. 1981), and crabs (Full 1987) use anaerobic metabolism, releasing energy by reactions such as the conversion of glucose to lactic acid (in vertebrates) or glucose and arginine to octopine (in squids). The anaerobic processes release far less energy from a given mass of food than the aerobic ones, but this does not mean that they are wasteful. After a burst of anaerobic activity, some of the lactic acid or other product is oxidized to release the energy needed to convert the rest back to glucose.

Aerobic and anaerobic fibers are generally mixed together in the same muscle, but fish have the aerobic muscle segregated as a red band running along the body on the outer surface of the white anaerobic muscle. The difference in color between the two types of muscle is very obvious in skinned fish, whether cooked or raw. This separation of the two types of muscle greatly aided the observations on fish that showed how the two types of muscle function (Bone 1966; Rome et al., 1984). Slow swimming at speeds that can be maintained for long times is powered by the aerobic fibers, but above a critical speed the white fibers are brought into use. The aerobic fibers contain many mitochondria, housing the enzymes of the Krebs cycle that play a central role in aerobic respiration, but the anaerobic fibers contain few mitochondria. The aerobic fibers have a good blood supply, needed to bring oxygen to them, but the anaerobic fibers have a much more restricted blood supply. Recent experiments have shown that white fish muscle may not be as strictly anaerobic as was previously believed; dogfish (*Scyliorhinus*) white muscle is capable of powering a short sequence of contractions aerobically (Lou et al., 2000). Squid have (yellow) aerobic muscle fibers and white anaerobic ones that differ from each other in the same sorts of ways as the red and white fibers of fishes (Bone et al., 1981).

Power output from aerobic muscle fibers is limited by the rate at which the respiratory and circulatory systems can supply them with oxygen or by the rate at which their mitochondria can use it. Anaerobic fibers are not limited in this way, but the energy that can be made available by anaerobic muscle in a single bout of activity is limited, because the body cannot tolerate excessive accumulation of lactic acid or (in squids) octopine. Also, anaerobic fibers can contain a higher proportion of myofibrils than aerobic ones, and so (other things being equal) can deliver more power, because less of their volume is occupied by mitochondria. Anaerobic respiration makes maximum sprint speeds possible, but sustained locomotion has to be powered aerobically. Human athletes exerting maximum effort for 10 s derive about 85% of their energy output from anaerobic metabolism; but in 10 min of maximum effort only 10–15% of the energy comes from anaerobic metabolism, and in 2 hours of maximum effort only 1% (Åstrand and Rodahl 1986). A 100-m sprint is powered predominantly by anaerobic metabolism, but the contribution of anaerobic metabolism to a marathon is very small. The change from mainly anaerobic metabolism for short races to mainly aerobic metabolism for longer ones may explain the change of gradient at 150 s in the graph of race speed against race duration (Fig. 1.3A).

Even aerobic respiration cannot be sustained at its maximum rate indefinitely, because the fuels that support it become depleted. This was evident in a study of red deer (*Cervus elaphus*) stags that had been killed by hunting with hounds (Bateson and Bradshaw 1997). The mean duration of hunts was 3 h and the mean distance covered was 19 km. Stags killed after long hunts had very much lower blood glucose concentrations than those killed after short hunts, indicating that their carbohydrate reserves had been reduced to low levels. It was also found that muscle enzymes had leaked into the blood in these animals, indicating that muscle damage had occurred.

2.6. SOME OTHER TYPES OF MUSCLE

This chapter has been mainly concerned with vertebrate striated muscle. Some invertebrates have very different muscles. Many muscles of molluscs, annelids, and nematode worms are of the type called obliquely striated, because the bands of thick and thin filaments do not run transversely across them, but are arranged helically. Their filaments are long (in some cases extremely long) and the fibers can exert high stresses, as already noted for a mussel muscle in Table 2.1. Because they are very long with large numbers of cross bridges, the thick filaments have to be strong enough to withstand large forces. They have a central core of the protein paramyosin, which makes them much thicker than the thick filaments of vertebrates. Some obliquely striated muscles are capable of exerting substantial stresses over remarkably wide ranges of sarcomere length (J.B. Miller 1975), but others have physiological properties more like vertebrate striated muscle (Milligan et al., 1997). Some of the obliquely striated muscles of bivalve molluscs have the property known as catch, which enables them to maintain tension for very long times with remarkably little expenditure of energy (Watabe and Hartshorne 1990). However, these muscles serve functions such as holding the shell closed, rather than locomotion, and so need not be discussed further in this book.

The flight muscles of advanced insects, including flies, beetles, and many bugs, bees, and wasps, have very different but equally remarkable properties (Josephson et al., 2000). They do not require a stimulus to elicit each contraction, as other muscles do. Instead, occasional action potentials in their nerves are enough to keep them in a state of oscillation in which they lengthen and shorten repeatedly. In one experiment, the wing muscles of a tethered fly beating its wings at a frequency of 120 Hz were found to be receiving action potentials at a frequency of only 3 Hz. To behave like this, the muscles must be connected to a resonant system. They then drive the system at its natural frequency of vibration (Machin and Pringle 1959). This behavior is due to properties that conventional muscles have to a lesser extent: stretching has an activating effect on an active muscle that persists briefly after stretching ends, and shortening has a deactivating effect that similarly persists.

These muscles are described as fibrillar muscles (because their myofibrils are thick and conspicuous) or asynchronous muscles (because their contractions are not synchronous with action potentials). All of them work at high frequencies, ranging from about 20 Hz in giant bugs to 1000 Hz in small midges. Fibrillar muscles have a large advantage for high-frequency operation, because they do not require a pulse of calcium ions to trigger each contraction, but only for the occasional action potentials. This greatly reduces the energy required for calcium pumping. It also greatly reduces the volume of sarcoplasmic reticulum required in the muscle fibers, leaving more space for myofibrils. Rattlesnake tail-shaker muscles, which operate at 90 Hz but are not fibrillar, contain 26% sarcoplasmic reticulum. In contrast, fibrillar flight muscle contains very little, for example 2% in a beetle that beats its wings at 80 Hz (Table 2.2).

The elastic properties of fibrillar flight muscle are unusual. Figure 2.2A showed that very little stress is needed to stretch resting frog muscle, until lengths at the upper end of its working range are reached. In contrast, unstimulated fibrillar flight muscle develops large passive elastic stresses when stretched only slightly beyond the length at which (when stimulated) it would develop maximum active stress (Fig. 2.8A).

Work loop experiments mimicking the conditions of flight have been performed on bumblebee (*Bombus*) flight muscles (Josephson 1997a). The frequency was about 150 Hz, the range of length change was about 3% of muscle length and the temperature was 40°C. The maximum power output obtained was 110 W/kg, which seems to be a little less than is possible in the living animal; calculations of aerodynamic power output of fully laden bumblebees in climbing flight have given values around 180 W per kilogram of muscle (Josephson 1997b). Even this higher value represents a work output in each wing beat cycle of only 180/150 = 1.2 J/kg, much lower than the outputs obtainable from vertebrate striated muscle. For example, as we saw in Section 2.3, work loop experiments with a mouse leg muscle gave a maximum power output of 94 W/kg at 5 Hz, representing a work output in each cycle of 19 J/kg (Askew and Marsh 1998). The net work output of fibrillar flight muscle is low because the range of length changes is small (3% compared to about 12% in the experiments on mice) and because the stress in the muscle remains quite high, during the lengthening phase of the cycle. Nevertheless, the power output is high, because the frequency is high.

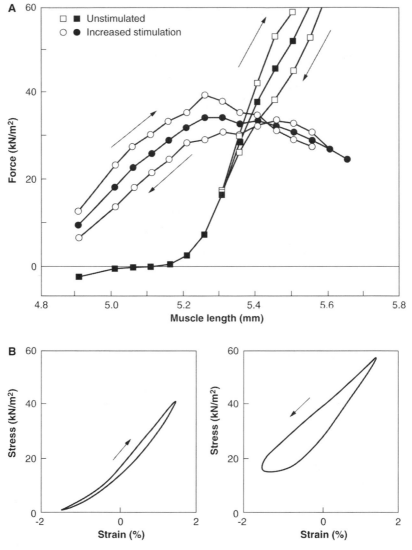

Fig. 2.8. Properties of bumblebee (*Bombus*) flight muscle. (A) A graph of stress against muscle length. Squares show stress in unstimulated muscle and circles show the additional stress that resulted from stimulation. Hollow symbols show stresses measured while length was increasing or decreasing, as indicated by the arrows. Filled symbols show the means of these values. (B) Examples of work loops obtained in experiments simulating the conditions of flight. These go clockwise for unstimulated muscle (above), showing that a little energy was being dissipated, and anticlockwise for stimulated muscle (below), showing that net work was being done. From Josephson (1997a b).

This chapter has shown that the muscles that power locomotion vary widely in structure and properties. Myofilament lengths are more or less constant among vertebrates, but longer myofilaments in some invertebrate muscles enable them to exert higher stresses. The stress that a muscle can exert depends both on the degree to which its sarcomeres are extended and on the rate at which they are shortening or being stretched. The properties of a muscle in a series of repetitive contractions and extensions cannot be predicted accurately from its behavior in single contractions. Aerobic muscles contain higher proportions of mitochondria (and correspondingly less contractile material) than anaerobic muscles. The peculiar oscillatory behavior of the fibrillar flight muscles of some insects enables them to work at very high frequencies without the high proportion of sarcoplasmic reticulum found in other fast muscles. The pennation patterns and moment arms of muscles, as well as the properties of their fibers, are important in matching them to their functions.

Though we know so much about muscle, scientists studying animal locomotion would like to know more. Equations 2.3a and 2.3b predict the force that a fully activated muscle will exert while contracting or being stretched at a constant rate. Equations 2.10a and 2.10b predict its metabolic rate in the same circumstances. However, during locomotion muscles do not simply shorten or stretch at constant rates. They generally undergo cycles of lengthening and shortening, and are activated to different extents at different stages of the cycle. In such circumstances, Equations 2.3 do not work well (Askew and Marsh 1998), and it seems likely that the same is true of Equations 2.10. If we had easily applicable equations that would predict the forces and metabolic rates of muscles in cyclic activity, we could greatly improve our understanding of the muscle forces and metabolic energy costs involved in locomotion. For example, we would be able to calculate what the forces and energy costs would be, if an animal's muscles had different properties or were differently arranged, or if it used a different pattern of movement.

Chapter Three

••

Energy Requirements for Locomotion

MUSCLES HAVE TO do work whenever an animal's body is moving against a resisting force, and whenever the mechanical energy of the body is increasing. This chapter discusses the forces and energies involved in a general way, in preparation for the discussions in later chapters of specific modes of locomotion. It also considers the relationship between the work that the muscles do and the metabolic rate of the body.

The major components of the mechanical energy of the body that are important for discussions of locomotion are kinetic energy, gravitational potential energy, and elastic strain energy. We will consider them in turn.

3.1. KINETIC ENERGY

Kinetic energy is the energy that a moving body has because it is moving. It is very simple to calculate the kinetic energy of a particle (an infinitely small body): the kinetic energy of a particle of mass m moving with velocity v is $\frac{1}{2}mv^2$. The calculation is more complicated for bodies of finite size because different parts of them may move with different velocities. Even if a body is rigid, different parts of it will be moving at different velocities if it is rotating. The kinetic energy U_{kin} of a rigid body is

$$U_{kin} = \tfrac{1}{2}\, mv_{cm}^2 + \tfrac{1}{2}\, I_{cm}\, \omega^2 \qquad (3.1)$$

where m is the mass of the body, v_{cm} is the velocity of its center of mass, I_{cm} is the moment of inertia of the body about its center of mass, and ω is its angular velocity (its rate of rotation, which must be measured in radians, not degrees, per unit time).

The center of mass is the point at which the mass of the body can be considered to act for purposes of calculations like this, for example the center of a uniform sphere or the midpoint of a uniform rod. Alexander (1983) describes a method for locating the centers of mass of other bodies. The moment of inertia is the constant needed to make this equation work. It is small if the mass of the body is all located close to the center of mass, and large if most of the mass is near the periphery. For example, it is larger for a hollow steel sphere than for a solid steel ball of the same mass. Further explanation, and descriptions of methods for measuring moments of inertia, can be found in textbooks of mechanics and in Alexander (1983).

The situation is even more complicated for bodies that are not rigid, for example, human bodies whose joints bend and extend as we walk or run. To calculate the kinetic energy of such a body, think of it as being composed of a very large number of tiny particles. Particle number 1 has mass m_1 and components of velocity u_1, v_1, w_1 in the X, Y, and Z directions; particle number 2 has mass m_2 and components of velocity u_2, v_2, w_2; and so on. The kinetic energy of the body is calculated by adding up the kinetic energies of all the individual particles.

$$U_{kin} = \tfrac{1}{2}\, m_1 \,(u_1^2 + v_1^2 + w_1^2) + \tfrac{1}{2}\, m_2 \,(u_2^2 + v_2^2 + w_2^2) + \ldots \quad (3.2)$$
$$= \tfrac{1}{2} \Sigma \,[\, m_i \,(u_i^2 + v_i^2 + w_i^2)\,]$$

The symbol Σ means "the sum of" and the subscripts i indicate that the total has to be calculated for $i = 1, 2, 3$, etc., to include all the particles. Rotational terms, like the term $\tfrac{1}{2}I_{cm}\omega^2$ in Equation 3.1, do not appear in this equation because it treats the body as an assembly of infinitely small particles.

It is often convenient to think of the kinetic energy of the body as having two components: energy due to movement of the body as a whole and energy due to movement of parts of the body (for example, limb segments) relative to each other. This is done by defining the external kinetic energy (the energy due to the motion of the center of mass) as $\tfrac{1}{2} m_{total} \,(u_{cm}^2 + v_{cm}^2 + w_{cm}^2)$ and the internal kinetic energy (due to motion of parts of the body relative to the center of mass) as $\tfrac{1}{2}\, m_1\, [(u_1 - u_{cm})^2 + (v_1 - v_{cm})^2 + (w_1 - w_{cm})^2] + \tfrac{1}{2}\, m_2\, [(u_2 - u_{cm})^2 + (v_2 - v_{cm})^2 + (w_2 - w_{cm})^2] + \ldots$ or $\tfrac{1}{2} \Sigma \,\{m_i[(u_i - u_{cm})^2 + (v_i - v_{cm})^2 + (w_i - w_{cm})^2]\}$. Here m_{total} is the total mass of the body, and u_{cm}, v_{cm}, w_{cm} are the components of the velocity of the center of mass:

$$U_{kin} = \tfrac{1}{2}\, m_{total} \,(u_{cm}^2 + v_{cm}^2 + w_{cm}^2) \qquad\qquad (3.3)$$
$$+ \tfrac{1}{2} \Sigma \,\{m_i \,[(u_i - u_{cm})^2 + (v_i - v_{cm})^2 + (w_i - w_{cm})^2]\}$$

It can be shown that Equation 3.3 gives the same result as Equation 3.2 (Chorlton 1967).

An unwelcome complication in calculations using Equation 3.3 is that the position of the center of mass is not fixed within the body. When I swing a leg forward, my center of mass moves forward a little, relative to my trunk, and when I raise an arm my center of mass rises a little.

3.2. GRAVITATIONAL POTENTIAL ENERGY

Gravity exerts a downward force mg on a body of mass m, where g is the gravitational acceleration (9.8 m s^{-2} at the surface of the earth). This force is known as the weight of the body. To raise the body through a height h, work must be done equal to the force multiplied by the distance, mgh.

This work can, in principle, be recovered if the body is allowed to fall back to its original level, so a body of mass m with its center of mass at a height h has gravitational potential energy mgh.

Gravitational potential energy and kinetic energy are interconvertible. A ball thrown into the air slows down as it rises and speeds up as it falls; it loses kinetic energy as it gains gravitational potential energy and vice versa. Similarly, a pendulum slows down as it rises at the end of a swing and speeds up as it falls at the beginning of the next swing. A frictionless pendulum in a perfect vacuum would go on swinging forever, swapping energy back and forth between the kinetic and the gravitational potential forms. Pendulumlike energy exchanges will figure largely in our discussion of walking in Section 7.3.

When you calculate the gravitational potential energy of a body, the result you get depends on the level from which you measure the height h. This does not matter, so long as you make it clear what your reference level is, because you will be interested in changes of gravitational potential energy rather than absolute values. It is often convenient to measure the height from ground level, because you cannot generally fall further than that.

3.3. Elastic Strain Energy

Energy is stored in an elastic structure when it is stretched or compressed. For example, energy is stored in the rubber of a catapult as it is stretched, and released, giving kinetic energy to the missile, when the catapult is fired. In the case of an ideal linear spring, the force F required to stretch it is proportional to the extension Δl, according to Hooke's law

$$F = S\,\Delta l = \frac{1}{C}\,\Delta l \tag{3.4}$$

where S is the stiffness and $C\,(= 1/S)$ is the compliance of the spring. A graph of force against extension is a line of gradient S, and the elastic strain energy U_{strain} is represented by the (triangular) area under it:

$$U_{\text{strain}} = \tfrac{1}{2}\,F\,\Delta l = \tfrac{1}{2}\,S\,\Delta l^2 = \frac{\tfrac{1}{2}\,F^2}{S} \tag{3.5}$$

Biological materials do not generally obey Hooke's law. The material of which the elastic properties will concern us most is tendon. Figure 3.1 shows the results of an experiment in which a tendon from a wallaby was repeatedly stretched and allowed to recoil. Its elastic behavior differs in two ways from that of an ideal, linear spring. First, contrary to Hooke's law, the graph is curved. Its gradient is shallower at small extensions than

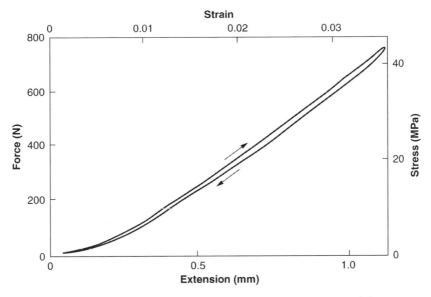

Fig. 3.1. A graph of force against extension obtained when a piece of the gastroc-nemius tendon of a wallaby (*Macropus*) was stretched and allowed to recoil repeat-edly at a frequency of 2.2 Hz (similar to the frequency with which the animal would have hopped). Additional scales show stress and strain. From Alexander (1988).

at large ones. Secondly, the graph for increasing force does not quite coin-cide with the graph for decreasing force. Instead, the upward and down-ward curves form a loop. In work loop experiments on muscle, graphs of force against length form anticlockwise loops, showing that the muscle is doing net work (Fig. 2.6). In Fig. 3.1, the loop is clockwise, showing that not all the work done stretching the tendon is returned in the elastic re-coil. Net work is being done on the tendon, and dissipated as heat. The area of the loop is small compared to the area under the graph, showing that the losses are small. About 7% of the work done stretching a tendon is lost in this way (Bennett et al., 1986).

A thick tendon has higher stiffness than a thin one of the same length, and a long tendon is less stiff than a short one of the same thickness. However, graphs of stress (= force / cross sectional area) against strain (= extension / length) are more or less the same for all tendons. Figure 3.1 shows scales of stress and strain as well as of force and extension. For materials that stretch according to Hooke's law, Young's modulus is the gradient of a graph of stress against strain. Because the graph for tendon is curved, its gradient is greater at high stresses than at low ones, and there is no unique Young's modulus. The gradient at any particular stress is known as the tangent Young's modulus at that stress. For tendon, it is low at low stresses and rises at high stresses to about 1500 N/mm². In later chapters, we will be concerned also with the elastic properties of resilin, a rubberlike protein found in insects that has a Young's modulus of about 2 N/mm² (Weis-Fogh 1960); and with those of a locust leg apodeme, which has a Young's modulus of 15,000 N/mm² (Bennet-Clark 1975).

The elastic strain energy stored in an ideal, linear elastic structure is proportional to the square of the applied force (Equation 3.5). Structures are broken by excessive forces, so the energy they can store is limited. Consider a cylinder of length l, cross-sectional area A, of a material of Young's modulus E, tensile strength τ, and density ρ. It can just withstand a force $A\tau$, which will stretch it to a strain τ/E, that is, to an extension $\tau l/E$. Equation 3.5 tells us that the strain energy stored in it will then be $\frac{1}{2}A\tau^2 l/E$. Its volume is Al and its mass is $Al\rho$, so the strain energy stored per unit mass is $\frac{1}{2}\tau^2/\rho E$. The ultimate tensile strength of tendon seems to be about 100 N/mm² or 1.0×10^8 N/m², its density is 1120 kg/m³ and its Young's modulus is about 1.5×10^9 N/m² (Bennett et al., 1986) Thus, the maximum strain energy that could be stored is $\frac{1}{2} \times (1.0 \times 10^8)^2/(1120 \times 1.5 \times 10^9) = 3000$ J/kg. This is a slight underestimate, because we have ignored the curvature of the stress–strain graph.

No tendon would store as much strain energy as that in a normal activity. Tendons grow thick enough to ensure that the forces their muscles can exert on them are well below the forces needed to break them, leaving a margin of safety. However, stresses of 50 N/mm² occur in some leg tendons of running mammals (Biewener 1998) and, by the same calculation, store 750 J/kg in them. Compare this with the maximum work that vertebrate striated muscle can do in a single contraction, 120 J/kg (Section 2.1). One gram of tendon is enough to store the work that 6 g of muscle can do.

3.4. WORK THAT DOES NOT INCREASE THE BODY'S MECHANICAL ENERGY

Muscles have to do work, not only when the body's mechanical energy is increased but also when the body moves against forces that transfer energy to the environment. For example, a crawling snake does work against friction as it slides forward over the ground. The energy used for this is degraded to heat and lost. A crawling snail does work against the viscosity of the mucus (slime) under its foot, and again the energy is degraded to heat. A swimming fish does work against drag, the hydrodynamic force that resists its movement. Some of this work is required to overcome the viscosity of the water in the boundary layer (the thin layer of water close

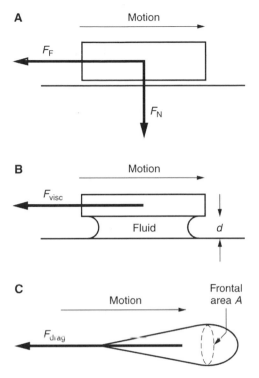

Fig. 3.2. Diagrams of situations in which motion is resisted by (A) friction, (B) viscosity, and (C) aerodynamic or hydrodynamic drag.

to its skin). The rest gives kinetic energy to the water that is inevitably set moving in the fish's wake, disturbed by the passage of the fish.

Here is how friction is calculated. Consider two surfaces that are pushed together by a force F_N (Fig. 3.2A). The subscript N indicates that this is a normal force, a force perpendicular to the surfaces. If one surface is made to slide over the other, the movement is resisted by a frictional force F_F,

$$F_F = \mu \, F_N \qquad (3.6)$$

where μ is the coefficient of dynamic friction for the surfaces in question. Coefficients of dynamic friction between dry metal surfaces are generally around 0.2. In a simple experiment in which I pressed my finger on a force plate, and slid my finger across the plate I found that the coefficient of dynamic friction between my dry finger and the smooth aluminium surface of the plate was about 1.0. Lubrication can reduce coefficients of friction greatly, for example, to around 0.01 between the metal surfaces of engineering bearings. The work done when a body moves a distance s against a frictional force F_F is sF_F.

The limb joints of mammals are very well lubricated, with coefficients of friction around 0.003. Consequently, the work animals have to do against friction in their own joints is so small that it can generally be ignored. For example, when an adult human walks the normal force in the hip joint of the supporting leg is about three times body weight, or just over 2000 N for a 70-kg man. (It is greater than body weight because of the forces exerted by hip muscles, see Crowninshield et al. 1978.) In the course of a step, the head of the femur slides about 20 mm (0.02 m) in the acetabulum. Hence, the work done during the step, against friction in the hip joint, is about $0.003 \times 2000 \times 0.02 = 0.1$ J. This is very small compared to the work of about 15 J needed to increase the sum of kinetic and gravitational potential energy in the course of the step (Cavagna et al. 1977).

Frictional forces are more or less the same for all speeds of sliding, but viscous forces increase with speed. Consider a plate of area A separated by a fluid layer of thickness d, viscosity η, from a fixed surface (Fig. 3.2B). If it slides over the fixed surface with velocity v, the viscous force that resists its movement is

$$F_{visc} = \frac{\eta \, Av}{d} \tag{3.7}$$

The viscosity of air at atmospheric pressure at 20°C is 1.8×10^{-5} N s/m^2. Freshwater has a viscosity of 1.8×10^{-3} N s/m^2 at 0°C and 1.0×10^{-3} N s/m^2 at 20°C, and seawater is slightly more viscous (Denny 1993). The mucus under the foot of a crawling slug has much higher viscosity, about 5–10 N s/m^2 (Denny 1980b).

The formula used for calculating aerodynamic and hydrodynamic drag depends on the size and speed of the moving object and on the properties of the fluid in which it is moving. For small objects moving slowly in highly viscous fluids, drag is due mainly to viscosity and is proportional to the velocity. For large objects moving fast in fluids of low viscosity, drag is due mainly to the kinetic energy given to the fluid and is proportional to the square of velocity. A more specific statement about this will be made in Chapter 4, when Reynolds number is defined. For the present, it is sufficient to know that drag is about proportional to velocity for swimming spermatozoa and microorganisms, and to velocity squared for swimming fishes. In this book we are concerned almost entirely with animals that are large and fast enough for drag to be calculated using the equation

$$F_{drag} = \frac{1}{2} \rho \, Av^2 C_D \tag{3.8}$$

where ρ is the density of the fluid; A is an area that can be defined in several alternative ways, as we shall see, v is the velocity, and C_D is the drag coefficient, a quantity that depends on the shape of the body. Confusingly, the drag coefficient is also affected by the size and speed of the body and by the properties of the fluid, but I will show in Section 10.1 how account can be taken of these effects, by the use of Reynolds numbers. Vogel (1981) gives a very good, clear explanation of drag.

Depending on the type of body being discussed, the area A can be defined in different ways. For bodies such as the fuselages of aircraft and the bodies (excluding the wings) of birds it is often defined as the frontal area, the area of the body as seen in front view (Fig. 3.2C). For the wings of birds and aircraft it is generally defined as the plan area, the area of a full-scale plan of the wings (Fig. 10.1A). Another possibility is the wetted area, the total surface area exposed to the fluid. Yet another is (body volume)$^{2/3}$; volume has dimensions (length)^3 and area has dimensions (length)^2, and so (volume)$^{2/3}$ has the dimensions of area. Different drag coefficients apply, depending on the definition of area that is used. For example, the drag coefficient based on frontal area of a sphere is about 0.5, over a wide range of Reynolds numbers; the drag coefficient based on wetted area is about 0.13, because the surface area of a sphere is four times the frontal area; and the drag coefficient based on (volume)$^{2/3}$ is about 0.6. All these drag coefficients give the same drag, when used in Equation 3.8, provided that the corresponding area is used with them.

Other shapes have different drag coefficients. The lowest coefficients are obtained with streamlined bodies, rounded in front and tapering to a point behind like the body shown in Fig. 3.2C. The drag coefficients of the streamlined body of a sea lion, in the range of Reynolds numbers at which sea lions swim, are about 0.08 (based on frontal area), 0.005 (based on wetted area), or 0.04 (based on (volume)$^{2/3}$) (Stelle et al., 2000). Comparison of the drag coefficients based on (volume)$^{2/3}$ shows that the drag on a sea lion's body is only one-fifteenth of the drag on a sphere of equal volume, moving at the same speed. Streamlining has a very large effect on drag.

The aerodynamic or hydrodynamic forces on bodies may have components at right angles to the direction of motion, in addition to the drag that acts backward along the direction of motion. These components of force are called lift. They are the forces that keep aircraft airborne. Work is done against drag but not against lift, because work is distance moved multiplied by the component of force *in the direction of motion*.

Different parts of an animal's body may move with different velocities and have different resistive forces acting on them. Suppose that in some interval of time, part number 1 of the body moves a distance s_1 against a force F_1, part number 2 moves a distance s_2 against a force F_2, and so on. The total work done against resistive forces is

$$W_{res} = F_1 s_1 + F_2 s_2 + \cdots = \Sigma\ (F_i s_i) \tag{3.9}$$

3.5. WORK REQUIREMENTS

Muscles do work when they shorten while exerting force. If the length of a muscle changes by Δl while it is exerting a force F, it does work $-F\,\Delta l$. The negative sign indicates that positive work is done when the length decreases (when Δl is negative). If the length increases (Δl is positive), $-F\Delta l$ is negative. A muscle that is stretched while exerting a force acts like a brake and is said to do negative work.

Suppose that in some interval of time the kinetic energy of a body increases by an amount ΔU_{kin}, the gravitational potential energy by ΔU_{grav}, and the elastic strain energy by ΔU_{strain}. Suppose that in the same interval of time work W_{res} is needed to move parts of the body against resistive forces such as friction or drag; that some of the muscles do positive work totaling W_{pos}; and that others do negative work totaling $-W_{neg}$. Then considerations of energy balance tell us that

$$W_{pos} - W_{neg} = \Delta U_{kin} + \Delta U_{grav} + \Delta U_{strain} + W_{res} \qquad (3.10)$$

Having some muscles do positive work while others do negative work is like driving a car with the brakes on. It may seem wasteful of energy but it is often necessary, for several reasons.

First, the changes of kinetic energy, gravitational potential energy, and elastic strain energy may be either positive or negative. The total mechanical energy of one part of the body may rise while another falls. In that case, muscles will have to do positive work on one part of the body while other muscles do negative work on the other, unless there is a mechanism that transfers energy from one part to the other.

As an example of an activity in which mechanical energy is transferred, consider kangaroo hopping. I will show in Chapter 7 that while the feet are on the ground, the kinetic energy of the body falls and then rises; at the same time tendons in the lower leg stretch (storing elastic strain energy) and then recoil (returning the energy). It seems clear in this case that kinetic energy of the body is converted to elastic strain energy in tendons, and back again. Similarly, when a ball bounces, its kinetic energy is converted to elastic strain energy, then returned in the rebound.

Now imagine yourself swinging your arms one-quarter of a cycle out of phase with each other, so that when one arm is moving fastest, in mid swing, the other is stationary at the end of its swing. The kinetic energy of the right arm falls while that of the left arm rises, and vice versa, but it seems unlikely that energy can be transferred from one arm to the other; instead, the muscles of one arm must do positive work while those of the other do negative work. This is rather an artificial example (the movement is hard to perform) and I use it only because it can be described briefly. At

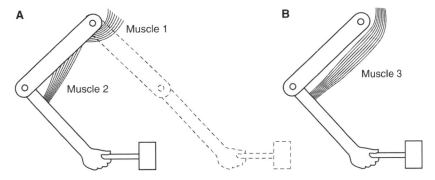

Fig. 3.3. (A) A diagram of an arm with one-joint muscles only, making a movement that is discussed in the text. (B) A two-joint muscle.

various stages of walking and running strides, the kinetic energy of one part of the body falls while that of another rises, and it is difficult to determine how much energy transfer occurs (Winter 1990).

A second reason why one muscle must do positive work while another is doing negative work depends on the arrangement of the muscles involved. Consider a person holding a heavy object, who moves it slowly forward from the position shown by the continuous line in Fig. 3.3A, to the position shown by the broken outline. This movement leaves the gravitational potential energies of arm and object unchanged. Throughout the movement, moments that can be balanced by the muscles shown in the diagram act about the shoulder and elbow. If these muscles are used for the movement, the shoulder muscle has to shorten, doing positive work, while the elbow muscle lengthens, doing negative work. The need for muscles to work against each other could be reduced if use were made of the muscle shown in Fig. 3.3B, which crosses both joints and exerts moments in the required direction about both. The movement involves 90° rotation about both joints, so if this muscle had equal moment arms about both joints there would be no need for it to change length during the movement. Though it would exert force, it need do neither positive nor negative work. Other muscles would have to be activated at appropriate stages of the movement to ensure that equilibrium was maintained at both joints, but the amounts of positive and negative work required for the movement would be reduced.

In the human arm, the brachialis muscle is arranged like the elbow muscle of Fig. 3.3A, and the biceps is arranged like the muscle of Fig. 3.3B. Alexander (1997a) presented a theoretical argument that seems to show that the metabolic energy cost of many arm movements would be greater, if two-joint muscles such as the biceps did not exist.

3.6. Oscillatory Movements

In this section we will ask how muscles and their tendons could be arranged to minimize metabolic energy costs in movements that are oscillatory in nature. Such movements are very common in locomotion. Flying insects and birds beat their wings up and down, swimming fishes beat their tails from side to side, and running mammals swing their legs backward and forward. In all these cases, a structure that has mass is oscillated in a fluid (either air or water), which resists its motion. We will have to consider two kinds of forces. Forces are required to accelerate and decelerate the moving body parts; we will call these inertial forces. In addition, forces are needed to move the body parts through the fluid against hydrodynamic (or aerodynamic) drag. In some cases the inertial forces may be much larger than the hydrodynamic forces. For example, the inertial forces needed to accelerate and decelerate the legs of a horse are much larger than the aerodynamic drag that resists their motion. In other cases the inertial forces may be much smaller than the hydrodynamic forces, or may even be canceled out by springs.

Any mass that is mounted on springs will vibrate if disturbed, at a natural frequency of vibration that depends on the ratio of the mass to the stiffness of the spring; high stiffness and low mass give a high frequency. When a mass–spring system is vibrating at its natural frequency, kinetic energy is being converted repeatedly to elastic strain energy and back again. The inertial forces are balanced by elastic forces in the spring. This seems to happen in scallops, which swim by repeatedly clapping their shells shut (squirting out a jet of water) and allowing them to spring open (see Section 16.2). The mass of the valves of the shell and the elastic stiffness of the hinge that joins the two valves of the shell together interact to give the animal a natural frequency of vibration that seems to match the swimming frequency (De Mont 1990). When the swimming scallop opens and closes its shell at this frequency, its muscles have only to work against hydrodynamic forces. They do not have to contend with the inertial forces, which are, however, relatively small.

Figure 3.4 shows a simple model of oscillatory movement that will help us to establish general principles applying to a wide range of styles of locomotion, including the leg movements of horses, the shell movements of scallops, and the wing movements of insects. Two muscles that have elastic tendons contract alternately, oscillating a plate in a fluid. The plate has mass and the fluid exerts drag on it. In addition, there may be springs in parallel with the muscles that supply, at least in part, the forces needed to overcome the inertia of the plate. The muscles are required to vibrate the

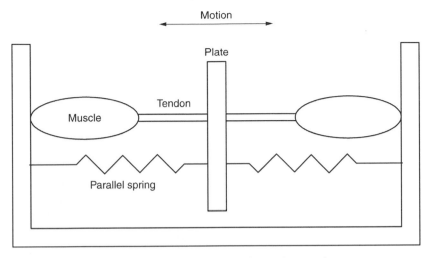

Fig. 3.4. A diagram of oscillatory movement driven by muscles.

plate at a specified frequency, with specified amplitude. Alexander (1997b) made calculations for this model, varying the unloaded shortening speed v_{max} of the muscles and the elastic compliance of their tendons. For each combination of speed and compliance I calculated the metabolic rate of the muscles, using Equations 2.3 and 2.10.

For each stage of a cycle of oscillation I calculated the forces that the muscles must exert. I also calculated the rates at which they would have to shorten or lengthen, taking account of the elastic stretching and recoil of their tendons. I put the force F and the rate of shortening v, obtained in this way, into Equations 2.3. This enabled me to calculate F_{iso}, an indicator of the cross-sectional area of muscle that would have to be active at that stage of the cycle to supply the required force. I then put the values of F_{iso} and v into Equations 2.10, which gave an estimate of the metabolic power needed at that stage of the cycle to power the movement. Finally, I averaged this power over a complete cycle to obtain the metabolic rate.

Results are shown in Fig. 3.5. Graph A refers to a negative ratio of inertial forces to drag. This is not impossible (it represents a spring–mass system driven below its natural frequency), but I know no examples in animal locomotion and will not discuss it further. Graph B represents a system with parallel springs driven at its natural frequency, such as a swimming scallop. The highest possible efficiency (star) is obtained with relatively fast muscles and no tendon compliance. In (C), peak inertial and drag forces are equal, as they are for the wing movements of some flying insects. The muscles should still be fairly fast, for maximum efficiency, but

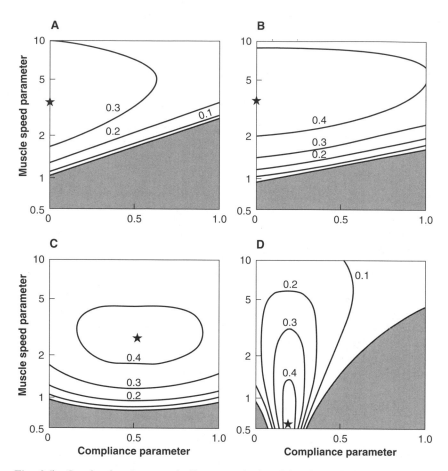

Fig. 3.5. Graphs showing metabolic rates calculated for the model of oscillatory movement (Fig. 3.4). In each graph, the vertical axis shows a muscle speed parameter, the v_{max} of the muscles divided by the peak speed of the plate. The horizontal axis shows a compliance parameter, the amount by which the peak hydrodynamic force would stretch the tendons, divided by the amplitude of the oscillation. The contours show efficiency calculated as (Work done against drag)/(Metabolic energy consumption). Efficiency is calculated in this way because the work against drag is the essential work that the muscle must do, whereas the need to do work against inertial forces can be eliminated by judicious choice of tendon compliance or of the compliance of parallel springs. Maximum efficiency is marked by stars. Shaded regions of the graphs are unattainable because the muscles would have to shorten faster than v_{max}, at some stage of the cycle. In (A), the ratio of peak inertial force to peak drag is −1; in (B) it is zero; in (C) it is +1; and in (D) it is +5. From Alexander (1997b).

there should in this case be some tendon compliance. In (D), inertial forces are five times greater than drag, just a little higher than has been estimated for some hovering hummingbirds. The optimum now requires a much slower muscle, but efficiency is in this case much less sensitive to muscle speed and much more sensitive to tendon compliance. Note the closely spaced contours, which show that a small change of compliance can result in a large change of efficiency.

We will refer back to Fig. 3.5 in later chapters, where we will discuss the various modes of oscillatory locomotion, but I have to admit a very serious gap in current knowledge. The swimming scallop seems to be the only case in which we know both the tendon compliance and the v_{max} of the muscle. Remember that inertial forces for the scallop seem to be zero, canceled out by elastic forces in its springlike hinge. The scallop's swimming muscle has no tendon, so tendon compliance is zero, as required for maximum efficiency (Fig. 3.5B). Also, v_{max} has approximately the value required for maximum efficiency. Thus, in the one case we know about, the properties of the muscle agree well with the prediction of the theory.

The highest attainable efficiency, marked by stars in Fig. 3.5, is between 0.43 and 0.45 in (B), (C), and (D), about the same as the maximum efficiency shown in Fig. 2.3D. It is only when the inertial forces are negative (Fig. 3.5A) that such high efficiencies cannot be attained.

Two words of caution seem necessary. First, we should not expect muscles to be adapted simply to maximize efficiency. Rather, we should expect muscle design to be a compromise between the requirements of efficiency, speed, acceleration, etc., as discussed in Chapter 1.

Secondly, the physiological data used to make the calculations for Fig. 3.5 came from experiments in which muscles made single contractions at constant speed. In oscillatory movements, muscles lengthen and shorten repeatedly at constantly changing speeds. We have noted in Section 2.3 that the forces exerted by muscles in work loop experiments can be significantly different from those predicted by Equations 2.3. Consequently, the results shown in Fig. 3.5 may be to some extent misleading. The approach used to calculate them seems to be the best currently available. Let us hope that advances in muscle physiology may soon make more reliable calculations possible.

In this chapter we have seen that muscles have to do work whenever the sum of the kinetic energy, the gravitational potential energy, and the elastic strain energy of the body are increased. They also have to do negative work, degrading mechanical energy to heat, when the sum of these energies is reduced. Additional work has to be done against friction when an animal slides over a solid surface, against viscous forces when it slides over a fluid layer, and against drag when it moves through air or water. Finally, we have seen how muscle and tendon properties can be adjusted to minimize the metabolic costs of oscillatory movements. I had to add a note of warning at the end of the discussion of oscillatory movements, because advances in muscle physiology are needed to enable us to do the calculations properly, as I explained at the end of Chapter 2.

· ·

Chapter Four

Consequences of Size Differences

D OMESTIC CATS and lions are very different in size, but they are similar in shape and move in similar ways. Both walk to go slowly, trot at intermediate speeds, and gallop to go fast. Small minnows and large salmon are similar in shape and make similar movements when they swim. Both hummingbirds and vultures fly by beating their wings.

There are important differences as well as similarities between the movements of animals of different sizes. In each of their gaits, lions run faster than cats and take longer strides, at a lower stride frequency. Minnows make more tail beats per second than salmon, and hummingbirds beat their wings at higher frequency than vultures. Hummingbirds hover but vultures cannot. Vultures soar and hummingbirds do not.

These examples suggest that we should want to understand how the structure of animals and their patterns of movement depend on body size. In this chapter I try to establish some of the basic principles that will be useful in the later chapters in which I discuss particular modes of locomotion. You will find further discussion of the consequences of size differences in McMahon and Bonner (1983), Schmidt-Nielsen (1984), and Brown and West (2000).

4.1. Geometric Similarity, Allometry, and the Pace of Life

Two shapes are geometrically similar if one could be made identical to the other by multiplying all length dimensions by the same factor. For example, a triangle with sides 3, 4, and 5 cm long is geometrically similar to one with sides of 6, 8, and 10 cm.

Imagine two animals that are geometrically similar to each other, one a precise half-scale model of the other. The larger one is twice as long as the smaller and has twice the circumference, so has $2 \times 2 = 4$ times the surface area. It is twice as long, twice as wide, and twice as high as the smaller one, so has $2 \times 2 \times 2 = 8$ times the volume and (if the animals are made of the same materials) eight times its mass. More generally, geometrically similar animals (or other objects) have areas proportional to

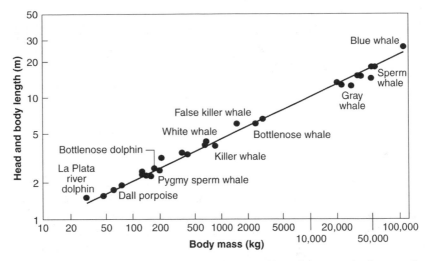

Fig. 4.1. A graph on logarithmic coordinates of length against body mass for whales. The slope of the regression line is 0.34. From Economos (1983).

(length)² and volumes proportional to (length)³; and if they are made of the same materials they have masses also proportional to (length)³. This implies that lengths are proportional to (mass)$^{1/3}$ and areas to (mass)$^{2/3}$.

This suggests that when we analyze relationships between body dimensions in animals of different sizes we may expect to find that our data can be approximated by equations of the form

$$y = ax^b \tag{4.1}$$

where y is a body dimension (perhaps length or area), x is another (perhaps volume or mass), and a and b are constants. Equations like this are called allometric equations. By taking logarithms of both sides of Equation 4.1, we get

$$\log y = \log a + b \log x \tag{4.2}$$

implying that a graph of log y against log x should be a straight line of gradient b. The allometric equation that best fits a set of data can be found by regression of the logarithms of the data. Least-squares (model 1) regression, or reduced major axis (model 2) regression may be the more appropriate, depending on the nature of the data and the purpose for which the equation is required (Rayner 1985a).

Figure 4.1 is a graph of length against body mass for whales. Distances along the axes are proportional to the logarithms of length and mass, not to length and mass themselves. Thus, the distance along the horizontal

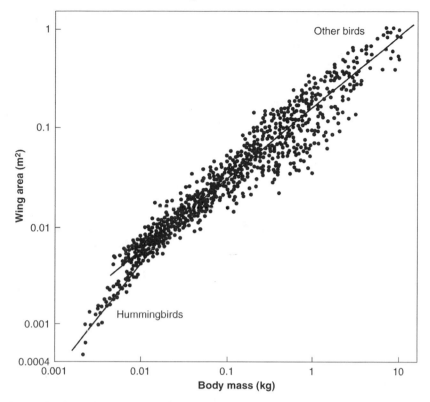

Fig. 4.2. A graph on logarithmic coordinates of wing area against body mass for birds. The slopes of the lines, fitted by reduced major axis regression, are 1.09 for hummingbirds and 0.72 for other birds. From Rayner (1987).

axis from 10 to 100 kg is the same as the distances from 100 to 1000 kg, and from 1000 to 10,000 kg. Drawn like this, the graph is equivalent to a graph of log(length) against log(mass). The equivalence of a graph on logarithmic coordinates, like this, and a graph of logarithms is made explicit in Fig. 4.3, which has a scale of logarithms at the bottom and a logarithmic scale at the top. If whales of different sizes were geometrically similar to one another, their lengths would be proportional to (body mass)$^{1/3}$, and all the points in Fig. 4.1 would lie on a line of slope 1/3. The slope of the line is actually 0.34, almost exactly as predicted.

In contrast, in Fig. 4.2 wing area is plotted against body mass for birds, again on logarithmic coordinates. If birds of different sizes were geometrically similar to each other, all the points would lie on a line of slope 2/3. In fact, the points form two lines of different slopes. The smallest birds are hummingbirds. The points for them (filled circles) are scattered around a

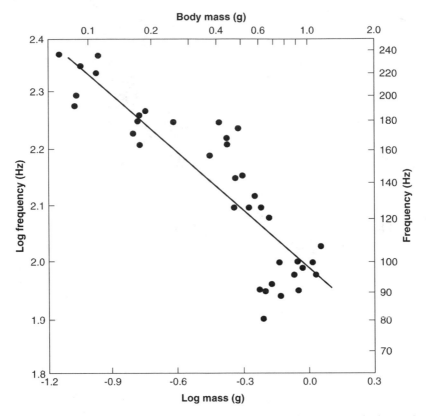

Fig. 4.3. A graph of the logarithm of wing beat frequency against the logarithm of body mass for euglossine bees. Scales of frequency and mass are also shown. From Casey et al. (1985).

line of slope 1.09, and the points for other birds around a line of slope 0.72. Both these slopes are significantly greater than the predicted slope of 2/3, showing that the wing areas of large birds are generally larger than they would be, if large birds were geometrically similar to small ones. Notice, however, that the points in Fig. 4.2 are quite widely scattered above and below the lines. This reflects differences between the birds' ways of life. For example, 10-kg vultures have wings of about twice the area of those of 10-kg albatrosses.

As well as being useful for describing how the dimensions of animals' bodies are related to body mass, Equation 4.1 is also often useful for describing how the rates of animal movements and of physiological processes are related to body size. Figure 4.3 shows that wing beat frequencies of a group of species of bees tend to be proportional to (body mass)$^{-0.35}$. Again there is a good deal of scatter about the line.

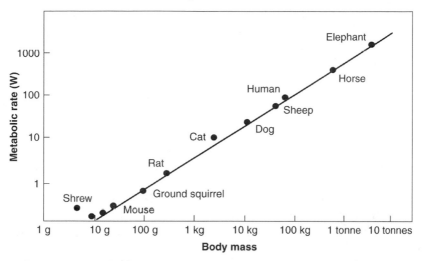

Fig. 4.4. A graph of logarithmic coordinates of resting metabolic rate against body mass for mammals. From Alexander (1999).

Figure 4.4 shows that the resting metabolic rates of mammals tend to be proportional to (body mass)$^{0.76}$. This is just one example of a general tendency: the resting metabolic rates of similar animals of different sizes tend to be about proportional to (body mass)$^{3/4}$ (Peters 1983). Maximum active metabolic rates of similar animals also tend to be roughly proportional to (body mass)$^{3/4}$. For example, Taylor et al. (1981) found that the maximum rates of oxygen consumption while running, of mammals ranging from mice to cattle and elands, were proportional to (body mass)$^{0.81}$.

If the metabolic rates of animals are proportional to (body mass)$^{3/4}$, metabolic rates per unit mass are proportional to (body mass)$^{-1/4}$. This fits in with a general tendency for the frequencies of animal movements to be about proportional to (body mass)$^{-1/4}$ and for the times required for biological processes to be about proportional to (body mass)$^{1/4}$. For example, the gestation periods of mammals are about proportional to (body mass)$^{0.24}$ and their heart beat frequencies to (body mass)$^{-0.25}$ (Peters 1983). If the energy used in each repetition of a movement (for example, a stride or a heart beat) is proportional to body mass, and the frequency of repetition is proportional to (body mass)$^{-1/4}$, the rate at which energy is used will be proportional to (body mass)$^{3/4}$.

However, there are some marked deviations from the general rule. Figure 4.3 has shown us that the wing beat frequencies of bees tend to be proportional to (body mass)$^{-0.35 \pm 0.06}$ (95% confidence limits), and Heglund et al. (1974) found that the galloping stride frequencies of mammals are proportional to (body mass)$^{-0.14}$ (confidence limits not calculated).

West et al. (1997) have tried to explain why metabolism and other bio-logical processes tend to proceed at rates proportional to (body mass)$^{-1/4}$. Their theory seems to have great explanatory power, but does not fully satisfy me, as I have explained elsewhere (Alexander 1999).

4.2. Dynamic Similarity

Lions are much larger than domestic cats and take fewer strides per second, but apart from that the movements of a galloping lion are very like those of a galloping cat. Large salmon beat their tails at lower frequencies than minnows, but may leave similar patterns of eddies in their wake. The con-cept of dynamic similarity will help us in comparisons like these and will enable us to make generalizations about the movements of animals of dif-ferent sizes.

Two shapes are geometrically similar if one could be made identical to the other by multiplying all lengths by some factor λ. By an extension of the same kind of thinking, two motions are dynamically similar if one could be made identical to the other by multiplying all lengths by a factor λ, all times by a factor τ and all forces by a factor ϕ. As an example of dynamically similar motion, think of two pendulums of different lengths swinging through the same angle.

What does dynamic similarity imply? If all lengths are multiplied by λ and all times by τ, all velocities must be multiplied by λ/τ and all accelera-tions by λ/τ^2. Newton's second law of motion tells us that force equals mass multiplied by acceleration, so if all forces are multiplied by ϕ, all masses must be multiplied by $\phi/(\lambda/\tau^2) = \phi\lambda/(\lambda/\tau)^2$. In other words,

$$\text{Ratio of masses} = \frac{\text{Ratio of forces} \times \text{Ratio of lengths}}{\text{Ratio of velocities}^2}$$

or

$$\frac{\text{Ratio of masses} \times (\text{Ratio of velocities})^2}{\text{Ratio of forces} \times \text{Ratio of lengths}} = 1 \qquad (4.3)$$

This tells us that for two motions to be dynamically similar, the following condition must be satisfied. Let m_1, m_2 be corresponding masses in the two motions (for example, the masses of corresponding parts of two animals' bodies); let v_1, v_2 be corresponding velocities (for example, the velocities of corresponding body parts at corresponding stages of the mo-tion); let F_1, F_2 be corresponding forces (for example, peak forces on the feet) and let l_1, l_2 be corresponding lengths (for example, stride lengths). If the motions are dynamically similar,

$$\frac{m_1 v_1^2}{F_1 l_1} = \frac{m_2 v_2^2}{F_2 l_2} \qquad (4.4)$$

Both motions must have the same value of mv^2/Fl.

This must be true for all the kinds of forces that are important for the motion. Suppose, for example, that gravitational forces are important, as they are for running mammals. The force F exerted by gravity on a mass m is mg, where g is the gravitational acceleration. Thus, $mv^2/Fl = v^2/gl$. When gravity is important, motions can be dynamically similar only if they have equal values of v^2/gl, a quantity that is called a Froude number.

This rule helps us to predict the speeds at which terrestrial animals change gaits. Quadrupeds walk at low speeds, trot at intermediate speeds, and gallop at high speeds. Walking, trotting, and galloping are markedly different patterns of movement, as every horse rider knows, and as the descriptions in Section 7.2 will show. Alexander and Jayes (1983) formulated the hypothesis that quadrupeds tend where possible to move in dynamically similar ways, which implies, among other things, that they will change gaits at equal Froude numbers. We took v to be running speed and l to be leg length (or, more precisely, the height of the hip joint from the ground in normal standing). We analyzed film of a wide variety of mammals ranging in size from small rodents to rhinoceros and concluded that in almost every case the change from trotting to galloping was made at a Froude number between 2 and 3. Our data seem also to show that the changes from walking to trotting in quadrupeds, from walking to running in humans, and from shuffling to hopping in kangaroos are all generally made at Froude numbers between about 0.3 and 0.8.

So far, we have assumed that gravitational forces are important. Now we will consider instead motions, such as swimming, in which viscous forces are important. Figure 3.2B showed a plate of area A moving with velocity v over a layer of thickness d of a fluid of viscosity η. Equation 3.7 told us that the force required to drive this motion is $\eta A v/d$. In dynamically similar motions, corresponding areas are proportional to the squares of corresponding lengths l^2, and corresponding thicknesses must be proportional to l. Thus, forces are proportional to $\eta l v$. Also, the masses of corresponding regions of fluid are proportional to ρl^3, where ρ is the density of the fluid. Thus, mv^2/Fl (Equation 4.4) is proportional to $(\rho l^3 v^2)/(\eta l v l) = \rho l v/\eta$. The quantity $\rho l v/\eta$ is called a Reynolds number. Motions in which viscosity is important can be dynamically similar only if their Reynolds numbers are equal.

The fluids that will concern us most in our discussions of locomotion are air and water. To calculate Reynolds numbers in them we need to know values of η/ρ, the quantity that is known as kinematic viscosity. For air at

20°C at a pressure of one atmosphere, it is 1.5×10^{-5} m²/s; and for fresh-water or seawater at 20°C it is 1.0×10^{-6} m²/s (Denny 1993).

Reynolds numbers will appear in our discussions both of swimming and of flight. For example, fluid flow in the boundary layer around a streamlined body of length l traveling at velocity v becomes turbulent if the Reynolds number $\rho l v / \eta$ rises above about 2×10^6, causing an abrupt increase of drag. We will need to know the range of Reynolds numbers involved when we discuss the drag that acts on swimming dolphins (Section 14.4).

We have considered motions in which gravitational forces are important and ones in which viscous forces are important, and turn now to elastic forces. A force $S \Delta l$ is needed to stretch a spring of stiffness S by an amount Δl (Equation 3.4). In dynamically similar motions, extensions Δl will be proportional to lengths l, so forces will be proportional to Sl. Thus, $mv^2/Fl = mv^2/Sl^2$, and the condition for dynamic similarity (Equation 4.4) is that the motions being compared must have equal values of mv^2/Sl^2. The natural frequencies of vibration of spring–mass systems are proportional to $(S/m)^{0.5}$, so this implies that the motions must have equal values of v^2/f^2l^2, where f is the natural frequency of the system; hence, they must have equal values of fl/v, which is called the Strouhal number. The reduced frequency referred to in many discussions of animal swimming and flight is simply 2π times the corresponding Strouhal number.

Strouhal numbers are applicable to cyclic motions in general, whether or not elastic forces are important. Consider two dynamically similar motions that repeat in regular cycles. All times in one of them are τ times corresponding times in the other, and all lengths are λ times corresponding lengths. Thus, frequencies are proportional to $1/\tau$ and velocities to λ/τ. It follows that fl/v is proportional to $(1/\tau)\lambda/(\lambda/\tau)$: in other words, it is constant. Any two dynamically similar cyclic motions must have equal Strouhal numbers. I will show in Section 11.1 that hovering humming-birds of different sizes beat their wings at frequencies that make their Strouhal numbers about equal.

Froude numbers, Reynolds numbers, and Strouhal numbers are all dimensionless; they have no units. However, it is essential when calculating them to use a consistent system of units. The easiest way to do this is to express everything in SI units: lengths in meters (not millimeters or kilometers), times in seconds, masses in kilograms, forces in newtons, etc.

4.3. ELASTIC SIMILARITY AND STRESS SIMILARITY

McMahon (1973) suggested that animals and plants of different sizes should be built in such a way as to deform under gravity in geometrically similar ways; gravity should cause equal strains in corresponding parts of

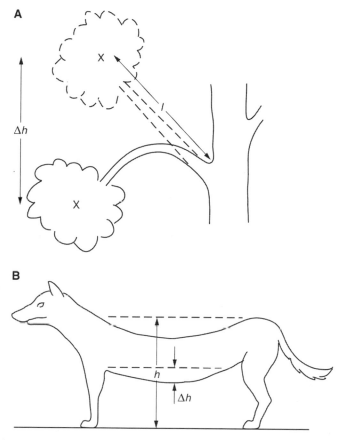

Fig. 4.5. Diagrams of (A) a branch and (B) a mammal, showing elastic deformations due to gravity. Broken outlines show their shapes as they would be in the absence of gravity.

their bodies. For example, Fig. 4.5A shows a branch of length l that bends elastically under its own weight, depressing its end by Δh. The theory of elastic similarity predicts that branches of different sizes will have equal values of $\Delta h/l$.

The theory is an attractive one for trees. Gravity is one of the principal forces they have to withstand (the other is drag exerted on them by wind). Less energy is needed to grow a thin branch than to grow a thick one of equal length, but too thin a branch will bend so much that its leaves are not well positioned to receive sunlight. It seems likely that the optimum compromise would result in branches of different sizes being elastically similar. McMahon and Kronauer (1976) published evidence that branches of different sizes are indeed proportioned so as to be, more or less, elastically similar.

The theory is less attractive for animals. McMahon (1975) applied it to the leg bones of mammals, but elastic deformation of leg bones is not a problem in any mammal known to me. McMahon (1973) also applied it to the trunks of mammals, arguing that these should be proportioned so as to sag under gravity in geometrically similar ways, with equal values of $\Delta h/h$ (Fig. 4.5B).

Geometrically similar structures have lengths and diameters proportional to (body mass)$^{1/3}$, but McMahon showed that his theory predicted that leg bones and trunks should have lengths proportional to (mass)$^{1/4}$ and diameters proportional to (mass)$^{3/8}$. This would make them relatively shorter and stouter in larger animals. He found good agreement with these predictions, both for the leg bones of Bovidae (cattle and antelopes [McMahon 1975]) and for the chests of Primates (McMahon 1973). However, in other groups of mammals (Alexander 1979a), and especially in smaller mammals (Economos 1983), leg bones scale more nearly as predicted for geometric similarity.

Because leg bones bend only by amounts that seem trivial, I am inclined to think their proportions more likely to depend on the need to be strong enough, than on the need to be stiff enough. We will examine a theory along these lines shortly.

The theory of elastic similarity seems more promising for structures that undergo substantial elastic deformations in life, for example, the leg tendons of mammals. Consider two mammals of different sizes that run in dynamically similar ways, and suppose that their leg tendons stretch to equal strains. Dynamic similarity implies forces proportional to body weight, so forces proportional to body weight would be causing similar elastic deformations; the animals would have to be elastically similar. However, we will see in Section 7.4 that the leg tendons of kangaroo rats do not stretch to the same strains as those of kangaroos, when they hop like kangaroos.

Biewener (1989, 1990) shifted the emphasis, in discussions of mammal leg design, from elastic strain to stress. He suggested that forces proportional to body weight should set up equal stresses in the skeletons and muscles of mammals of different sizes; mammals of different sizes should show stress similarity. In animals built of the same materials, equal stresses imply equal elastic strains, so stress similarity and elastic similarity are two aspects of the same design principle. The difference of viewpoint is nevertheless important. There was no obvious reason why the tiny elastic strains that occur in leg bones should be expected to be the same in animals of different sizes, but there is a clear reason why we might expect to find equal stresses. Bones must be strong enough to withstand the forces that act on them. Leg bones of different sized mammals are built of essentially the same material, capable of withstanding the same stress.

Figure 4.6A represents an imaginary one-legged terrestrial animal that will help us to work out how stress similarity might be possible. A vertical force equal to the animal's weight mg acts on the foot, exerting a moment $mgl \sin \theta$ about the joint halfway up the leg. The muscle has cross-sectional area A and has a moment arm r about the joint, so when it exerts a stress σ the force is $A\sigma$ and the moment about the joint is $A\sigma r$. Balancing the moments about the joint gives

$$A\sigma r = mgl \sin \theta \tag{4.5}$$

$$\sigma = \frac{mgl}{Ar} \sin \theta$$

If animals of different sizes were geometrically similar, A would be proportional to (body mass)$^{2/3}$, l and r would be proportional to (body mass)$^{1/3}$, and θ would be constant; stresses in a 3000-kg elephant would be ten times as high as in a 3-kg rabbit.

How could animals be built to avoid this unacceptable result? I will make two assumptions. The first of these is that muscle mass is the same proportion of body mass in animals of all sizes. I assume this because muscles make up so large a proportion of body mass in mammals of all sizes that there can be little scope for increase. For example, a blacktail jackrabbit (*Lepus californicus*) was found to have 46% muscle in its body (Grand 1977). Secondly, I assume that the fascicles of the muscles have lengths proportional to the moment arm r. This implies that if the fascicles of different-sized animals shorten by the same fraction of their length, they will move the joint through the same angle. Together, these two assumptions imply that Ar is proportional to the volume, and hence the mass, of the body. Hence, from Equation 4.5, the stress in the muscle is proportional to $l \sin \theta$.

This tells us that one way of making muscle stress the same in animals of different sizes would be to keep l constant, but that would not be feasible; the legs of an elephant could not be made as short as those of a rabbit. Alternatively, θ could be made smaller in larger animals. This seems to be the case. Larger mammals generally do stand and run on straighter legs than small ones. Elephants hold their legs straighter than rabbits.

Let us look at the terms in Equation 4.5 and see how they actually scale. Biewener (1989) made measurements of the extensor muscles of the principal limb joints of mammals ranging in size from mice to horses. He described $r/(l \sin \theta)$ as the effective mechanical advantage of a muscle and found that it was about proportional to $m^{0.26}$, where m is body mass. Alexander et al. (1981) dissected mammals ranging from shrews to an elephant and found that limb muscles generally had cross-sectional areas about proportional to $m^{0.8}$. Hence, by Equation 4.5, muscle stresses should

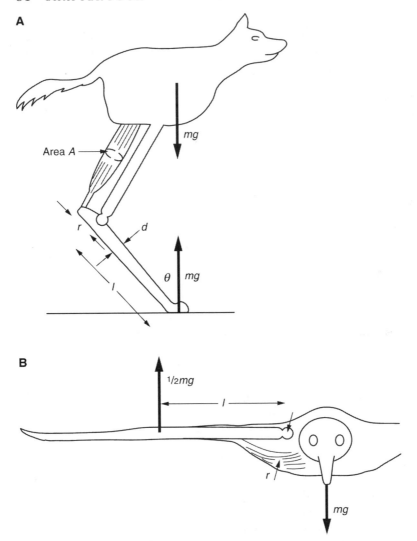

Fig. 4.6. Diagrams of (A) a terrestrial animal, represented as having only one leg, and (B) a flying animal seen in front view. These diagrams are used in a discussion of the stresses in bones and muscles of animals of different sizes.

be about proportional to $m/(m^{0.8}m^{0.26}) = m^{-0.06}$; they should be somewhat smaller in larger animals. The straightness of their legs is more than adequate to explain how large mammals can be supported by their leg muscles.

Now consider how bones should be built to ensure that they are strong enough to support animals of all sizes. The force mg on the foot (Fig. 4.6A) has an axial component $mg \cos \theta$ along the length of the lower leg bone and a transverse component $mg \sin \theta$ at right angles to it. The axial component compresses the bone along its length and the transverse component bends it. Calculations of the stresses in limb bones in locomotion tell us that the transverse components of force generally give rise to much larger stresses than the axial components (Rubin and Lanyon 1982). This should not surprise us; long, slender structures such as bones or sticks are more easily broken by bending than by axial compression. For this reason, we will consider only stresses due to transverse forces.

The transverse component of force in Fig. 4.6A exerts bending moments on the lower leg bone, which increase from zero at the distal end of the bone to a maximum value of $mgl \sin \theta$ close to the joint. The peak stress in a cross section of a bent beam (or a bent bone) is Bending moment/Section modulus. Section modulus is a quantity that depends on the diameter and shape of the cross section; for cross sections of the same shape, it is proportional to $(\text{diameter})^3$ (Alexander 1983). Thus, the peak stress in the bone is proportional to $(mgl/d^3) \sin \theta$. If animals of different sizes were geometrically similar to each other, l and d would both be proportional to $m^{0.33}$ and θ would be constant, so the stress would be proportional to $m^{0.33}$. Either rabbits would have bones far stronger than necessary, or elephants would fracture their bones when they tried to stand.

Just as we found for muscles, the tendency for larger mammals to stand and run on straighter legs helps to avoid excessive stresses in bones. We have already seen that $r/(l \sin \theta)$ is about proportional to $(\text{body mass})^{0.26}$. Biewener (1990) found that muscle moment arms r tend to be proportional to $m^{0.44}$, making $l \sin \theta$ proportional to $m^{0.44}/m^{0.26} - m^{0.18}$. Alexander, Jayes et al. (1979a) found that the diameters of mammal leg bones are generally about proportional to $m^{0.36}$. Thus, $(mgl/d^3) \sin \theta$ is expected to be proportional to $mm^{0.18}/(m^{0.36})^3 = m^{0.10}$. Bone stress should be proportional to $m^{0.10}$, implying that it should increase with animal size, but not by nearly as much as if mammals of different sizes were geometrically similar.

These arguments suggest that muscle stresses in standing animals should decrease slowly with increasing body size, in proportion to $m^{-0.06}$, and that bone stresses should increase slowly, in proportion to $m^{0.10}$. There is too much uncertainty about both exponents for us to be confident that either is different from zero, so our conclusion should probably be simply that bone and muscle stress change far less with changing body size than they would if mammals of different sizes were geometrically similar.

We have been thinking of the weight of the body as the load on an animal's legs. This is correct for standing, but much larger forces act on feet in running and jumping. For example, the peak forces on the feet of a galloping greyhound were four times as high as when it was standing still (Bryant et al., 1987). The two hind feet of bushbabies (*Galago moholi*) taking off for a jump together exerted a force of up to 13 times body weight (Günther et al. 1989).

The faster an animal runs, the lower the duty factor (the fraction of the duration of the stride for which each foot is on the ground). The force exerted on the ground, averaged over a complete stride, must match body weight, so the lower the duty factor, the larger the forces that must be exerted while the foot is on the ground; fast running requires large forces. Alexander et al. (1977) filmed African ungulates, ranging from small gazelles to giraffes, galloping fast in their natural habitat. We found that duty factors were about proportional to (body mass)$^{0.14}$, implying that the forces exerted by the larger animals were smaller multiples of body weight. In contrast, Bennett (1987) filmed kangaroos of different sizes hopping fast and found that duty factors were proportional to (body mass)$^{-0.10}$, implying that the forces on the feet of larger kangaroos were larger multiples of body weight. I do not know which group is more typical of mammals in general. However, the very largest terrestrial mammals, rhinoceros and elephants, are less athletic than smaller ones, presumably because their legs are not strong enough to exert such large multiples of body weight. Elephants neither gallop nor jump. The lowest duty factor that I have observed for an elephant (0.49 [Alexander 1979b]) is much higher than the duty factors of around 0.2 that Alexander et al. (1977) observed for small antelopes galloping fast.

Flying animals as well as running ones have to support their own weight. A flying bird of weight mg requires an upward lift $mg/2$ on each wing (Fig. 4.4B). This lift is distributed along the length of the wing, but the moment it exerts about the shoulder is the same as if the whole force acted at the center of pressure, at a distance l from the wing base. Thus, the moment is $mgl/2$. It must be balanced by the pectoralis muscle, which has cross-sectional area A and moment arm r, and exerts stress σ. Hence,

$$\sigma = \frac{mgl}{2Ar} \qquad (4.6)$$

The argument that we used in our discussion of Equation 4.5 tells us that if birds of different sizes were geometrically similar, muscle stress would be proportional to $m^{0.33}$; stresses would be ten times as high in a 10-kg swan as in a 10-g tit. Birds of different sizes are not geometrically similar. Wingspan, and therefore the length l, tends to be proportional to $m^{0.39}$ (Rayner 1987), but that deviation from geometric similarity makes the problem worse rather than better. Not enough seems to be known about the scaling of bird wings and wing muscles to tell us how muscle stresses actually scale in birds of different sizes.

This chapter has introduced the concepts of geometric similarity, dynamic similarity, elastic similarity, and stress similarity. These concepts will be very helpful when we compare the structure and movements of animals of different sizes, but it is important to remember that no animal is a precisely scaled model of another. We may discuss what the consequences would be if animals were, for example, geometrically similar or moved in dynamically similar ways, but the similarity is never exact. The chapter has also explained some dimensionless numbers that will be important when we compare the locomotion of animals of different sizes.

· ·

Chapter Five

Methods for the Study of Locomotion

*T*HIS CHAPTER outlines some of the principal techniques that have been used in the research that is described in later chapters. Many of them have been used in research on several different modes of locomotion, described in separate chapters. My aim is to help readers to understand what was done in the experiments that I will describe, not to supply the practical details they will need if they wish to perform similar experiments themselves.

5.1. CINEMATOGRAPHY AND VIDEO RECORDING

Cinematography or video recording is often the best way of recording movement. Cinematography has had a great deal of use in the past, but film and processing are expensive, and processing takes time. Video recording has become by far the more common means of making moving images of locomotion, especially since high-speed video cameras have become available. Standard video cameras record only 30 frames per second (in North America) or 25 frames per second (in Europe), so are much too slow to record fast animal movements such as the wing beats of insects. Weis-Fogh (1973) took cine film at 7150 frames per second to capture the details of the wing movements of a tiny wasp (*Encarsia*) that beats its wings with a frequency of 400 Hz. That gave him 18 frames for each wing beat cycle. As a rough general rule, about 20 frames per cycle of movement will usually provide the information that a researcher needs.

That example was of a very fast movement that required an exceptionally fast framing rate. High-speed video cameras taking up to 500 frames per second are readily available, though expensive. Faster cameras are seldom needed in research on animal movement.

A single view can provide information only about movement in two dimensions. For three-dimensional information, two views are needed. Sometimes it is convenient to record the two views with one camera; for example, a side view and a top view of an animal can be obtained in the same frame if a mirror is set at 45° in the field of view. Alternatively, two synchronized cameras may be used. It may seem convenient to set these up at 90° to each other so that one shows a side view and the other a front or top view of the animal, but this is not essential. Software is available for extracting three-dimensional information from two cameras set at any angle to each other. It is sometimes necessary to have more than two cameras if there is a danger of the points of interest getting hidden at some stages of the movement behind other parts of the body.

It is often helpful to put marks on the points of interest on the animal's body. For example, in many studies of running, marks have been put on the skin over the principal joints of the legs. In some cases these marks have been painted on; in others, adhesive markers have been used. One of the disadvantages of using skin markers is that a marker that is directly over the center of a joint in one frame may be well away from it in another, if the skin moves relative to the underlying skeleton.

Quantitative analysis of a film generally depends on determining the coordinates of a number of points on the body, in successive frames of the film. This can be done by displaying each frame in turn on a video monitor, and placing a digitizing cursor over each point of interest in turn. This can be very laborious, if the numbers of frames and of points of interest are large. However, several commercially available systems will do the job automatically, finding the markers in each frame of the film and recording their coordinates.

Data obtained by film analysis can be used to calculate the forces acting on the body and the body's kinetic and gravitational potential energy at successive stages of the movement. The body is treated as an assembly of rigid segments (thigh, lower leg, foot, etc.). The mass and moment of inertia of each segment must be known, and the position of the center of mass within the segment. The coordinates of the joints between segments, measured from the film, can be used to calculate the velocity and angular velocity of each body segment at each stage of the motion, and so its kinetic energy. Further calculation gives the acceleration and angular acceleration of each segment, from which forces can be calculated. For animals with many body segments that can be moved relative to each other, the procedure requires collecting and processing a great deal of information (see Winter [1990] for more detail). It is prone to error, because the process of differentiation that gets velocities from successive positions, and accelerations from successive velocities, magnifies random errors. This kind of analysis should not be undertaken lightly.

X-ray cinematography systems are available, and are sometimes very useful. For example, Jenkins et al. (1988) took X-ray cine film of starlings in flight, and were able to show how the wishbone bends in each wing beat cycle. They took their film at 200 frames per second, obtaining about 15 frames for each wing beat cycle. Some of the limitations of X-ray cinematography are that very high framing rates are not available, resolution is not very good (at least in the films that I have seen), and only a small field of view can be filmed. Pictures can be obtained of the whole body of a starling, or of the thorax of a labrador dog (Bramble and Jenkins 1993), but not of the dog's whole body.

5.2. STATIONARY LOCOMOTION

Research on animal locomotion often involves connecting the animal to large items of equipment. For example, a researcher may want to collect the animal's breath and pass it along a tube to gas analysis equipment, so that its rate of oxygen consumption can be measured. Alternatively, he or she may want to have wires connecting electrodes in the animal's muscles to recording equipment, to show when the muscles are active. For experiments like these, it is often inconvenient to have the animal moving around. However, Langman et al. (1995) collected the breath of an elephant as it walked round a zoo, by placing the gas analyzer on a golf cart that was driven alongside the animal.

Figures 5.1 and 5.2 show how a running, swimming, or flying animal can be kept stationary relative to the laboratory. In each case the diagram shows provision for measuring the rate at which the animal is using oxygen, but discussion of this is deferred to the next section. In Fig. 5.1A the animal is running on a treadmill (a moving belt), matching its speed to the speed of the belt so as to stay on the belt. A wide variety of mammals, birds, and reptiles have been trained (usually without difficulty) to do this, but it is sometimes advisable to have a safety rope to prevent the animal from falling off the belt. With a sufficiently large, fast treadmill it is possible to have a racehorse galloping in a laboratory (Young et al., 1992). Much smaller treadmills have been used for running insects (Full et al., 1990).

When an animal is running on fixed ground out of doors, its movement and any wind together ensure that air is moving around its body. There is no such air movement around an animal running on a treadmill indoors, unless a fan is used, as shown in the diagram. The air movement may be important to ensure that the animal does not overheat. It will generally have little effect on the energy cost of locomotion, unless a strong wind is simulated.

Figure 5.1B shows how a flying animal can be kept stationary in a wind tunnel. A powerful fan draws a current of air through the tunnel, and the animal matches its speed to the air flow, so as to remain stationary in the tunnel. Many birds and bats have been trained to do this. The grid

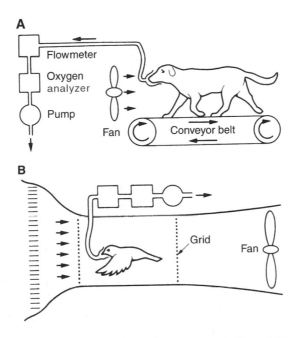

Fig. 5.1. Diagrams of (A) a mammal running on a treadmill, and (B) a bird flying in a wind tunnel. In both cases the air that the animal breathes out is drawn through an oxygen analyzer, so that the rate at which the animal is using oxygen can be measured. From Alexander (1975a).

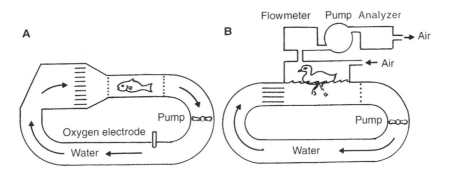

Fig. 5.2. Diagrams of (A) a fish swimming in a water tunnel and (B) a duck swimming in a flume. Provision is made, in both cases, for measuring the animal's rate of oxygen consumption. From Alexander (1999).

behind the animal ensures that it cannot be blown backward and chopped up by the fan.

The aim in experiments like this is generally to simulate flight through still air. For air flow round the stationary bird to be the same as if it were moving through still air, the flow in the working section of the tunnel must be smooth and steady and have the same velocity throughout the cross section of the tunnel. Wind tunnels are carefully designed to achieve this as nearly as possible. If they were parallel-sided throughout, the air would flow more slowly near the walls than at the center. The marked narrowing at the entrance to the working section largely eliminates this problem. The honeycomb (represented by a band of parallel lines across the entrance to the tunnel) is a grid of metal plates that helps to make the air flow parallel to the axis of the tunnel, without swirling around.

The cross section of the tunnel must be large compared to the animal's wingspan. Otherwise, the aerodynamic forces on the animal are affected by interaction with the tunnel walls (Rayner 1994). The effects are small if the diameter of the working section is at least 2.5 wingspans and the animal flies near the center of the tunnel, but large wind tunnels are expensive, and many experiments have been done in tunnels much smaller than the ideal. An excellent, unusually large wind tunnel at Lund (Pennycuick et al., 1997) has a working section 1.2 m wide and is used for birds up to the size of small ducks. Ducks fly fast, but the fan motor of this tunnel is powerful enough to blow air through it at speeds up to 38 m/s, which is sufficient. Very much smaller, slower tunnels suffice for experiments with insects (Dudley and Ellington 1990a).

Figure 5.2A shows a fish swimming against a current of water in a water tunnel. Fish quickly learn to match their swimming speed to the flow. The principle is the same as for the wind tunnels used in experiments on flight, but the water is recirculated instead of being allowed to flow away. (Similarly in some wind tunnels, the same air is used repeatedly.) As in wind tunnels, a marked reduction of diameter at the entrance to the working section helps to make the velocity uniform across the cross section, and a honeycomb reduces swirling. Also as in wind tunnels, it is important to have the diameter of the working section large enough to avoid serious hydrodynamic interactions between the animal and the tunnel wall. Much research on fish swimming has been done in undesirably narrow tunnels, in which significant artifacts are inevitable (Webb 1993).

Figure 5.2B shows the equivalent experiment for a duck swimming on the water surface. In the working section, the water has a free surface exposed to air, so this is a flume rather than a water tunnel.

5.3. Measurement of Energy Consumption

Researchers often want to know how much energy is being used for loco-motion. By far the commonest way of doing this is by measuring the rate at which the animal is using oxygen. This works, of course, only when locomotion is being powered by aerobic metabolism. For example, it can be used to measure energy consumption by humans running at speeds up to about 6 m/s, the highest speed that athletes can sustain aerobically. It is not suitable for measuring the energy cost of sprinting, which can involve speeds exceeding 10 m/s.

Conveniently for physiologists, a liter of oxygen releases almost the same amount of energy, whatever food is being oxidized. For example, it releases 20.9 kJ if glucose is oxidized, and 19.6 kJ if palmitic acid (a fatty acid) is oxidized. More than twice as much oxygen is needed to oxidize a gram of fatty acid as to oxidize a gram of glucose, but more than twice as much energy is released, and the energy per unit volume of oxygen is almost the same.

In Fig. 5.1, the dog and the bird each has a mask attached over its face. The masks fit loosely, so air can flow in freely round their edges. A pump draws air through the mask and on through a flowmeter and an oxygen analyzer. The flow must be fast enough to ensure that it carries with it all the air that the animal breathes out; it does not matter that air that the animal has not breathed also passes through the analyzer. The flowmeter records the volume of air that is analyzed, and the analyzer measures the concentration of oxygen in it. Hence, knowing the composition of fresh air, it is possible to calculate how much oxygen has been removed. This is the oxygen that has been used in the animal's metabolism.

In Fig. 5.2B, the same principle is used to measure the oxygen consump-tion of the duck, but instead of wearing a mask the whole animal is covered by a hood. Hoods covering the whole animal are also used in some experi-ments with animals running on treadmills. It is generally convenient to use masks for large animals (such as horses) and whole-body hoods for small ones (such as mice).

In Fig. 5.2A the water is being recirculated, so the concentration of dissolved oxygen in it is gradually reduced by the animal's metabolism. Plainly, it must not be allowed to fall too far. An oxygen electrode measures the falling oxygen concentration. If the volume of water in the system is known, the volume of oxygen used can be calculated.

The methods that have been described so far are suitable for laboratory experiments, but not for measuring the oxygen consumption of animals moving around in the field. This can be done by the doubly labeled water technique. The animal is given an injection of water labeled with isotopes both of hydrogen and of oxygen, that is, of water to which both ^2H_2O and $H_2^{18}O$ have been added. The animal is released for a period of hours or days and then recaptured, and the concentrations of the isotopes in a sample of its blood are measured. While free it will have lost both ^2H and ^{18}O in the water that has left its body by evaporation and as urine. It will also have lost ^{18}O in the carbon dioxide produced by respiration. If the changes in concentration of both isotopes are known, the quantity of carbon dioxide produced can be calculated. This is not quite as useful as it would be to be able to calculate oxygen consumption, because it does not give as reliable an indication of the metabolic rate. Production of a liter of carbon dioxide is accompanied by the release of 28 kJ of energy if fat is being oxidized, but only 21 kJ if it comes from oxidation of carbohydrate. Nevertheless, a fairly good estimate of the metabolic rate is generally possible. A disadvantage of this method is that it gives only the total amount of energy used over a period of hours or days, with no indication of fluctuations of metabolic rate within that period.

It is generally easier in field experiments to measure heart beat frequency than to measure oxygen consumption. Animals' hearts generally beat faster while their metabolic rates are high, and oxygen consumption can be calculated from heart beat frequency if the two variables can be shown to be well correlated. This method has been used, for example, to calculate metabolic rates for free-ranging albatrosses (Bevan et al., 1995). Signals from electrocardiograph electrodes were recorded as explained in Section 5.7.

Another method that can be used to determine an animal's metabolic rate is to measure its heat output. If the animal is in a steady state, with constant body temperature, the metabolic rate equals the rate of heat loss plus the rate at which the animal is doing mechanical work on the environment. Because muscles generally work with efficiencies of 25% or less, the work output is much smaller than the heat output. Ward et al. (1999) used an infrared thermograph to make thermal images showing the distribution of surface temperature on flying starlings. From these they calculated the rate of heat loss from the body.

5.4. Observing Flow

Flying and swimming animals set the air or water around them moving. Observation of the fluid movements can provide valuable information. For example, observation of the moving fluid in the wakes that flying birds and bats leave behind them showed that they use different gaits at different flying speeds (Section 12.1). Observation of the flow of air over the wings of flying moths revealed the aerodynamic mechanism that enables them to fly (Section 11.1). Much of the work required for flight and swimming is used giving kinetic energy to the wake, so if that energy can be measured it provides information about the work that the animal has to do. For these reasons, it is often useful to be able to observe or measure the flow in the fluid around the animal.

One approach is to use flowmeters, such as electrically heated thermistors. A thermistor is a small bead of metal oxide or oxide mixture with electrical leads attached. The electrical resistance of thermistors falls as the temperature rises. When an electrically heated thermistor is placed in moving air or water, the flow cools it and its resistance increases. The change in resistance gives a measure of the rate of flow. Wood (1970) used a flowmeter of this kind to measure the velocity of air flow in the wake of the beating wings of tethered flies. A serious disadvantage of this approach is that a flowmeter can record the flow in only one place at a time.

It is often more useful to make the flow visible, by suspending particles in the fluid. Plainly, particles should be chosen that will move with (as nearly as possible) the velocity of the surrounding fluid, neither sinking nor rising through the fluid. Ideally, the particles should have the same density as the fluid. Airflow in the wake of flying birds and bats has been made visible by filling the room with a cloud of tiny soap bubbles filled with a mixture of helium and air. Ordinary (air-filled) soap bubbles sink, and helium-filled bubbles rise, but the mixture was adjusted to match the density of the bubbles to the air (Spedding et al., 1984). For making water movements visible, polystyrene or nylon spheres are available that have densities sufficiently close to that of water, with diameters ranging from a few micrometers upward.

The tiny water droplets that form fog sink only exceedingly slowly in air, because they are so small. Similarly, if the particles used to make flow visible are small enough it does not matter what their density is. A smoke-like suspension of minute oil droplets has been used to make visible the flow of air around the wings of moths in tethered flight (Willmott et al., 1997).

The flow may be observed in three dimensions by using two cameras viewing it from different directions. Alternatively, a plane in the field of view may be illuminated by a thin sheet of light, so that only particles in the sheet of light are visible. A picture of the whole field of flow can be built up from repeated observations with the sheet of light vertical or horizontal, and at different positions in the field.

Identifying the same particles and measuring their positions in successive frames of a film is difficult and very time-consuming. Fortunately, it can be done automatically, by the technique of particle image velocimetry. This produces plans of a plane in the field of flow, showing the speed and direction of flow at many different points in the plane. It has been used, for example, to analyze the flow in the wake of swimming fish (Müller et al., 1997).

5.5. FORCES AND PRESSURES

Transducers are devices that translate a signal into a different modality; for example, force transducers give an electrical output in response to applied forces. Many of the force transducers used by biologists incorporate strain gauges, which do not measure forces directly but register stretching and compression. For example, a horizontal beam is bent by vertical forces acting on its end. If strain gauges are glued to its upper and lower surfaces, one of them will be stretched and the other compressed when the beam is bent. The electrical output of the strain gauges can be used to measure the force on the beam.

A strain gauge is a strip of thin metal foil, or a thin slice of semiconductor material, mounted on a plastic backing. Stretching increases the electrical resistance of metal foil strain gauges, but decreases the resistance of semiconductor gauges.

The forces that animals exert on the ground can be measured by means of force plates, which are instrumented panels set into the floor. They are usually rectangular, with a force transducer under each corner. These may be beams with strain gauges bonded to them, or piezoelectric devices. Ideally, each transducer should register components of force along three mutually perpendicular axes; these are usually forces parallel to the direction of locomotion, transverse forces, and vertical forces. The force plate is then capable of reporting not only the magnitude and direction of any force that acts on it, but also the coordinates of the point on the platform at which the force acts. The vertical component of force on the plate is the sum of the vertical forces registered at the four corners, and similarly for the forward (or backward) component and the transverse component. If the point of action of the force is near the front of the plate, the transducers at the front corners will register larger forces than the ones at the rear; and if the point of action is on the left side of the plate, the transducers on the left will register larger forces than those on the right. Thus, the coordinates of the point of action on the plate can be calculated. In addition, the horizontal components of the forces at the four corners can be used to calculate the moment exerted by the forces on the plate, about a vertical axis. All these calculations are performed automatically by a computer connected to the force plate.

Figure 5.3 is a force plate record of a sheep walking across a force plate. Only four of the six outputs of the plate are shown; the transverse compo-

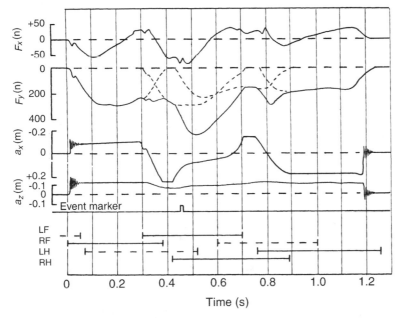

Fig. 5.3. A record of the output from a force plate, as a sheep walked over it. F_x is the component of force in the forward direction; F_y is the vertical component of force; a_x and a_z are the coordinates in the forward and transverse directions of the point of application of the force on the plate; and the event marker is a signal used to synchronize the record with a cine film taken simultaneously. Bars below the record show when each foot was on the platform (continuous bars) or on the floor off the platform (broken bars). LF, left fore; RF, right fore; RH, right hind; LH, left hind. From Jayes and Alexander (1978).

nent of force and the moment of the force about a vertical axis are omitted. The right fore foot was set down 0.1 m behind the center of the plate; then the left fore foot was placed 0.2 m in front of the center of the plate; then the right hind foot was placed almost 0.2 m behind the center of the plate; and finally the left hind foot was placed almost 0.2 m in front of the center. At times when two feet were on the plate simultaneously, the force records indicate the total force and the a_x record indicates the position between the two feet at which the resultant of the forces on them acted. This has made it possible to calculate separately the vertical component of force exerted by each foot (shown as broken lines on the F_y record).

A force plate should be stiff enough to deflect very little, under the forces that are to be recorded. Otherwise, it will alter the forces it is designed to measure; running across a platform that deflected a lot would be like running on mattresses. Also, the platform should be capable of registering faithfully the most rapid fluctuations of force that it is desired to record. This demands a high natural frequency of vibration; the significance of this will become apparent when we discuss pressure transducers. Further details of the construction and use of force plates are given by Biewener and Full (1992).

Pressure is the force acting per unit area in a fluid. There is sometimes a need to measure it, for example, in research on animals swimming by jet propulsion (Section 16.1). The instruments generally used are pressure transducers incorporating a small metal diaphragm, one side of which is exposed to the pressure. Distortion of the diaphragm is registered by strain gauges, so that the output of the strain gauges indicates the pressure. Large pressure transducers generally have to be connected through a fluid-filled tube to the point where pressure is to be measured. However, miniature pressure transducers are available with diameters less than 2 mm, which can be placed within the bodies of all but the smallest animals.

Figure 5.4 illustrates basic principles of design that apply not only to pressure transducers, but to transducers generally. The essential point is that any mass mounted on a spring will vibrate at a natural frequency; the smaller the mass and the stiffer the spring, the higher the frequency. Lauder (1980) wanted to ensure that his pressure transducer was capable of recording the fast changes of pressure inside the mouths of feeding fish. It was too large to implant in the fishes' bodies, so it had to be connected through a water-filled cannula. To test it, he connected it through a water-filled tube to an inflated balloon. He burst the balloon to obtain a very rapid pressure change, and obtained the records shown. In Fig. 5.4A the connection was made through a short (20-mm) metal tube. Oscillations follow the bursting of the balloon, at the natural frequency determined by the stiffness of the transducer diaphragm and the mass of water in the connecting tube. Their frequency is 250 Hz, and they continue for about 70 ms. This is not a faithful record of the pressure change in the balloon. For the remaining records, a much longer (500-mm) plastic tube was used. In Fig. 5.4B, as in A, the tube was filled with water that had been boiled to ensure that no air bubbles formed in it. Oscillations again continue for about 70 ms (note that the time scale has been changed). The frequency of the oscillations is lower (65 Hz) because the tube is longer (increasing the vibrating mass) and because its plastic wall is more compliant than the metal wall of the other tube (reducing the effective spring stiffness of the system). If there had been air bubbles in the tube, they would have reduced the stiffness and natural frequency further. In Figs. 5.4C and D, the tube has been filled with solutions of glycerol in water, instead of pure water. The viscosity of the glycerol damps out the vibrations, more completely in (D), in which the concentration is slightly higher, than in (C). Figure 5.4D shows critical damping, the degree of damping that gives the transducer the fastest possible response. With more

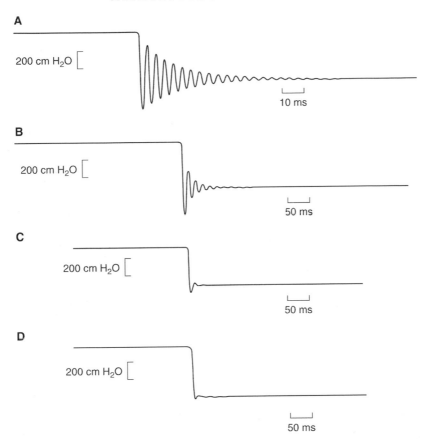

Fig. 5.4. The pressure change in a balloon, as it was burst, recorded by means of a pressure transducer. In (A) the transducer was connected to the balloon by a water-filled metal tube 20 mm long. In (B–D) the tube was 500 mm long. In (B) it was filled with water, in (C) with a 50% aqueous solution of glycerol, and in (D) with a 55% glycerol solution. From Lauder (1980).

severe damping, there would be no oscillation, but the signal would move more slowly to its final level. Following an instantaneous change of pressure or whatever else it is designed to measure, the output of a critically damped transducer reaches 99% of its final value in a time about equal to the period of vibration it would have had, if there had been no damping. Thus, transducers of all kinds should have undamped natural frequencies that are high, compared to the highest frequency component of the event that is to be recorded, and they should be critically damped. The system represented in Fig. 5.4D, with an undamped natural frequency of 65 Hz, is suitable only for recording pressure changes that take times substantially

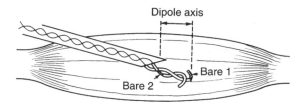

Fig. 5.5. A diagram of an electrode being inserted into a muscle for electromyography. From Gans (1992).

longer than 1/65 s. Lauder (1980) would not have had to take so much trouble, to ensure that his system's frequency response was adequate, if pressure transducers small enough to put inside the fish had been available at the time. Miniature pressure transducers are now readily available, but it remains important for workers in biomechanics to be aware of the frequency responses of their transducers.

5.6. Recording Muscle Action

When muscle fibers are activated, action potentials travel along them, similar to the action potentials that transmit information along nerves. In an action potential, the potential difference across the cell membrane is briefly reversed. This reversal of potential travels along the muscle fiber or axon. While it is traveling there are differences of electrical potential between one part and another of the fiber's outer surface. By recording these electrical events, we can find out when each muscle is active as the animal moves. The technique is called electromyography. It is explained, with many practical details, by Basmajian and de Luca (1985) and by Loeb and Gans (1986).

Figure 5.5 shows one of several types of electrode that are used for electromyography. It consists of two very fine insulated wires, twisted and glued together, with the insulation removed only from the regions labeled "bare 1" and "bare 2." The wires are threaded down a hypodermic needle, and their ends bent back as shown. The needle is inserted into a muscle in a living animal and then withdrawn, leaving the electrode in place. If the needle and wires are fine enough, this causes little discomfort, as many humans who have had electrodes placed in their muscles can testify. The electrode should be placed so that the dipole axis defined by the two bare patches is parallel to the muscle fibers. The leads are connected to an amplifier and recording equipment. Figure 5.6 is an example of a record obtained in this way. It shows electromyographic records from the principal wing muscles of a flying pigeon; there were two electrodes in the pectoralis

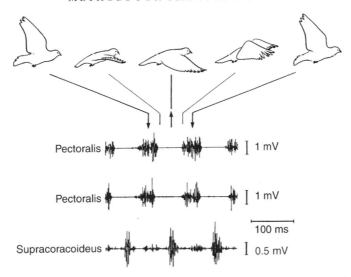

Fig. 5.6. Electromyographic records from the principal wing muscles of a flying pigeon (*Columba livia*). The outlines above, drawn from film taken simultaneously, show the stages of the wing beat cycle corresponding to selected points on the record. From Dial et al. (1988).

muscle and one in the supracoracoideus. In some of the experiments, long (18-m) wires trailed behind the birds, carrying the electrical signals to the amplifiers. In others, radio transmitters were used, as explained in Section 5.7. The record shows that the pectoralis muscle was active at the beginning of the downstroke of the wings, and the supracoracoideus at the beginning of the upstroke.

The electrical potentials recorded in this way are small, of the order of 1 mV. The changes of potential difference across the cell membrane are very much larger, of the order of 100 mV, but can be recorded only by inserting electrodes into individual muscle fibers. Electrodes of the kind shown in Fig. 5.5 record potential differences between two points in the extracellular space, and pick up action potentials from all the muscle fibers in the immediate vicinity. They will generally not detect action potentials from all parts of the muscle, and they may pick up action potentials from adjacent muscles. The very small spikes in the record from the supracoracoideus muscle, in Fig. 5.6, are probably action potentials in the pectoralis.

The length changes of muscle fibers can be recorded by sonomicrography (see Griffiths 1991). This requires insertion into the muscles of piezoelectric crystals, connected by fine wires to recording equipment. They are inserted by means of hypodermic needles, in the same way as electromyographic electrodes. One crystal has to be inserted at each end of the same muscle fascicle. This is difficult to do, in an intact animal, so it is

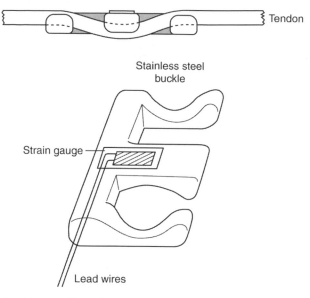

Fig. 5.7. A buckle transducer for measuring the forces acting in tendons. From Biewener (1992).

advisable to kill and dissect the animal after the experiment to make sure that the crystals were satisfactorily placed. Pulses of ultrasound are emitted by one crystal and received by the other. The time that the ultrasound takes to travel between them indicates the distance.

The forces that individual muscles exert in locomotion can sometimes be measured by fitting buckle transducers to their tendons. This requires a surgical operation, and is feasible only if there is an adequate length of tendon outside the muscle belly. Biewener (1992) explains the procedure. A buckle transducer (Fig. 5.7) is an E-shaped piece of metal with a strain gauge bonded to it. Fitting it onto a tendon leaves the tendon slightly bent where it winds round the arms of the E. Tension in the tendon straightens the tendon and distorts the E, and the distortion is detected by the strain gauge. The amount of distortion and so the strength of the signal from the strain gauge depend on the dimensions of the buckle and the thickness of the tendon, so it is necessary to calibrate the transducer for every tendon that it is used on. This is done after the experiment by killing the animal, dissecting out the tendon with the buckle still in place, and then stretching it in a testing machine of the kind described in Section 5.8. In that way a graph can be obtained of the output of the transducer against the force in the tendon.

The forces exerted by muscles can also sometimes be calculated from measurements of the force that the animal is exerting on the environment, for example, from force plate records. Calculations of this kind often involve assumptions about how the effort is being shared by several muscles that have similar effects. For a discussion of the difficulties involved see Herzog and Leonard (1991).

5.7. RECORDING MOVEMENT AT A DISTANCE

Many of the methods described so far require wires connecting the animal to recording equipment. This is often not convenient. For example, in the electromyographic study of pigeon flight (Fig. 5.6) the bird had to fly trailing 18-m wires behind it, when it was connected directly to the recording equipment. There are two alternative methods of getting recordings without connecting wires. Both of them require the animal to carry a piece of equipment, which must be made appropriately small and light.

One possibility is for the animal to carry a small radio transmitter, which will transmit the signals from whatever electrodes or transducers may be being used. This method was used for some of the electromyographic recording from pigeons. The transmitter must be small enough to have no appreciable effect on the animal's movements. In particular, transmitters attached to birds that are feeding nestlings should be light in weight, compared to the food loads carried in foraging flights (Pennycuick et al., 1989). The weight of the transmitter is less critical if the animal is not going to fly, but should still be kept reasonably small. A variant of the method that has sometimes been applied to aquatic animals uses an ultrasonic transmitter (which will work satisfactorily when submerged) instead of a radio transmitter (which will not). Both types of transmitter were used by Thompson and Fedak (1993) in their research on gray seals (*Halichoerus grypus*) in their natural habitat off the west coast of Scotland. They attached to the seals a pair of electrodes to record the electrocardiogram, a pressure transducer to measure the depth at which the animal was swimming, and a paddlewheel flowmeter to measure the swimming speed. The outputs from all of these were transmitted continuously as ultrasonic signals, which were received by a hydrophone on a yacht a few hundred meters away. In addition, they glued a radio transmitter to the seal's head, which transmitted effectively only when the seal was at the surface. This told them when the seal's head was out of water, and was also useful for locating it if it moved out of range of the hydrophone.

The alternative to a transmitter is a data logger (a miniature computer) attached to the animal to record the output of the transducers. The animal is recaptured and the data logger recovered (if all goes well) at the end of the experiment. This method was used by Bevan et al. (1995) in a study of black-browed albatrosses (*Diomedea melanophrys*) in the South Atlantic. The birds were nesting, so could be recaptured easily when they returned to their nests from foraging trips. The sensors attached to them included electrocardiograph electrodes, and a thermistor that measured the temperature in the abdominal cavity. The function of the thermistor was to show when the bird was feeding; the temperature in the abdomen fell briefly, whenever it swallowed a cold squid. The data logger recorded the heartbeat frequency every 30 s, and the abdominal temperature every minute.

The animals mentioned in the examples in this section are all moderate or large in size: pigeons, seals, and albatrosses. It does not at present seem feasible to build radio transmitters or data loggers small enough for attachment to small animals such as ants or flatworms.

5.8. PROPERTIES OF MATERIALS

The skeletons and tendons of animals must be strong enough to withstand the forces that act on them in locomotion, and their elastic properties may be important; for example, tendons function as springs in running mammals (Section 7.4). Dynamic testing machines of the kind that engineers use for measuring the strength and elasticity of metals and plastics are often the most suitable machines for mechanical tests on animal materials. Figure 5.8A is a diagram of one of these machines. It has a rigid steel frame. At the top is a load cell, a transducer that gives an electrical output proportional to any vertical force that acts on it. At the bottom is a hydraulically driven actuator that can be made to move up and down. The diagram shows the machine set up to stretch a specimen, perhaps a tendon. One end of the specimen is clamped to the load cell and the other to the actuator, so that when the actuator moves down the specimen is stretched and the force is registered by the load cell. An electrical output from the machine indicates the movements of the actuator, but this may give a misleading impression of the length changes of the specimen if it is severely distorted, in and near the clamps, by the pressure required to grip its ends firmly. An alternative is to use an extensometer like the one shown in Fig. 5.8B. This is a transducer that registers the length changes of an undistorted section of the specimen. Further practical details are explained by Ker (1992).

This chapter has reviewed the range of methods available for research on animal locomotion. Many of them will be referred to repeatedly in later chapters. Though recently introduced methods enable us to make observations that would have seemed impossible ten or twenty years ago, there are still a great many things that we would like to do, but cannot.

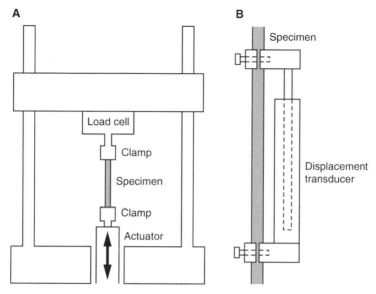

Fig. 5.8. Diagrams of (A) a dynamic testing machine and (B) an extensometer. From Alexander (1988).

It would be very useful to have an imaging system that could record the movements of the whole skeleton of large animals in fast locomotion. At present, we can take whole body cine X-ray of small mammals and birds, but if we want to study the movements of large ones we have to rely on optical systems that record the movements of markers on the skin, which may move relative to the underlying skeleton. Treadmills and wind and water tunnels are excellent for studying locomotion at constant velocity, but we need better facilities for studying acceleration and turning. There are well-established methods for measuring oxygen consumption, but we cannot easily measure the rates of anaerobic metabolism that make fast locomotion possible. Force plates that record the vertical component of force are being incorporated into treadmills, but it would be very helpful to be able to record the horizontal components as well. Tendon buckles and sonomicrography crystals enable us to investigate the forces and length changes in muscles and tendons during locomotion, but the surgery involved in implanting them restricts the number of muscles that can be investigated simultaneously. Less invasive alternatives would be very helpful.

Chapter Six

....

Alternative Techniques for Locomotion on Land

T HERE ARE many possible ways of traveling over land, some using legs and some not. This chapter reviews the possibilities, using very simple models to introduce the principles and to make rough estimates of energy costs. Descriptions and explanations of the movements of real animals and actual measurements of energy costs follow in later chapters.

A few definitions are needed. A stride is a complete cycle of movement; for example, from the setting down of a foot to the next setting down of the same foot. The stride length is the distance traveled in one stride, and the stride frequency is the number of strides taken in unit time. The mechanical cost of transport is the work required to move unit mass of animal unit distance. The metabolic cost of transport is the metabolic energy used when unit mass of animal travels unit distance.

6.1. TWO-ANCHOR CRAWLING

Figure 6.1A shows an imaginary animal, crawling in a very simple way. Bristles are shown on the underside of its body that slope backward so that the animal can slide forward over the ground much more easily than it can slide back. The animal alternately lengthens its body and shortens it. When it is lengthening, the bristles prevent the hind part of the body from sliding back, so the fore part is pushed forward. When it is shortening, the bristles prevent the fore part from sliding back, so the hind part is pulled forward. Thus, the simple action of lengthening and shortening the body moves the animal forward over the ground. If the animal lengthens and shortens by λ, it advances in each cycle of lengthening and shortening by a distance λ, so λ is the stride length.

Let the mass of the animal be m, let the gravitational acceleration be g, and let the coefficient of friction between the animal and the ground for forward sliding be $\mu_{forward}$. The coefficient of friction for backward sliding (μ_{back}) is greater, because of the bristles. The weight of the animal is mg,

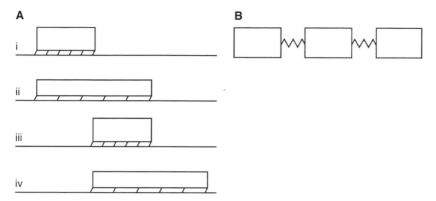

Fig. 6.1. (A) Successive stages of crawling by a simple model designed to illustrate the two-anchor principle. (B) The three-part crawler discussed in the text.

and the frictional force resisting its forward motion is $\mu_{forward} mg$. The work that must be done against friction to move the animal forward by one stride length λ is $\mu_{forward} mg\lambda$. The mechanical cost of transport T is the work per unit mass and per unit distance

$$T_{friction} = \mu_{forward}\, g \tag{6.1}$$

This is a good estimate of the cost of transport if the animal is moving slowly, but not if it travels fast. As the animal moves, each end of its body is repeatedly accelerated and halted. When it is accelerated, work is needed to give it kinetic energy. When it is halted, the kinetic energy is lost. To keep our argument simple, we can think of the body as three equal parts (Fig. 6.1B). In steady crawling at velocity v, the middle third of the body moves forward with constant velocity v, but the front and rear thirds are each stationary for half the time, and move forward at velocity $2v$ for the other half. The kinetic energy gained and lost in the course of a stride, by the two thirds of the body that are accelerated and halted, is $\frac{1}{2}(2m/3)(2v)^2 = 4mv^2/3$. This amount of work must be done in each stride, while the animal advances by a distance λ. Thus, the inertial cost of transport (work divided by mass and by distance) is

$$T_{inertia} = \frac{4v^2}{3\lambda} \tag{6.2}$$

At low speeds, the frictional cost of transport (Equation 6.1) is greater than the inertial cost (Equation 6.2), and at high speeds the reverse is the case. They are equal when $v^2/\lambda g = 0.75\mu_{forward}$. Notice that $v^2/\lambda g$ is a Froude number (Section 4.2) that uses the stride length as the scale of length. At Froude numbers lower than this, Equation 6.1 gives the better

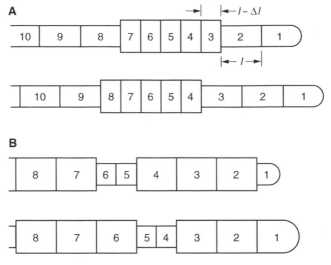

Fig. 6.2. Successive stages of crawling by peristalsis. In (A) the waves of contraction travel backward along the body, as in earthworms. In (B) they travel forward, as in such worms as *Polyphysia*. From Alexander (1982).

estimate of the cost of transport, and at higher Froude numbers Equation 6.2 gives the better estimate. I have not added the two costs together, because the kinetic energy that is lost when one end of the body is halted can be used to do work against friction.

Maggots (larval Diptera) crawl more or less like this simple model, advancing 0.15–0.25 body lengths in each stride (Berrigan and Pepin 1995). However, they lack the bristles shown in Fig. 6.1. Instead, they anchor the anterior end when shortening the body by hooking the head downward.

6.2. Crawling by Peristalsis

Figure 6.2A shows how earthworms crawl. Instead of the body lengthening and shortening as a whole, successive segments lengthen and shorten in turn, each segment changing length a little after the one in front. Thus, waves of lengthening and shortening travel backward along the body. Each segment remains stationary while it is short, and moves forward while it is long; we will discuss the conditions in which this will happen in the next paragraph. The diagram shows two successive positions of the animal. In the interval between them, segment 3 lengthens. The segments immediately behind it are prevented from sliding back, so the segments in front are pushed forward. Meanwhile, segment 8 has shortened. The segments in front of it cannot be pulled back, so the segments behind are pulled forward.

The number of segments that can be in motion at any instant is limited. Let the coefficients of friction for forward and backward sliding be $\mu_{forward}$ and μ_{back}. These coefficients may be equal, but may be made different, for example by backward-sloping bristles such as earthworms have on the ventral surface of the body. Let a fraction q of the segments be moving forward at any time. For this to be possible, the frictional force potentially available to stop the short segments from sliding backward, $(1 - q)mg\mu_{back}$, must be greater than the frictional force on the long segments, $qmg\mu_{forward}$. This implies

$$q < \frac{\mu_{back}}{\mu_{back} + \mu_{forward}} \tag{6.3}$$

That condition applies if the animal is crawling slowly enough for inertial forces to be negligible. If it were crawling fast enough for inertial forces to be important, friction on the stationary segments would have to balance the inertial forces as well as the frictional forces on the moving segments, and fewer segments could be simultaneously in motion.

Now we will estimate the work required for crawling. The frictional cost of transport is the same as for the two-anchor model (Equation 6 1) Let one wavelength of the animal's motion extend over n segments, so that when segment 3 is the first of one group of short segments, segment $n + 3$ is the first of the next. Let m be the mass of this group of n segments. Let each segment have length l when it is long, and length $(l - \Delta l)$ when it is short, as shown in the diagram. Let it be long and moving forward for a fraction q of the time, so that each wavelength consists of qn long segments and $(1 - q)n$ short ones. This means that to move the worm forward with a mean speed v, the moving segments must have speed v/q. In the interval between the two instants represented in Fig. 6.2A, mass m/n of animal is given velocity v/q. Thus, the work done giving it kinetic energy is $mv^2/2nq^2$. While this work is being done, the animal advances through a distance Δl. This equals λ/n, where λ is the stride length, the distance traveled while every segment makes a complete cycle of lengthening and shortening. The inertial cost of transport is the work divided by the mass and by the distance

$$T_{inertia} = \frac{v^2}{2q^2\lambda} \tag{6.4}$$

As for the two-anchor model, the frictional cost is dominant at low speeds and the inertial one at high speeds. They are equal when the Froude number based on stride length, $v^2/\lambda g$, equals $2\mu_{forward}q^2$.

A   **B**

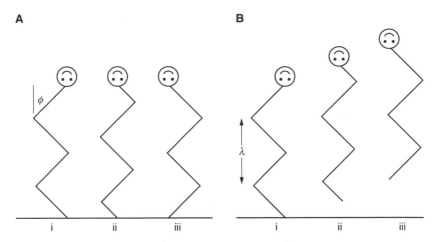

Fig. 6.3. Diagrams of an animal attempting to crawl by passing waves of bending posteriorly along its body. In (A) the body slides easily sideways, and no progress is made. In (B) the body slides more easily along its long axis than at right angles to it, and the animal crawls successfully. Each sketch is displaced to the right of the previous one, to avoid overlap.

6.3. SERPENTINE CRAWLING

Figure 6.3 shows a simple model of another technique of crawling. It resembles serpentine crawling of snakes, but the body is shown as a zigzag instead of a smooth curve, to simplify the mathematical analysis (Bekker 1956). Waves of bending travel backward along the body, pushing the animal forward. This could have either of two effects, depending on how easily the body can slide over the ground in different directions. In Fig. 6.3A, the body easily slides sideways. Waves pass backward along it, but the animal makes no progress. In Fig. 6.3B, the body slides more easily along its own length than at right angles to it, and the animal moves forward, the hind parts of the body traveling along the track made by the fore parts.

Let the coefficients of friction for the body be μ_{axial} for sliding along its own axis and $\mu_{\text{transverse}}$ for sliding at right angles to it. To slide a segment of the body of mass δm forward along its own axis requires a force $\delta m g \mu_{\text{axial}}$, and to slide it sideways at right angles to its axis requires a force $\delta m g \mu_{\text{transverse}}$. For the animal to move as in Fig. 6.3B, the forward component of the available transverse frictional force must exceed the backward component of the axial frictional force

$$\mu_{\text{transverse}} \sin \phi > \mu_{\text{axial}} \cos \phi \qquad (6.5)$$

$$\tan \phi > \frac{\mu_{\text{axial}}}{\mu_{\text{transverse}}}$$

In this style of crawling, the stride length λ is equal to the wavelength of the waves formed by the body (Fig. 6.3B). We will calculate the work required to move the body (mass m) forward by one wavelength. To advance a distance λ, the body must slide a distance $\lambda/\cos \phi$ along its zigzag path, so the work that has to be done against friction is $\mu_{\text{axial}} mg\lambda/\cos \phi$. When the animal crawls at speed v, each segment of the body slides with speed $v/\cos \phi$ at an angle $\pm\phi$ to the direction of travel, so its velocity has a transverse component $\pm v \tan \phi$. Each time the segment passes a bend in the zigzag, this transverse component of velocity is lost and regained. The associated kinetic energy $\frac{1}{2}\delta mv^2 \tan^2 \phi$ is lost and has to be replaced. In a complete stride, this happens twice to every part of the body, so the inertial work that has to be done by the whole animal is $mv^2\tan^2\phi$.

The frictional cost of transport is the frictional work divided by the mass of the body and by the stride length

$$T_{\text{friction}} = \frac{\mu_{\text{axial}} g}{\cos \phi} \qquad (6.6)$$

The inertial cost of transport is

$$T_{\text{inertia}} = \frac{v^2 \tan^2 \phi}{\lambda} \qquad (6.7)$$

The frictional cost is the larger at low speeds, and the inertial cost at high speeds. They are equal when the Froude number based on stride length, $v^2/\lambda g$ equals $\mu_{\text{axial}}/(\sin \phi \tan \phi)$.

6.4. FROGLIKE HOPPING

Frogs travel in a series of jumps, coming to a halt between each jump and the next. No work is done against friction with the ground unless the feet skid while pushing off, but work is required to give the body kinetic energy for each jump. Consider the model shown in Fig. 6.4A, which jumps like a frog. Its total mass is m. It travels in a series of jumps at speed v, taking off for each jump as soon as it has landed from the previous one. It takes off at an angle α to the horizontal. The horizontal component of its velocity at takeoff is v, so the resultant velocity is $v/\cos \alpha$, and the kinetic energy it has to be given for the jump is $\frac{1}{2}m(v/\cos \alpha)^2$. To calculate the jump length, note that the vertical component of the animal's velocity is

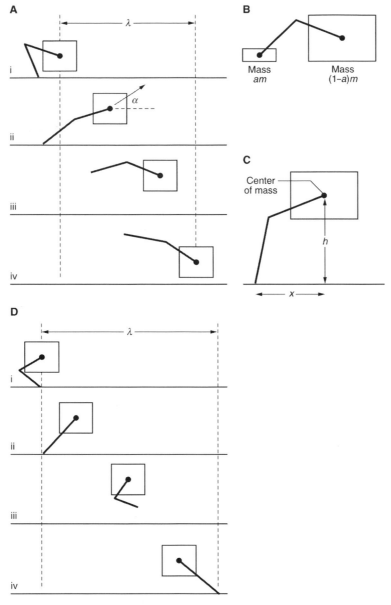

Fig. 6.4. Diagrams representing (A) successive positions of an animal hopping like a frog; (B) an animal with mass only in its body and feet; (C) an animal with its foot on the ground, showing how two dimensions are defined; and (D) successive positions of an animal hopping like a kangaroo.

$+v$ tan α at takeoff and $-v$ tan α when the animal lands. The time for which the animal remains airborne is the time required for the gravitational acceleration g to change the velocity by this amount; it is $2v$ tan α/g. Since the horizontal component of the velocity is v, the animal travels a distance $2v^2$tan α/g in this time. The cost of transport is the kinetic energy divided by the mass and by this distance

$$T = \frac{\tfrac{1}{2}\, m \, (v/\cos \alpha)^2}{2mv^2 \tan \alpha/g} = \frac{g}{2 \sin 2\alpha} \qquad (6.8)$$

The cost of transport will have its lowest possible value, $g/2$, when the takeoff angle α is 45°.

Two assumptions are hidden in that very simple argument. First, we have assumed that the time taken by the animal to accelerate and decelerate is small enough to be ignored. If the animal jumps like a frog by extending its legs, the distance over which it accelerates is approximately equal to the length of the legs, and this assumption is in effect that the length of the jump is much greater than the length of the legs. Secondly, we have assumed that the mass of the legs is small enough to be ignored. We will investigate the effect of this second assumption.

Real animals have mass distributed all along the length of the leg, but to keep the mathematics simple we will imagine an animal that has mass only in its body and its feet (Fig. 6.4B). Let the feet have mass am and the body $(1 - a)m$. The length of the jump depends on the velocity of the animal's center of mass at takeoff. The feet remain stationary on the ground while the body is being accelerated, so to give the center of mass the required velocity $v/\cos \alpha$ the body must be accelerated to velocity $v/[(1 - a)\cos \alpha]$. A body of mass $(1 - a)m$ moving with this velocity has kinetic energy $\tfrac{1}{2}mv^2/[(1 - a)\cos^2 \alpha]$. The cost of transport is this kinetic energy divided by the mass of the animal and the length of the jump

$$T = \frac{g}{2\,(1 - a)\sin 2\alpha} \qquad (6.9)$$

Thus, the greater the fraction of body mass that is in the feet, the greater the cost of transport. It will be an advantage for an animal that moves in this way to have the lightest possible legs and feet.

6.5. AN INELASTIC KANGAROO

Our next model travels in a series of hops like the last, but with an important difference; it does not come to a halt between one hop and the next. Thus, it moves more like a kangaroo than a frog. However, it differs from kangaroos in having no springs in its legs. (I will show in Section 7.4 that tendons in the legs of kangaroos function as energy-saving springs.)

This model is shown in Fig. 6.4D. Its mass is m and it travels with velocity v, taking strides of length λ. Its feet are on the ground for a fraction β of the time. The vertical component of the force on the ground, averaged over a complete stride, must equal the animal's weight mg. Thus, the mean vertical force while the feet are on the ground is mg/β. We will make the simple assumption that the vertical component of the force on the ground is constant and equal to mg/β, throughout the time that the foot is on the ground.

We will assume that while the feet are on the ground, the force they exert on the ground is always in line with the animal's center of mass. There are two reasons for this assumption. First, if the force is not in line with the center of mass, it will make the animal rock. Secondly, it will be shown in Section 7.3 that the energy cost of hopping would be greater, if the force were not kept more or less in line with the center of mass. If the vertical component of the force is mg/β and the center of mass is at a height h and at a distance x in front of the foot (Fig. 6.4C), the horizontal component of force needed to keep the resultant in line with the center of mass is $mgx/\beta h$. As the animal moves forward by a small amount δx, this component of force does work amounting to $(mgx/\beta h)\,\delta x$. The foot remains on the ground until $x \approx \beta\lambda/2$. By integrating the work we have just calculated from $x = 0$ to $x = \beta\lambda/2$, we find that the work done as the center of mass moves forward over this distance is $mg\beta\lambda^2/8h$.

The animal moves forward from $x = 0$ to $x = \beta\lambda/2$ in time $\beta\lambda/2v$. During this time, the vertical component of force mg/β gives it an upward acceleration $(g/\beta) - g = (1 - \beta)g/\beta$. A body accelerating from rest with acceleration a for time t moves a distance $\frac{1}{2}at^2$, so the animal rises by $\frac{1}{2}[(1 - \beta)g/\beta](\beta\lambda/2v)^2 = (1 - \beta)g\beta\lambda^2/8v^2$. The work done equals the force multiplied by the distance, $(1 - \beta)mg^2\lambda^2/8v^2$. At high speeds (when the Froude number v^2/gh is large) this work done against vertical forces is small, compared to the work done against horizontal forces.

The total work done during the stride is the sum of the work done by the horizontal and vertical components of force. It is $(mg\lambda^2/8)[(\beta/h) + (1 - \beta)(g/v^2)]$. The cost of transport is obtained by dividing this work by the mass of the animal and the stride length

$$T = \frac{g\lambda}{8}\left[\frac{\beta}{h} + (1 - \beta)\frac{g}{v^2}\right] \tag{6.10}$$

In deriving that equation, we assumed that the vertical force on the ground was constant, throughout the time that the feet are on the ground. Alexander (1977a) made the more realistic assumption that the vertical

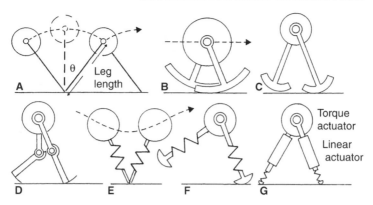

Fig. 6.5. Models of walking and running. (A) the minimal bipedal walker; (B) the synthetic wheel; (C, D) walking models with heavy legs; (E, F) mass–spring models of running; and (G) a model discussed in Section 7.7. From Alexander (1995a).

component of force rises and then falls while the feet are on the ground, like half a cycle of a cosine curve. This led to the equation

$$T = \frac{g\lambda}{8} \left[\left(\frac{0.73\beta}{h} \right) + (1 - 0.73\beta) \frac{g}{v^2} \right] \qquad (6.11)$$

At high speeds, the second term in the square brackets would be small, and T would be approximately $0.09g\beta\lambda/h$.

These arguments assume that the feet leave the ground when $x = \beta\lambda/2$. This is not exactly true, because the horizontal components of the forces on the feet make the animal travel a little more slowly while the feet are on the ground than when they are off it. Alexander (1977a) showed how the velocity fluctuations can be calculated.

This section has been concerned with kangaroo-like hopping, but similar arguments can be applied to bipedal running. To obtain equations for running, it is necessary only to substitute $\lambda/2$ for λ and 2β for β, in Equations 6.10 and 6.11. We will return to hopping and running in Section 6.9, where we will consider the possible role of structures that function as springs.

6.6. A MINIMAL MODEL OF WALKING

Figure 6.5A shows the simplest of all models of walking (Alexander 1976). It consists of a point mass mounted on rigid legs of negligible mass. Each foot is set down as the other is lifted. The body moves forward in a series of arcs of circles, of radius equal to the length of the legs. The body rises and falls in each step. Similarly, when humans walk we keep each leg straight, while its foot is on the ground. Consequently, our centers of mass are about 40 mm higher in midstep, when the supporting leg is vertical, than at the stage of the stride when we have both feet on the ground (Margaria 1976).

This style of walking becomes impossible at high speeds, because gravity cannot give the body a downward acceleration greater than g. A body moving at speed v along an arc of a circle of radius r has an acceleration v^2/r toward the center of the circle. Let the walker's body have velocity v_0, at the stage of the stride when the leg is vertical. Then at this stage the body has a downward acceleration v_0^2/h, where h is the length of the legs:

$$g \geq \frac{v_0^2}{h} \tag{6.12}$$

and the Froude number v_0^2/gh cannot be greater than 1. Adult humans have legs about 0.9 m long, and the gravitational acceleration is 9.8 m/s², so this argument predicts a speed limit for human walking of 3.0 m/s.

Adult humans generally walk at speeds below about 2 m/s, and run at higher speeds (Thorstensson and Roberthson 1987). However, we are capable of walking faster. In my experience, more or less normal walking is possible at speeds up to 2.7 m/s (data of Alexander and Jayes 1980). Athletes in walking races attain much higher speeds; the world records for the 10-km race imply mean speeds of 4.4 m/s for men, and 4.0 m/s for women. These speeds are made possible by peculiar hip movements that flatten the arc traversed by the center of mass, so that its radius of curvature is greater than the length of the leg (see Alexander 1984).

Now we will calculate the work required for walking, for the model shown in Fig. 6.5A. While only one foot is on the ground, and the body is moving along a circular arc, no work is required. The body slows down a little as it rises, and speeds up again as it falls. Kinetic energy is converted to gravitational potential energy and back again, as in a swinging pendulum. Work is required only at the instant when one foot hits the ground and the other leaves it, at which stage both legs make angles θ with the vertical. Immediately before this instant, the center of mass is traveling with velocity v_θ at an angle $-\theta$ to the horizontal (v_θ is a little greater than v_0, because some potential energy has been converted to kinetic energy). The vertical component of its velocity is $-v_\theta \sin \theta$. Immediately after this instant, the center of mass is traveling with velocity v_θ at an angle $+\theta$ to the horizontal, and the vertical component of its velocity is $+v_\theta \sin \theta$. The mass of the body is m, so it loses and regains kinetic energy amounting to

$\frac{1}{2}mv_\theta^2\sin^2\theta$. This amount of work has to be done twice in each stride, once for each foot, so the cost of transport is $v_\theta^2\sin^2\theta/\lambda$. The stride length λ equals $4h\sin\theta$, so the cost of transport is

$$T = \frac{v_\theta^2\lambda}{16h^2} \tag{6.13}$$

McGeer (1990a) presented an argument that predicted a cost of transport $4\cos^2\theta$ times this. The difference, as he pointed out, is that his model was entirely passive, with no control over the forces its feet exerted on the ground. In my model, only the vertical component of velocity is affected by the impact of the foot with the ground. In his, the horizontal component of velocity is reduced by the impact. To avoid this, the forces exerted by the feet must be controlled so that the two feet exert equal forces at the instant when both are on the ground.

We have noted that v_θ is a little greater than v_0, but we have not worked out by how much. The difference in height between the body's highest point and its lowest is $h(1 - \cos\theta)$. The kinetic energy gained equals the gravitational potential energy lost

$$\frac{1}{2}m\left(v_\theta^2 - v_0^2\right) = mgh\left(1 - \cos\theta\right) \tag{6.14}$$

$$v_\theta^2 = v_0^2\left[1 + \left(\frac{2gh}{v_0^2}\right)(1 - \cos\theta)\right]$$

For an adult human walking at a comfortable speed of 1.5 m/s, v_0^2/gh is about 0.25 and θ is about $25°$, whence v_θ is approximately $1.3v_0$.

6.7. THE SYNTHETIC WHEEL

The model just discussed had legs of negligible mass, but real legs are quite heavy. Inertial work must be done to swing them forward and back in walking, unless their swinging can be driven by gravity like the swinging of pendulums. Figure 6.5B shows a model of walking devised by McGeer (1990a), in which the legs swing like pendulums.

In this model, the legs have mass, but this mass is too small for their swinging to have an appreciable effect on the motion of the trunk. Each leg is a sector of a wheel. The model rolls forward at constant velocity, first on one foot and then on the other, as the legs swing forward and back. While its rim is on the ground, each wheel rolls with constant angular velocity. When it leaves the ground, it swings freely as a pendulum. It completes its backward swing, swings forward, and starts swinging back again, before it is set down on the ground for its next step. It is set down at the instant when the foot is once again stationary relative to the ground. McGeer (1990a) shows that this occurs 0.65 of a pendulum period after it was lifted. Once the mathematical model is set in motion on level ground, no further work is needed to keep it moving. A physical model would of course need a little work, to overcome air resistance and friction in its joints.

Electromyographic records of human walking show very little activity in the leg muscles, while the foot is off the ground. This suggests that the legs may swing like pendulums. However, this model cannot explain the cadence of human walking at normal speeds. If each foot is off the ground for 0.65 of a pendulum period, the duration of a stride is 1.3 periods. An adult human leg, pivoted at the hip and swinging like a pendulum, would have a period of about 1.5 s (Mochon and McMahon 1980). Thus, the model suggests that the stride period for human walking should be 1.3 × 1.5 = 2.0 s. Observed stride periods range from about 2.0 s in extremely slow walking at 0.4 m/s, to 1.0 s in brisk walking at 1.6 m/s (Zarrugh et al., 1974).

An important difference between this model and real human walking is that people rise and fall as they walk, but the model does not. This is not a realistic model of human walking. It nevertheless seems interesting to ask whether it would be advantageous for animals to walk in the manner of the model. At first sight, the very low cost of transport seems most attractive.

This synthetic wheel model shares the advantages and disadvantages of real wheels. A wheeled vehicle can travel at constant velocity on smooth level ground at very low energy cost. Cycling on roads is far more economical of energy than walking or running, as Section 18.2 shows. However, wheels are much less satisfactory on rough terrain. Indeed, many places that are accessible to legged animals are inaccessible to wheeled vehicles. For example, wheels cannot climb vertical steps that are higher than about half their radius (Bekker 1956; LaBarbera 1983).

6.8. WALKERS WITH HEAVY LEGS

Figure 6.5C shows a model of walking that is more realistic than the previous one. The legs are allowed to be heavier. The feet are again arcs of circles, but their radius is much less than leg length. Consequently, the model rises and falls as it walks, much like Fig. 6.5A. McGeer (1990a) analyzed the motion of this model by computer simulation. He showed that the rise and fall of the body affected the movement of the legs, making them swing slightly faster than they otherwise would. The model would walk passively down a slope with a stride period of about 1.2 times the pendulum period of the legs, corresponding to 1.8 s for adult humans.

As well as analyzing the mathematical model, McGeer (1990a) built a physical one. He found that it would walk down gradients as small as 2.5%. A model of mass m walking a distance λ down a 2.5% gradient loses potential energy $0.025mg\lambda$, so the cost of transport is $0.025g$. Kinetic energy is lost at each impact of a foot with the ground, as in the minimal biped (Fig. 6.5A).

Figure 6.5D shows another mathematical model that walks on heavy legs. The knee bends passively as the foot leaves the ground, but straightens (again passively) as it swings forward. Mochon and McMahon (1980) and McGeer (1990a) analyzed the motion of this model. The gait is more realistic, in that the bending of the knee makes the foot clear the ground, but the cost of transport is a little higher because energy is lost when the knee hits the stop that limits its extension. The stride duration is realistic for human walking, when the segments of the legs are given realistic lengths and masses (Mochon and McMahon 1980).

6.9. Spring–Mass Models of Running

Figure 6.5E shows a model in which the legs are springs of negligible mass. It bounces along like a bouncing ball. Each time it lands on the ground, a spring is compressed. The animal loses kinetic energy, which is stored as elastic strain energy in the spring. Then the spring recoils, restoring the kinetic energy. If the springs were perfect and there were no air resistance, the model would continue bouncing forever, once set in motion. The cost of transport would be zero. The motion of this model has been studied by computer simulation, by Blickhan (1989) and by McMahon and Cheng (1990).

Figure 6.5F is a slightly more elaborate model (McGeer 1990b). The upper parts of the legs have appreciable mass, but the springs, and the rounded feet on their ends, have negligible mass. As well as compression springs in the legs there are torsion springs at the hips, which make the legs oscillate backward and forward in a scissorlike action. McGeer showed by computer simulation that the model would run with a stride frequency very close to the frequency that the scissorlike oscillations would have if the feet were kept off the ground. Speed could be increased by increasing the amplitude of scissoring. If the stiffness of the springs is kept constant, increasing the amplitude leaves the stride frequency little changed but increases the length of the strides. If the springs were perfect, if there were no friction in the joints, and if there were no air resistance, the cost of transport would be zero.

Table 6.1.
Mechanical costs of transport for the models of terrestrial locomotion analyzed in this chapter

Model	Frictional cost	Inertial cost	Conditions
Two-anchor crawling	$\mu_{forward}\,g$	$4v^2/3\lambda$	$\mu_{back} > \mu_{forward}$
Peristaltic crawling	$\mu_{forward}\,g$	$v^2/2q^2\lambda$	$q < \mu_{back}/(\mu_{back} + \mu_{forward})$
Serpentine crawling	$\mu_{axial}\,g/\cos\phi$	$v^2\tan^2\phi/\lambda$	$\tan\phi > \mu_{axial}/\mu_{transverse}$
Froglike hopping	0	$g/2\sin 2\alpha$	
Inelastic kangaroo	0	$(g\lambda/8)\,[(0.73\beta/h) + (1 - 0.73\beta)\,(g/v^2)]$ $\approx 0.09g\beta\lambda/h$ at high speeds	
Inelastic runner	0	$(g\lambda/16)\,[(1.46\beta/h) + (1 - 1.46\beta)\,(g/v^2)]$ $\approx 0.09g\beta\lambda/h$ at high speeds	
Minimal walker	0	$v_\theta^2\lambda/16h^2$	$1 \geq v_0^2/gh$
Synthetic wheel	0	0	
Walker with heavy legs	0	$\approx 0.025g$	
Spring–mass running	0	0	

Note. Conditions that must be satisfied, for each technique of locomotion, are also shown. Symbols: g, gravitational acceleration; h, leg length; q, fraction of segments in motion; v, velocity; α, takeoff angle; β, duty factor; ϕ, angle shown in Fig. 6.3; λ, stride length; μ, coefficient of friction for sliding in the direction indicated by the subscript.

6.10. COMPARISONS

Table 6.1 shows the costs of transport that we have calculated. The most striking of the points shown by the table are the following:

1. Work is done against friction only in the three crawling techniques, at the top of the table. If the coefficients of friction are the same, this work is greater for serpentine crawling than for the other techniques, because a serpentine crawler takes a longer, sinuous route.

2. Neither frictional nor inertial work is required for the synthetic wheel, or for the spring–mass models of running.

3. The inertial cost of transport is proportional to v^2, and so can be expected to increase rapidly with increasing speed, for the three crawling techniques and for the minimal walker. The techniques of locomotion represented by these models are unlikely to be favored at high speeds. The minimal walker is in any case incapable of traveling fast.

We will now compare the inertial costs shown in the table using rough estimates for the values of the variables. For two-anchor crawling, the cost is $1.33v^2/\lambda$. For peristaltic crawling, if the backward and forward coefficients of friction are equal, q cannot exceed 0.5 and the inertial cost cannot be less than $2.0v^2/\lambda$. However, if the backward coefficient is much higher than the forward one, q can approach 1.0 and the cost may be little more than $0.5v^2/\lambda$. For serpentine crawling, if the axial and transverse coefficients of friction are equal, $\tan \phi$ cannot be less than 1 and the lowest possible inertial cost is $1.0v^2/\lambda$. In human walking at moderate speeds, stride length λ is typically about 1.5 times leg length h, making the cost of transport for the minimal walker $0.14\ v_\theta^2/\lambda$. This is less than for any of the crawling techniques. However, it may not be reasonable to assume that a peristaltic crawler, serpentine crawler, and a legged walker traveling at the same speed would take strides of equal length.

Froglike hopping has no frictional cost of transport, and the inertial cost, with the optimal takeoff angle of 45°, is $0.5g$. This is less than $0.5v^2/\lambda$, if the Froude number based on stride length, $v^2/g\lambda$, is greater than 1.0. Thus, froglike hopping is more economical than any of the crawling techniques, at high speeds. However, inelastic kangaroolike hopping is much more economical than frog hopping, because the animal does not lose all its kinetic energy every time it lands. A kangaroo hopping very slowly, with a Froude number based on leg length (v^2/gh) equal to 1, would have a stride length of about 2.1 leg lengths (Hayes and Alexander 1983). The cost of transport for an inelastic kangaroo hopping like this would be $0.26g$. At very high speeds, the second term in the square brackets in Equation 6.11 would be small, and the cost would be little more than $0.09g\beta\lambda/h$. For a kangaroo hopping very fast, β would be about 0.25 and λ/h would be about 12 (Bennett 1987), so the cost of transport calculated from the inelastic model would be little more than $0.3g$. Springs could make the cost of transport even less, but the zero cost of transport predicted by the mass–spring models is an unattainable ideal.

This analysis has shown that crawling techniques like those of earthworms and snakes would be expensive of energy at high values of the Froude number based on stride length, $v^2/g\lambda$. Appropriately, they are used only at relatively low Froude numbers. Earthworms with masses around 0.8 g crawl at speeds up to 6 mm/s, with stride lengths up to 20 mm (Quillin 1999). At this speed and stride length, $v^2/g\lambda = 0.0002$. Cockroaches of the same mass run at speeds up to 1.5 m/s taking 60-mm strides (Full and Tu 1991); for them, $v^2/g\lambda = 3.8$. Black racer snakes (*Coluber constrictor*) with masses of about 100 g attain crawling speeds up to 1.5 m/s in short bursts (Walton et al., 1990). If their stride length at that speed is the same as at lower speeds (0.3 m), $v^2/g\lambda = 0.75$. Desert iguanas (*Dipsosaurus dorsalis*) of slightly lower mass (70 g) go much faster on legs. They can run for short distances at 5 m/s taking 0.4-m strides, which makes $v^2/g\lambda = 6$ (Marsh 1988).

This chapter has presented extremely simple models of various means of traveling overland. We cannot expect the costs of transport that we have calculated for these models to be accurate for real animals. The value of the calculations is that they reveal general principles that might be obscured by the complexity of more realistic models.

··

Chapter Seven

Walking, Running, and Hopping

*I*N CHAPTER 6, we used simple models to compare locomotion on
legs with other means of moving over land. In this chapter, we will
examine walking, running, and hopping in much more detail, referring much more to observations and experiments. We will be concerned
with all legged animals, both tetrapods (amphibians, reptiles, birds, and
mammals) and arthropods (insects, crabs, etc.).

7.1. SPEED

In almost every case, legged animals can move faster over land than animals of similar size that lack legs. Figure 7.1 shows some data for mammals. Figure 7.1A shows maximum sprint speeds, which cannot be sustained for long because they are powered largely by anaerobic metabolism;
and Figure 7.1B shows the lower speeds that can be sustained by aerobic
metabolism.

It is very difficult to get reliable maximum sprint speeds for large mammals. Small mammals have been chased and filmed in the laboratory, but
sprint speeds of large animals have to be measured in the field. A very
large proportion of published speeds are estimates, based probably on the
observer's experience of motor traffic (see Garland 1983). These are highly
unreliable. Even speeds read from the speedometer of a vehicle driving
alongside the animal may be unreliable, because if the animal swerves
away from the vehicle, the vehicle will have to travel a longer path, on the
outside of the bend. The most reliable data are for human athletes and for
racehorses and greyhounds. Elite human athletes run 100-m races in about
10 s, at a mean speed of 10 m/s. They continue accelerating throughout the first half of the race, and reach a peak speed of over 11 m/s (Reilly
et al., 1990). The results of experiments on fast treadmills indicate that
80% or more of the energy for the race is liberated by anaerobic metabolism
(Margaria 1976). Times given in the sporting pages of newspapers
show that most greyhound races are won at 15–16 m/s, and most horse
races over distances up to 1600 m (one mile) at 16–17 m/s. It seems
possible that horses could go even faster without a rider, and also that they
might attain higher peak speeds over shorter distances. Eaton et al.

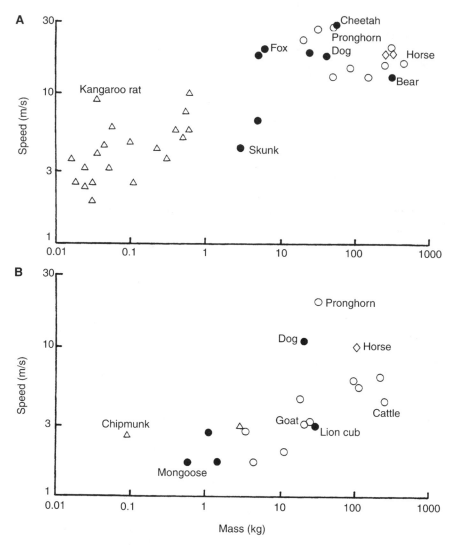

Fig. 7.1. Maximum speeds of mammals, plotted against body mass: (A) maximum sprint speeds; (B) the maximum speeds that can be sustained by aerobic metabolism. Filled circles, Carnivora; hollow circles, Artiodactyla; diamonds, Perissodactyla; triangles, Rodentia. From Alexander (1993a), with the point for the cheetah moved to take account of Sharp's (1997) record.

(1995a) measured the oxygen consumption of thoroughbred horses galloping on a treadmill at speeds that they could not sustain for long, and
calculated that anaerobic metabolism may supply less than 30% of the energy for races.

Cheetahs (*Acinonyx jubatus*) seem to be the fastest of all runners. A
widely quoted record of one running at 32 m/s has been discredited, but
Sharp's (1997) measured speed of 29 m/s can be relied on. Sharp is an
experienced athletics timer. His record is of an animal that had been
tamed, but had been allowed to return to the wild. Sharp measured the
times it took to cover a 200-m course from a flying start, in both directions. The animal was induced to run by towing a piece of meat behind a
vehicle. Cheetahs have remarkably large thigh muscles; Alexander (1993a)
weighed the thigh muscles of a cheetah and found that they were 50%
heavier than predicted by allometric equations for a typical quadrupedal
mammal of the same body mass. A lion and a tiger, however, had thigh
muscles very close to the allometric predictions for their body masses.

Figure 7.1A shows that, as a general rule, larger mammals can run faster
than small ones. However, the very largest mammals seem to be relatively
slow. Alexander and Pond (1992) were unable to induce white rhinoceros
(*Ceratotherium simum*; adult mass around 2 tonnes) to run faster than 7.5
m/s. I know no reliable measurements of high speeds for elephants. The
graph also shows substantial differences in speed between mammals of
similar mass. For example, horses are faster than bears, and kangaroo rats
are faster than other mammals of similar mass.

Ostriches (*Struthio camelus*) seem to be as fast as horses. I have driven
alongside one with the speedometer reading 60 km/h (17 m/s) (Alexander et al., 1979c). I have already pointed out that speeds measured in this
way are not always reliable, and quote this speed only because no better
measurements are available. Birds that can fly are much slower runners
than ostriches, which should not surprise us, because their leg muscles
make up a much smaller proportion of the mass of the body. Many lizards
are as fast as mammals of similar mass. Bonine and Garland (1999) recorded a 16-g *Cnemidophorus* running at 6 m/s, which is slower than the
kangaroo rat in Figure 7.1A, but faster than the other small mammals.
Some arthropods run at speeds that seem high for animals of their size.
Ghost crabs (*Ocypode quadrata*, about 50 g) can run short distances at 1.6
m/s (Blickhan and Full 1987). Cockroaches (*Periplaneta americana*, 0.8
g) sprint at up to 1.5 m/s (Full and Tu 1991).

Speeds that can be sustained by aerobic metabolism are generally much
lower than maximum sprinting speeds. For example, elite human sprinters
run 100-m races at 10 m/s, but the (different) elite athletes who race over
10,000 m achieve only 6 m/s over the longer distance. Figure 7.1B shows
some data for animals, most of it obtained by running them on treadmills.

In the cases of dogs, horses, and elands (*Taurotragus*) the treadmill could not be driven as fast as the animal's maximum aerobic speed on level ground. Accordingly, the animal's oxygen consumption was measured when it was running as fast as possible uphill on a sloping treadmill, and used to estimate the maximum aerobic speed on level ground. Maximum aerobic speeds, like sprint speeds, are generally higher for larger animals. However, species of similar size may have very different maximum aerobic speeds. For example, dogs are faster than goats of equal mass, and horses are faster than cattle. These particular differences may be to a large extent the result of selective breeding; dogs and horses have been bred for tasks that involve running substantial distances, whereas it is probably an advantage to have goats and cattle that do not run too fast.

The pronghorn antelope (*Antilocapra americana*) of the North American prairies seems to be able to maintain higher speeds than any other mammal. There is a field observation of pronghorn running 11 km in 10 min, a speed of 18 m/s. The maximum rate of oxygen consumption has been measured for captive pronghorn galloping uphill on a sloping treadmill. It is five times as high as the maximum rate of oxygen consumption of a goat of equal mass, and it has been estimated from it that the maximum aerobic speed may be 20 m/s (Lindstedt et al., 1991). Differences between pronghorn and goats, which help to explain the difference in athletic performance, include the following. Pronghorn have lungs more than twice the volume of the lungs of goats of the same mass. Pronghorn hearts can pump blood three times as fast as goat hearts, and the hemoglobin concentration is higher than in goat blood; consequently, the rate at which hemoglobin can be pumped round the body is almost five times as high as in goats. Pronghorn have only a little more muscle in their bodies than goats, but the concentration of mitochondria in the muscles is much higher than in goats, making the total volume of muscle mitochondria 2.5 times as high as in goats. Maximum sprinting speed may depend largely on the mass of muscle in the body, as a proportion of body mass, but maximum aerobic speed depends on the capacity of the lungs and blood system to supply oxygen to the muscles, and on the capacity of the mitochondria to use it.

The maximum aerobic speeds of ghost crabs are one-tenth or less of their maximum sprinting speeds (Full 1987).

The speed of an animal is stride length multiplied by stride frequency, so speed can be increased by taking longer strides, by increasing stride frequency, or by a combination of the two. Mammals generally increase both stride length and stride frequency to increase speed within the walking and trotting range; but once they have reached the high speeds at which they gallop, further increases of speed depend on increased stride length, with stride frequency kept more or less constant (Heglund et al.

1974). The lizard *Dipsosaurus dorsalis* and the cockroach *Periplaneta americana* increase both stride length and stride frequency as they increase speed (Fieler and Jayne 1998; Full and Tu 1991).

Relative stride length is stride length divided by the height of the hip joint from the ground. Animals of different sizes moving in dynamically similar fashion (see Section 4.2) would have equal relative stride lengths. Dynamically similar movement is possible only when they are traveling with equal Froude numbers, (Speed)^2/Hip height x Gravitational acceleration). Figure 7.2A is a graph of relative stride length against Froude number for eleven species of mammal ranging from small rodents to rhinoceros. Data are shown for each species walking or running at several different speeds. Open and filled symbols distinguish between the two groups of mammals that Jenkins (1971) described as noncursorial and cursorial. Noncursorial mammals, such as rodents and small carnivores, stand and run on strongly bent legs, with the humerus and femur more nearly horizontal than vertical. Cursorial mammals, such as ungulates and large carnivores, keep their legs much straighter, with the femur and often also the humerus more nearly vertical than horizontal. Most mammals of less than 3 kg mass are noncursorial, and most mammals of more than 3 kg are cursorial. Domestic cats are unusually small cursorial mammals, and coypu (*Myocastor coypus*, about 7 kg) are unusually large noncursorial ones. Figure 7.2A shows that each species of mammal increases stride length as it increases speed, except for a drop in stride length at the transition from walking to trotting. At any given Froude number, noncursorial mammals use larger relative stride lengths than cursorial ones.

Figure 7.2A includes no primates and no bipeds. At the same Froude number, nonhuman primates use even larger relative stride lengths than noncursorial mammals (Alexander and Maloiy 1984). Humans and kangaroos use about the same relative stride lengths as cursorial mammals at the same Froude numbers (Fig. 6 of Alexander 1989c), and hopping birds use relative stride lengths within the range for noncursorial running mammals (Hayes and Alexander 1983). When running fast, the lizard *Dipsosaurus* takes longer strides than noncursorial mammals of equal hip height, running at the same speed (Fieler and Jayne 1998). Cockroaches run with higher stride frequencies and shorter strides than predicted for mammals of the same mass, running at the same speed (Full and Tu 1991).

Step length is the distance traveled while a particular foot is on the ground. Duty factor is defined as the fraction of the duration of a stride, for which a foot is on the ground, but is approximately equal to step length divided by stride length. As an animal increases its speed and takes longer strides, it must increase step length, or reduce duty factor, or both. Mammals and birds increase step length a little as they increase speed, but achieve longer strides mainly by reducing the duty factor. Step length is

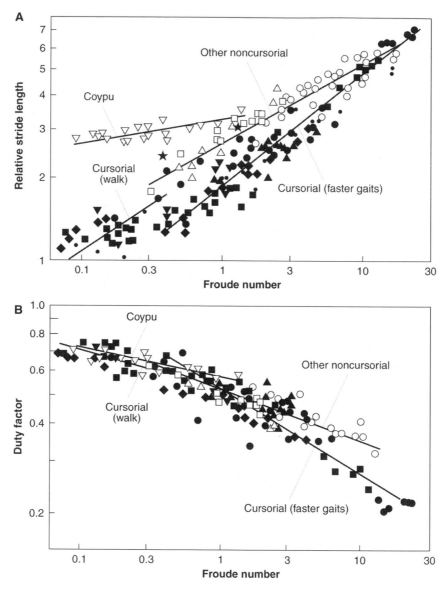

Fig. 7.2. Stride lengths and duty factors of mammals, related to speed. In (A) relative stride length and in (B) duty factor is plotted against Froude number. Different symbols are used for different species. Hollow symbols refer to noncursorial species (rodents and ferret) and filled symbols to cursorial species (dog, cat, sheep, camel, horse, and two species of rhinoceros). From Alexander and Jayes (1983).

limited by leg length, but higher relative step lengths (step length/hip height) are possible in birds and rodents that run with their legs strongly bent than in cursorial mammals such as horses that run with their legs straighter (Hoyt et al., 2000). Figure 7.2B shows that quadrupedal mammals reduce their duty factors from around 0.7 in slow walking to 0.2 (in dogs) or 0.3 (in ferrets, *Putorius putorius*) in fast galloping. Duty factors of humans, kangaroos, and crows fall through similar ranges as speed increases (Alexander and Maloiy 1984; Hayes and Alexander 1983). An ostrich running fast had a duty factor of 0.29 (Alexander et al., 1979c). The lizard *Dipsosaurus* reduces its duty factor from 0.63 to 0.34 as it increases its speed from a slow walk to a fast run (Fieler and Jayne 1998). In contrast, the cockroach *Blaberus discoidalis* reduces the duty factors of its feet only a little as it increases speed (Ting et al. 1994). It seldom uses duty factors outside the range 0.50 to 0.75.

7.2. Gaits

Gaits in which the duty factors are greater than 0.5 are generally described as walks, and those with smaller duty factors as runs. By this criterion, there are inevitably stages in a running stride at which both feet of a pair are off the ground. Quadrupedal mammals and humans generally change from walking to running at Froude numbers close to 0.5 (Alexander and Jayes 1983; Alexander 1989a). The change is often abrupt, but in sheep the two gaits merge into each other (Jayes and Alexander 1978). At about the same Froude number, kangaroos change from a shuffling gait (using all four feet and the tail) to a bipedal hop, and crows change from walking to hopping (Hayes and Alexander 1983). If runs are defined as gaits involving duty factors less than 0.5, cockroaches seldom run, even at high speeds.

The Froude number is $(Speed)^2/(Gravitational\ acceleration \times Hip\ height)$, so if animals of different sizes change gaits at equal Froude numbers, they do so at speeds proportional to the square root of hip height. For example, dogs change gaits at about half the speeds at which horses with legs four times as long make the corresponding changes. In normal life, the gravitational acceleration is almost constant, but if the speed of gait change depends on the Froude number, we should expect an animal that is tested in different gravitational environments to change gaits at speeds proportional to the square root of the gravitational acceleration. For example, the gravitational acceleration on the surface of Mars is 0.4 of the value on earth, so a person walking on Mars could be expected to change from walking to running at $\sqrt{0.4} = 0.63$ of the speed at which the change is made on earth. Other mechanically equivalent speeds, for example, the speed at which the mechanical cost of transport is least, should be reduced by the same factor.

Kram et al. (1997) have performed ingenious experiments in which they simulated low gravity by having people walk while suspended by an elastic rope that partially counteracted their weight. They found that this reduced the speed of the walk/run transition, but left the Froude number (calculated using the simulated gravitational acceleration) unchanged. However, the simulation of low gravity in these experiments was not entirely satisfactory. Though the gravitational force on the body as a whole was partially counteracted, the gravitational forces on limb segments were unaltered. Thus, the periods with which the limbs would swing passively, as pendulums, were unaltered. If the speed at which running becomes more economical than walking depends in part on the energy needed to swing the limbs, this speed might not be the same in the simulation as in real reduced gravity.

A more satisfactory (but much more expensive) experiment was performed by Cavagna et al. (1998). Their subjects walked across a force plate fixed to the floor of an aircraft, while it executed parabolic flights. As the aircraft flew to the top of the parabola and down again, it had a large downward acceleration that remained more or less constant for 30 s. During this time, the gravitational acceleration for bodies in the aircraft, relative to the floor of the aircraft, matched the gravitational acceleration on Mars. We saw in Section 6.6 that walking involves pendulumlike energy exchanges between kinetic energy and gravitational potential energy. Cavagna and his colleagues used the force plate records to calculate how well kinetic energy losses were matched by potential energy gains, and vice versa. They found that the speed at which kinetic and potential energy were most effectively exchanged in simulated Martian gravity was about 0.63 of the speed of most effective exchange on earth. In other words, the energy exchange was most effective at the same Froude number in the two gravitational environments.

Animals use four bipedal gaits. In walking and running, the left and right legs move alternately, half a stride out of phase with each other. In in-phase hopping they move in phase with each other. Finally, in out-of-phase hopping gaits (called skipping gaits by Minetti [1998]) the two legs are clearly out of phase with each other, but by less than half a stride. Humans and some birds walk at low speeds and run at high ones. Kangaroos and some rodents and birds use the in-phase hop. Crows walk at Froude numbers less than about 0.5, and use the out-of-phase hop at higher Froude numbers (Hayes and Alexander 1983).

Quadrupedal gaits are described as symmetrical if the left and right legs of each pair move half a stride out of phase with each other. Other gaits are described as asymmetrical. The walking gaits used by quadrupedal

Fig. 7.3. Horses (A) trotting and (B) galloping. From Gambaryan (1974).

mammals are symmetrical. The feet move in the order left fore, right hind, right fore, left hind, left fore again, and so on; the significance of this order will become apparent when we discuss the stability of walking, in Section 7.9. Hildebrand (1976) has shown that different groups of mammals regularly use different variants of the quadrupedal walk. Some move the feet of the same side of the body in rapid succession, with a longer interval before the feet of the other side move; some move diagonally opposite feet in rapid succession; and some move all four feet at approximately equal intervals.

The trot, the pace, and the amble are symmetrical quadrupedal runs. In trotting (Fig. 7.3A), diagonally opposite feet move in phase with each other; in pacing, the two feet on the same side of the body move in phase with each other; and in ambling, the four feet move at roughly equal intervals, in the same order as in walking. Most mammals trot when traveling at moderate speeds, but camels pace and elephants sometimes amble (Gambaryan 1974).

At high speeds, quadrupedal mammals use asymmetrical gaits, generally making the change from a trot or pace to an (asymmetrical) gallop at a Froude number of about 2.5 (Alexander and Jayes 1983). Several variants of galloping are recognized (Gambaryan 1974; Hildebrand 1977). The canter, which is recognized by horse riders as a distinct gait, is a form of gallop used at low speeds. Galloping mammals arch their backs during the part of the stride when the fore feet are on the ground, and straighten them while the hind feet are on the ground (Fig. 7.3B). This enables them to take advantage of the elastic properties of a sheet of tendon in the back, as described in Section 7.5.

Cursorial mammals and birds move their legs more or less in parasagittal planes (i.e., in vertical planes, parallel to the long axis of the body [Jenkins 1971]). At the midpoint of a step, their feet pass vertically under the hip and (in quadrupeds) the shoulder joints. In a set of footprints, the line of left footprints is close to the line of right footprints (see, for example, tracks of fox and deer in Lawrence and Brown [1974]). Noncursorial mammals hold the femur and humerus at an angle to the sagittal plane, and the separation between the lines of their left and right footprints is a larger fraction of stride length (see, for example, tracks of weasels and rodents). Modern reptiles and amphibians walk and run with the feet far lateral to the hip and shoulder joints, but fossil footprints show that dinosaurs walked more like birds and mammals (Thulborn 1990). If the feet are lateral to the body, bending the body to the left moves the right fore foot forward relative to the left fore foot, and the left hind foot forward relative to the right hind foot (Fig. 7.4A). Bending to the right has the reverse effect. Lizards and salamanders take advantage of this to extend their steps, by bending their backs from side to side as they walk or run quadrupedally.

This side-to-side bending throws the body into waves. These may be standing waves or traveling waves; Fig. 7.4B and C shows the difference. Lizards with well-developed legs use standing waves at low speeds, and traveling waves that move posteriorly along the body at high speeds (Ritter 1992). Lizards that have reduced or rudimentary legs use traveling waves at all speeds. Section 9.4 describes how snakes crawl, using traveling waves.

Whether standing or traveling waves are used, the wavelength is twice the distance from the shoulders to the hips. Thus, the shoulders are turned to the left while the hips are turned to the right, and vice versa. For this to lengthen the step, the left fore foot must be on the ground while the pectoral girdle is turning to the left, and the right hind foot must be on the ground while the pelvic girdle is turning to the right. Lizards use symmetrical gaits in which diagonally opposite feet commonly move more or less simultaneously, as in the mammalian trot (White and Anderson 1994). Tortoises and turtles, which do not (and cannot) bend their backs from side to side, move diagonally opposite feet slightly out of phase with each other, as described in Section 7.9. At high speeds, many lizards run bipedally on their hind legs (Irschick and Jayne 1999). Bipedal running is possible because the long tail brings the center of mass of the body back to a position near the hip joints.

Cockroaches and other insects, running on all six legs, move them in two groups of three. Each group consists of the front and hind legs of one side, and the middle legs of the other. The two groups move half a stride out of phase with each other, so the animal is supported alternately by the

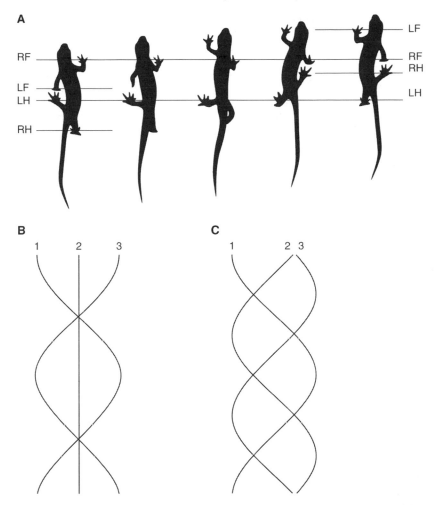

Fig. 7.4. (A) Successive stages in the stride of a newt (*Triturus*). RF, right fore; LF, left fore; LH, left hind; RH, right hind. From Roos (1964). (B, C) Successive positions of (B) a standing wave and (C) a traveling wave.

two tripods of legs (Fig. 7.5A: see, for example, Ting et al., [1994]). The cockroach *Periplaneta americana* runs on all six legs at speeds up to 1.0 m/s, but at higher speeds uses only the middle and hind legs, or the hind legs alone (Full and Tu 1991).

Short-legged centipedes such as *Scolopendra* move each leg slightly after the leg immediately in front, so waves of leg movement travel backward along the body. In *Scutigera* (a centipede with very long legs) and in millipedes waves of leg movement travel forward along the body (Manton

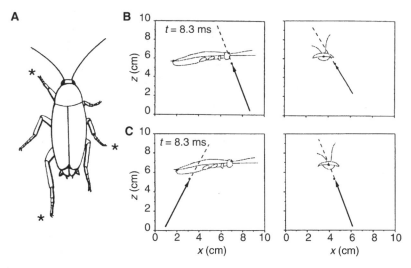

Fig. 7.5. (A) A cockroach. The three legs marked by asterisks move together, and then the other three. (B, C) Forces on a fore (prothoracic) and a hind (meta-thoracic) foot, respectively, of a walking cockroach. Small circles mark the position of the foot, in each case. (B) and (C) are from Full et al. (1991).

1973). Waves of lateral bending travel posteriorly along the body of *Scolopendra* as it runs. Anderson et al. (1995) showed that this bending is caused actively by contractions of longitudinal muscles; it is not the passive consequence of forces exerted by the legs.

7.3. FORCES AND ENERGY

When an animal walks or runs, little energy is dissipated by friction (Section 3.4) or air resistance (unless there is a strong wind [Pugh 1971]). Therefore, we will have to concern ourselves in this section only with kinetic, gravitational, and elastic strain energy. Each of these energies has the same value at corresponding stages of successive strides, if the animal is traveling at constant speed on level ground. Therefore, the positive work and the negative work that the muscles have to do are almost exactly equal.

Figure 7.6A shows the forces on the feet of a hopping kangaroo rat, recorded by means of a force plate. Both hind feet landed on the plate, followed 0.1s later by the fore feet, which exerted much smaller forces. We will look in detail only at the forces on the hind feet. The vertical component of force rises to a peak of 2.5 N (3.2 times the animal's weight) and then falls. The mean value of this component of force over a complete stride must equal body weight. It is because the feet are off the ground

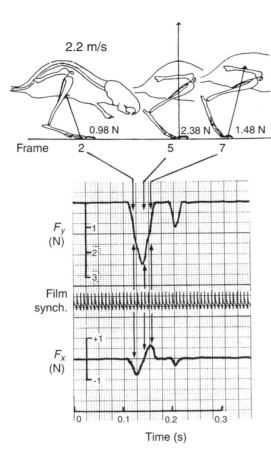

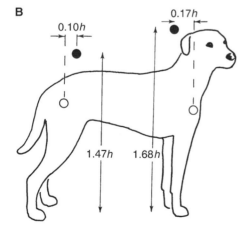

Fig. 7.6. (A) Records of the vertical (F_y) and horizontal (F_x) components of the forces on the feet of a kangaroo rat (*Dipodomys spectabilis*) as it hopped across a force plate. Above the force record are tracings from an X-ray cine film taken simultaneously, with arrows representing the forces. From Biewener et al. (1981).

(B) An outline traced from a photograph of a dog, showing (hollow circles) the hip and shoulder joints and (filled circles) the points through which the forces on the feet tend to act throughout a step. Distances are expressed as multiples of the hip height h. From Jayes and Alexander (1978).

for much of the stride that the force while they are on the ground must exceed body weight. The record of the horizontal component shows a decelerating force followed by an accelerating one. If the animal is hopping at a steady speed, the mean horizontal component of force, over a complete stride, must be zero. The outlines at the top of the figure show three stages of the step, traced from an X-ray cine film. Superimposed on them are arrows showing the resultant forces on the feet, calculated from the force records below. Throughout the step, the line of action of the force passes through the central part of the body, close to the presumed position of the animal's center of mass. The mean moment of the force about the center of mass, over a complete stride, must be zero; otherwise, the animal would somersault at an ever increasing rate. There is no requirement for the moment to be zero throughout the step, but if the line of action is always close to the center of mass, keeping the moments small, the animal will rock very little as it hops.

Other bipeds similarly keep the lines of action of the forces on their feet close to the center of mass of the body when they hop or run or walk; see, for example, Alexander and Vernon (1975a) on kangaroos, Clark and Alexander (1975) on quail (*Coturnix*), and Alexander and Vernon (1975b) on humans. In quadrupedal mammals, as in bipeds, each foot exerts a decelerating followed by an accelerating horizontal component of force in each step. The forces on the fore feet tend to keep in line with a point above the shoulder, and the forces on the hind feet tend to keep in line with a point above the hips. Figure 7.6B shows the positions of these points for dogs. The positions for sheep are very similar (Jayes and Alexander 1978).

Force plate records of lizards show that each foot exerts a decelerating horizontal component of force, followed by an accelerating one, while on the ground (Farley and Ko 1997). The feet also exert transverse components of force, the right feet pushing toward the right and the left feet toward the left, making the lines of action of the ground forces slope inward toward the shoulder and hip (Ritter 1995). The feet of walking turtles exert near-vertical forces on the ground, so that much larger moments act about the shoulder and hip than about the elbow and knee (Jayes and Alexander 1980; van Leeuwen et al., 1981). The feet of cockroaches exert transverse as well as vertical and longitudinal components of force, so that the resultant force on each foot is more or less aligned with the leg (Fig. 7.6B) (Full et al., 1991, 1995).

In Section 6.5 we discussed an inelastic kangaroo model of hopping or running. Our analysis of it showed that at high speeds most of the work required for locomotion was due to the horizontal components of force on the feet. This suggests that the energy cost of locomotion might be less if the forces on the feet were kept vertical throughout the step. However, if

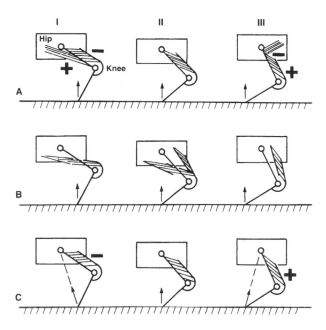

Fig. 7.7. Diagrams of a running animal, showing the trunk and one leg. (A–C) Three possible patterns of muscle action. (I), (II), and (III), in each case, represent successive stages of a step. Muscles doing positive and negative work are labeled + and −, respectively. The mass of the leg is ignored, in the discussion of this diagram. From Alexander (1988).

the force were kept vertical, energy would be wasted by muscles doing work against each other, in the same sort of way as arm muscles work against each other in Fig. 3.3A. In Fig. 7.7A the force on the foot remains vertical throughout the step. At all three stages, the force on the foot exerts a clockwise moment about the knee, so the knee extensor muscle (illustrated) must be active. Between stages I and II the knee is bending, so the extensor muscle is being stretched, doing negative work. Between stages II and III, however, it is extending, and the muscle must be doing positive work. Between stages I and II the force exerts an anticlockwise moment about the hip, so the hip extensor muscle must be active. The hip joint is extending, so the muscle is doing positive work. Similarly, between stages II and III the hip flexor muscle is active and is doing negative work, resisting the continued extension of the hip. Throughout the stride, one muscle is doing positive and another negative work. Energy is being wasted by muscles doing work against each other.

In Fig. 7.7B, the forces are still vertical throughout the step. In this case, however, each muscle crosses both the hip joint and the knee, ex-

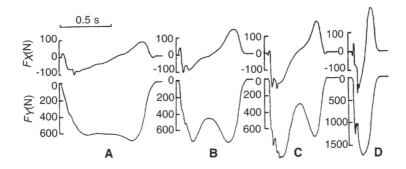

Fig. 7.8. Records of the force on one foot of a 70-kg man walking at (A) 0.9 m/s; (B) 1.5 m/s, and (C) 2.1 m/s; and (D) running at 3.6 m/s. From Alexander and Jayes (1980).

erting moments about both. The muscles shown could balance the moments about both joints, if the ratios of their moment arms about the two joints were right. In that case, their lengths would remain constant while they were active; they would have to exert forces, but they would do no work. This is a possible design for economical locomotion, but evolution does not seem to have adopted it. The muscle shown in (III) resembles the rectus femoris muscle of mammals, but kangaroos are the only mammals I know to have a muscle like the one shown in (I) (the femorococcygeus muscle).

In Fig. 7.7C, the force on the foot is kept in line with the hip joint. It exerts no moment about the hip, so no hip muscle need be active. The knee extensor muscle must be active throughout, doing negative work as the knee bends followed by positive work as it extends. Work is done, but it is not work against other muscles.

Alexander (1991b) considered a leg like the one in Fig. 7.7A and C, with muscles that each crossed only one of the two joints. I calculated the work done by the muscles, as the hip moved forward horizontally, for different directions of the force on the foot. I showed that the work (both positive and negative) was least if the force on the foot was kept in line with the hip joint as in (C). I assumed that the hip moved precisely horizontally, and I did not consider how the optimum force direction might change if some of the muscles crossed both joints. However, this simple analysis suggests that animals may be able to save energy by exerting horizontal components of force with their feet as they walk and run.

Now we will consider the vertical components of force in more detail. Figure 7.8 shows force plate records of a man walking at three different speeds, and running. In slow running (A) the vertical component of the force on a foot rises, plateaus, and falls. In faster walking (B) the force record has two peaks, and in very fast walking (C) the peaks are higher. In running (D), however, there is only a single main peak of force (I am ignoring a small initial peak that occurs in walking as well as running and is due to the foot being brought suddenly to rest when it hits the ground). As speed increases, starting with a slow walk, the force pattern becomes more and more two-peaked, then changes abruptly to the single-peaked pattern of running. The duty factor is between 0.55 and 0.70 throughout the range of walking speeds, then drops abruptly to 0.3–0.4 when running commences (Fig. 1.5A).

To progress further, we need a more precise means of describing the one- and two-peaked patterns of force. Let a foot be on the ground from time $-\tau/2$ to time $+\tau/2$. At any time t within this interval, let the vertical component of force on the foot F_y be

$$F_y = A \left[\cos\left(\frac{\pi t}{\tau}\right) - q \cos\left(\frac{3\pi t}{\tau}\right) \right] \qquad (7.1)$$

where A is the constant needed to make the mean value of F_y, over a complete stride, equal to body weight; and q is a parameter that is called the shape factor because it determines the one- or two-humped shape of the force record (Alexander and Jayes 1978). Obviously, when $q = 0$ a graph of force against time is a half cycle of a cosine curve. As q increases, the graph becomes flatter-topped and eventually, for values exceeding 0.15, two-humped. Negative values of q make the graph bell-shaped. Equation 7.1 successfully imitates the patterns of vertical force observed in human walking and running. The shape factor q is 0.22 in Fig. 7.8a, 0.36 in (B), 0.55 in (C), and −0.07 in (D). More generally, it increases gradually from about 0.2 in slow walking to 0.7 in fast walking, then falls abruptly to about −0.1 when running starts (Fig. 1.5B).

Figure 7.9 shows the consequences of different combinations of shape factor and duty factor. In (i) the shape factor is 0.4, so the force curves are markedly two-peaked; and the duty factor is 0.75, so there are quite long periods of overlap, when both feet are on the ground. The total force (broken curves) is greater than body weight when both feet are on the ground, and less than body weight when only one foot is on the ground, with the supporting leg vertical. That implies that the body must have an upward acceleration when both feet are on the ground, and a downward acceleration when the supporting leg is vertical as in the stiff-legged models of walking shown in Fig. 6.5A and C. It must be lowest when both feet are on the ground, and highest when the supporting leg is vertical.

In Fig. 7.9(iv), the shape factor is 0, so the force curves are half-cycles of cosine curves. The duty factor is only 0.55, so the times when both feet

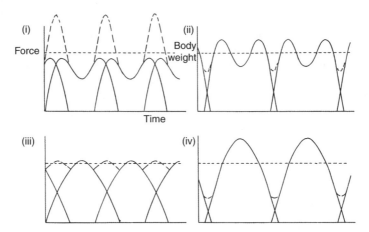

Fig. 7.9. Schematic graphs of the vertical components of force acting on the feet of a biped, with different shape factors q and duty factors β. Each graph shows the forces exerted by the left and right feet in several successive steps, and (by broken lines) the total force when both feet are on the ground. (i) $q = 0.4$, $\beta = 0.75$; (ii) $q = 0.4$, $\beta = 0.55$; in (iii) $q = 0$, $\beta = 0.75$ and (iv) $q = 0$, $\beta = 0.55$. From Alexander and Jayes (1978).

are on the ground are short. In this case, the total force is less than body weight when both feet are on the ground, and more than body weight when only one foot is on the ground, with the supporting leg vertical. This is the opposite of the situation in (i). The body must be highest when both feet are on the ground, and lowest when the supporting leg is vertical, which is how a mass–spring model (Fig. 6.5E and F) would behave. In Fig. 7.9(iv) the duty factor is 0.55, so the gait is a walk; it can be described as a compliant walk, to distinguish it from the stiff-legged walk of Fig. 7.9(i). However, the motion resembles running, in which the force is zero and the body is highest while both feet are off the ground.

In Fig. 7.9, parts ii and iii are graphs for two other combinations of shape factor and duty factor, which make the force rise above body weight twice in each step.

Both in walking and in running, each foot exerts horizontal components of force that tend to decelerate the animal in the first half of a step and accelerate it in the second half, as we have already noted. Consequently, the kinetic energy of a biped is least when the supporting leg is vertical and greatest when both feet are on the ground (in walking) or neither foot is on the ground (in running). The biped's gravitational potential energy is least when its body is lowest. In stiff-legged walking, this is when both feet are on the ground, but in compliant walking and in running it is when the supporting leg is vertical. In stiff-legged walking, kinetic energy is high when gravitational potential energy is low, and vice versa; but in running kinetic and gravitational potential energy fluctuate in phase with each other. Some authors use this to define walking and running (McMahon 1985). They regard a gait as a walk if kinetic and potential energy are out of phase, and as a run if they are in phase.

Dogs, sheep, and horses, like humans, use higher shape factors when walking than when running (Alexander and Jayes 1978, 1983; in the latter paper, the quantity a_3/a_1 equals $-q$).

Cavagna et al. (1977) studied the fluctuations of kinetic and gravitational potential energy during walking and running of various mammals and birds. They used a very long force plate, long enough to record several successive footfalls. From force records of the animals traveling over this plate they were able to calculate the fluctuations of external kinetic energy and of gravitational potential energy. (The distinction between external and internal kinetic energy is explained in Section 3.1. All references to kinetic energy so far in this chapter refer to external kinetic energy.) Cavagna and his colleagues found that when turkeys and a rhea walked slowly, kinetic and potential energy fluctuations were out of phase with each other, as they were also in human walking. Their records of quadrupedal walking of monkeys, dogs, and sheep are hard to interpret, because the fore and hind legs do not move simultaneously. A much clearer picture would probably have emerged if it had been possible to measure the energy changes of the fore and hind quarters separately. They got very different results for running by humans, turkeys, and the rhea; for hopping by kangaroos and a spring hare; and for trotting by the monkeys, dogs, and sheep. In all these cases, the fluctuations of kinetic and potential energy were in phase. There was no difficulty in interpreting the records of the quadrupeds trotting, because in the trot fore and hind feet are set down simultaneously.

Farley and Ko (1997) performed similar experiments with two species of lizard. At low speeds, gravitational potential energy fluctuations were sometimes in phase with external kinetic energy fluctuations, and sometimes 180° out of phase with them. The gait was sometimes a stiff-legged walk, and sometimes a compliant walk. At higher speeds, however, the kinetic and potential energy fluctuations were always in phase. The highest speed at which stiff-legged walking was observed corresponded to a Froude number of about 1.

Full and Tu (1990) ran cockroaches (*Blaberus discoidalis*) over a force plate and showed that external kinetic energy fluctuations and gravitational potential energy fluctuations were in phase with each other at all speeds. We have already noted that this species almost always uses duty factors greater than 0.5 at all speeds. Thus, its gait at all speeds is a compliant walk.

7.4. ENERGY-SAVING SPRINGS

Simple models in Chapter 6 showed how springs can reduce the work that muscles have to do in hopping and running. If the kinetic and potential energy that the animal loses in the first half of a step can be stored as elastic strain energy, and returned in the second half of the step, the muscles will be relieved of the necessity to do work. The inelastic kangaroo model (Section 6.5) required a substantial input of muscular work in each step, but the mass–spring model required none during steady locomotion.

Early evidence that springs might play an important energy-saving role in running came from measurements of human oxygen consumption. Cavagna et al. (1964) used their force plate to determine the fluctuations of external kinetic energy and of gravitational potential energy that occur in running. Using these data and measurements of oxygen consumption, they calculated the efficiency of the muscles, assuming that every increase in the mechanical energy of the body was supplied by muscular work. The calculated efficiency was much higher than the efficiencies of 25% or less that had been measured in experiments on bicycle ergometers. The paradox would be resolved if about half of the positive and negative work were done by springs.

At that time, anatomists were taught dogmatically that tendon is not elastic. However, Alexander (1974a) used a force plate to record the forces on the feet of a dog taking off for large jumps. I took films simultaneously that enabled me to calculate the moments of the measured forces about the ankle. Hence, I calculated the forces that must act in the gastrocnemius and plantaris tendons (the tendons of the principal extensor muscles of the ankle), and the amount by which these forces could be expected to stretch the tendons. These calculations led to the unexpected conclusion that the bending and reextension of the ankle that occurred in takeoff was largely due to passive elastic stretching and recoil of the tendons, with very little change in the lengths of the gastrocnemius and plantaris muscle fibers. A similar force plate study of kangaroo hopping (Alexander and Vernon 1975a) led to a similar conclusion; the gastrocnemius and plantaris tendons were stretching substantially, and recoiling elastically, in each step.

Subsequent developments in technology have made possible much more direct investigation of the muscles and tendons in kangaroo hopping. Biewener et al. (1998b) put tendon buckles on the tendons of wallabies (*Macropus eugenii*) to measure the forces acting in them, and implanted sonomicrometry crystals in the muscles to measure the length changes of their fascicles. They recorded the forces and length changes while the wallabies hopped on a treadmill, at various speeds. Afterward they killed the animals, calibrated the tendon buckles, checked that the sonomi-crometry crystals had been correctly placed and measured the elastic prop-erties of the tendons. They found that the plantaris muscle fascicles length-ened and shortened by 0.5 mm or less, and the gastrocnemius fascicles 2.2 mm or less, during the part of the stride while the feet were on the ground and the muscles were exerting substantial forces. These length changes are so small that they may have been due to elastic distortion of the muscle cross bridges, with no need for cross bridges to detach and reattach. While the muscles were changing length very little, the tendons were stretching and recoiling far more. Biewener and his colleagues calculated that work done by elastic recoil of tendons averaged 20 times as much as the work done by the muscle fascicles. They investigated only the muscles of the ankle joint. Tendon elasticity probably contributes much less to the work done by knee and hip muscles, but the elastic extension and recoil of the tendons of ankle extensor muscles is enough to reduce the total work re-quired of the leg muscles by 45% at 6 m/s, the fastest speed of hopping that was investigated.

Roberts et al. (1997) had performed similar experiments on turkeys. Instead of using a tendon buckle, they measured the force in the gastroc-nemius tendon by means of a strain gauge glued to an ossified part of the tendon. They found that the gastrocnemius muscle fascicles changed length very little while the foot was on the ground (but much more while it was off the ground, when the muscle was exerting very little force). Most of the work done by the muscle came from elastic recoil of the tendon.

Elastic stretching and recoil of tendons has been demonstrated by much simpler means in camels and horses (Alexander et al., 1982; Dimery et al., 1986). In these species, some of the distal leg muscles have such extremely short muscle fascicles that their changes of length in locomotion must be almost entirely due to elastic extension and recoil of their tendons. For example, the plantaris muscle of the camel is a stout tendon 1.3 m long, extending from the distal part of the femur to the phalanges, with the muscle belly represented only by a small quantity of muscle fascicles, 1 to 3 mm long. The digital flexor muscles of the forefoot of a pony had tendons about 0.7 m long, and muscle fascicles only 3 mm long in the superficial muscle and 6 mm in the deep one. In the investigations both of camels and of horses, the animals were filmed walking and running at various speeds, and the angular movements of the leg joints were measured. Ex-periments were performed on carcasses to discover the combinations of joint angles that each short-fascicled muscle allowed if its tendon was kept taut but not stretched (each of the muscles in question crosses several joints). This made it possible to calculate the length changes of the ten-dons that were occurring in running. It was found, for example, that the superficial digital flexor of the horse was extended 60 mm beyond its unstretched length, while the foot was on the ground in galloping. It is inconceivable that 3-mm muscle fascicles could extend as much as that. The extension must have been very predominantly elastic stretching of the tendons. The tendons investigated in the horse were stretched while the foot was on the ground by 3–6% in walking, 3–7% in trotting, and 4–9% in galloping. While the foot was off the ground, the tendons became slack and apparently folded. Biewener (1998) made force plate records of horses, and calculated the forces in the tendons of distal leg muscles, making plausible assumptions. He calculated that, in galloping, stresses of 40–50 MPa occurred in several tendons. This would be enough to stretch the tendons by about 5%. He estimated that elastic recoil of tendons contributed up to 40% of the positive work required for trotting and galloping.

Antelopes also have long tendons and very short fascicles in distal leg muscles that function as springs (Alexander et al., 1981). Metabolic energy is needed to develop tension in muscles, even when the contraction is isometric and no work is done. Consequently, energy can be saved by having short muscle fascicles where long ones are not needed. Monkeys have much longer fascicles than antelopes, in the homologous muscles, a difference that may be related to their arboreal habits. A climbing animal may need to be able to exert large forces with its limbs bent or extended, depending on the position of the next foothold. In contrast, for an animal that travels only on the ground, every step is much like the one before. The climbing animal needs to be able to adjust its limb length over a much wider range, and needs fairly long muscle fascicles to make the adjustments.

In humans, the plantaris muscle is rudimentary, but the tendon of the gastrocnemius and soleus muscles (the Achilles tendon) is an important energy-saving spring. The peak force exerted by the muscles in running at 4 m/s (4.9 kN [Thorpe et al., 1998]) divided by the cross-sectional area of the tendon (89 mm^2 [Ker et al., 1987]) gives a stress in the tendon of 55 MPa. The free part of the tendon, outside the muscle bellies, is quite short. However, the tendon continues on and in the muscle bellies as aponeuroses, which also stretch under load (Maganaris and Paul 2000). In addition to the Achilles tendon, ligaments in the arch of the foot serve as springs. Films of barefoot runners show that at the stage of a running step at which the force on the foot is greatest, the ankle joint is about 10 mm nearer the ground than when the foot is rested lightly on the ground (Alexander 1987). This is due to flattening of the arch of the foot, stretching its ligaments. Ker et al. (1987) compressed amputated feet in a dynamic testing machine, in a rig that imitated the pattern of forces that acts in running. We showed that the foot behaves as a passive spring, returning in its elastic recoil 78% of the work done flattening the arch. We estimated that of the external kinetic energy and gravitational potential energy lost and regained by the body in each running step, 35% is stored and returned by stretching and recoil of the Achilles tendon, and a further 17% by flattening and recoil of the arch of the foot.

Though tendon springs play so important an energy-saving role in large mammals, they may be much less important in small ones. Biewener and Blickhan (1988) assessed the role of tendon springs in kangaroo rats, which hop like kangaroos but are much smaller. The body masses of their specimens averaged 107 g, far less than the 3.6- to 5.8-kg wallabies studied by Biewener et al. (1998b). In earlier experiments, they had used a tendon buckle to measure the total force in the tendons of the two principal extensor muscles of the ankle, the gastrocnemius and plantaris. They found that in steady hopping, stresses of 5–10 MPa were developed in these tendons, far less than the 20–30 MPa that can be calculated from the data of Biewener et al. (1998b) for wallabies. Because the stresses are low, the tendons must stretch very little and can store only a little strain energy. Much larger stresses act in the large jumps that the animals make to escape from predators. The penalty for having tendons strong enough for the large jumps is that the energy savings in steady hopping are small.

We saw in Section 3.3 that about 7% of the work done stretching a tendon is not returned in its elastic recoil, but is dissipated as heat. This raises the temperature of the tendon, to as much as 45°C in the case of the superficial digital flexor tendons of galloping horses. This may damage the tendon (Birch et al., 1997).

7.5. Internal Kinetic Energy

So far we have ignored internal kinetic energy, the kinetic energy due to movements of parts of the body relative to its center of mass (Section 3.1). The internal kinetic energy of the animal changes whenever a limb is accelerated or decelerated as it swings forward and back. The following argument shows that internal kinetic energy fluctuations must contribute more to the energy cost of running at high speeds than at low ones. To keep the argument simple, we will ignore the masses of the upper parts of the legs, and consider only the mass of the feet. Consider an animal of mass m running at speed v, taking strides of length λ. Let the total of the masses of its feet be km. When the feet are on the ground, they have velocity $-v$ relative to the body. Every foot is given this velocity once in each stride, so internal kinetic energy $kmv^2/2$ has to be supplied in each stride. The contribution to the cost of transport, obtained by dividing this by body mass and stride length, is $kv^2/2\lambda$. Something should be added to this for the internal kinetic energy associated with the forward swing of the feet while they are off the ground, but we will ignore that. Stride frequency f equals v/λ, so the contribution of internal kinetic energy requirements to the cost of transport can be written $kvf/2$. Stride frequencies of mammals remain more or less constant once they have reached galloping speeds (Heglund et al., 1974), making this contribution to the cost of transport proportional to speed. In contrast, the cost of transport for the inelastic kangaroo model, which ignores internal kinetic energy, decreases as speed increases, approaching an asymptotic value at high speeds (Equation 6.10). If the contribution of internal kinetic energy increases in proportion to speed while other components of the cost of transport decrease, internal kinetic energy becomes increasingly important as speed increases.

Fedak et al. (1982) calculated the internal kinetic energy changes in running of four species of birds and three of mammals, ranging from a 44-g quail (*Excalfactoria chinensis*) to a 99-kg pony. Each animal was filmed running at various speeds on a treadmill, using high-speed X-ray cinematography for the small species and light cinematography for the large ones. Each film was analyzed to determine the coordinates of the joints between body segments in successive frames over several strides. The animals were killed and cut up into segments (foot, lower leg, thigh, etc.). The mass of each segment and the position of its center of mass were determined. From these data, Fedak and his colleagues were able to calculate the velocities of the segments and of the center of mass of the whole body, and hence the internal kinetic energy (the second term on the right-hand side of Equation 3.3). They expressed their result as the power requirement per unit body mass. From it we can calculate that the contribution of internal kinetic energy requirements to the mechanical cost of transport is $0.5v^{0.5}$ J/kg m (where the speed v is expressed in meters per second) for all the species. It increases with speed, but not as rapidly as our simple argument suggested.

Heglund et al. (1982) used force plate data to calculate the fluctuations of gravitational potential energy and external kinetic energy for 13 species of birds and mammals, ranging from kangaroo rats and quails to humans and sheep. From their results it can be calculated that the contribution to the cost of transport of the work needed to supply these energies is about 0.7 J/kg m for all species and all speeds. This agrees well with the cost predicted for high speeds by Equation 6.11 if $\beta\lambda/h$ (= step length/hip height) is a little less than 1. As speed increases, this component of the cost of transport remains more or less constant while the component associated with internal kinetic energy increases. Internal kinetic energy demands the majority of the work of running when $0.5v^{0.5}$ is greater than 0.7, that is, at speeds greater than 2 m/s.

The elastic mechanisms that we have been discussing save energy that would otherwise have had to be provided by muscles, to supply external kinetic energy and gravitational potential energy. They do nothing for

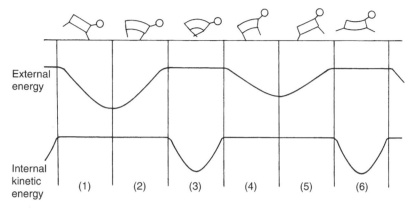

Fig. 7.10. Diagrams of successive stages in a galloping stride, with schematic graphs showing energy changes. "External energy" means external kinetic energy plus gravitational potential energy. From Alexander et al. (1985).

internal kinetic energy, but springs could help there too. The model of running shown in Fig. 6.5F had legs of appreciable mass, so internal kinetic energy fluctuated as it ran. However, it had torsion springs at the hips as well as the compression springs in its legs. The function of the torsion springs was to store up the kinetic energy lost when a leg was halted at the end of its forward or backward swing, and to return the energy by elastic recoil, to provide kinetic energy for swinging the leg in the opposite direction. No animal that I know has springs like this at the tops of its legs, but mammals do seem to take advantage of springs in their backs when they gallop. It is at the high speeds, at which galloping is the preferred gait, that it is most useful to have springs to save and return internal kinetic energy.

The springs are the aponeurosis (tendon sheet) of the principal extensor muscle of the back, and the lumbar part of the vertebral column (Alexander et al., 1985). The back bends and extends in the course of a galloping stride, as shown by the diagrams at the top of Fig. 7.10. At stage 3, when the back is most bent, the fore legs have been swinging back and are about to swing forward; and the hind legs have been swinging forward and are about to swing back. All four legs are halted and set swinging in the reverse direction; internal kinetic energy is lost and immediately regained. The forces needed to reverse the directions of swinging exert bending moments on the back, requiring the extensor muscle to be active. The force exerted by the muscle stretches its aponeurosis, storing up energy that is returned in an elastic recoil. We measured the elastic properties of strips of aponeurosis from dogs and deer, in a dynamic testing machine. From the results, we calculated that the aponeurosis was capable of storing and returning a substantial proportion of the internal kinetic energy lost and regained in each stride.

As well as stretching the aponeurosis, tension in the extensor muscle compresses the lumbar part of the vertebral column, especially the intervertebral disks. We compressed vertebrae and disks in the dynamic testing machine, and showed they could supplement the spring action of the aponeurosis. They could store less energy than the aponeurosis, but enough to seem useful.

The direction of swing of the legs is also reversed, and internal kinetic energy lost and regained, at the stage of the stride shown in Fig. 7.10(6). There is no structure that seems likely to serve as effectively as a spring, at this stage of the stride, as the aponeurosis of the back does at stage (3). However, the tendon of a muscle in the hind leg (tensor fasciae latae) may make a useful contribution (Bennett 1989).

We will see in the next section that galloping requires less metabolic energy than trotting at high speeds. The energy-saving role of the springs in the back may explain why. They cannot function in trotting, in which each left leg swings forward while the corresponding right leg swings back, and vice versa, and bending moments on the back are therefore small.

7.6. METABOLIC COST OF TRANSPORT

Hoyt and Taylor (1981) trained ponies to walk, trot, or gallop on command, so that they could, for example, make them trot at speeds at which they would have preferred to gallop. They measured the rates of oxygen consumption of these ponies, as they walked, trotted, and galloped on a treadmill at various speeds. An example of their results is shown in Fig. 7.11. Notice how the graphs for the three gaits cross. Below 1.7 m/s, walking is more economical of energy than trotting, but above 1.7 m/s the reverse is the case. Below 4.6 m/s, trotting is more economical than galloping, but above that speed the reverse is the case. Accordingly, the pony should walk when traveling at less than 1.7 m/s, trot between that speed and 4.6 m/s, and gallop at higher speeds. That is what it did.

The graph shows energy per unit time plotted against distance per unit time. Therefore, the slope of a line from the origin to a point on the curve, like the broken line in Fig. 7.11, gives the energy cost of traveling unit distance at the speed represented by the chosen point on the line. The broken line is the tangent to the walking curve, and has a lower gradient than any other line joining the curve to the origin. That tells us that the energy cost of walking unit distance at the speed it represents (1.2 m/s)

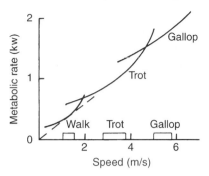

Fig. 7.11. Graphs of metabolic rate against speed for a 140-kg pony walking, trotting and galloping on a treadmill, from the data of Hoyt and Taylor (1981). Boxes on the speed axis show the speeds at which it and other ponies used each gait when moving spontaneously around their paddock. The tangent is explained in the text. From Alexander (2000a).

is less than at any other walking speed. A tangent to the trotting curve would show that the most economical speed for trotting is 3.2 m/s. A pony wishing to travel at a mean speed of 2 m/s would do best to walk part of the way at 1.2 m/s and trot the rest at 3.2 m/s. The boxes on the speed axis show the ranges of speed in which the pony used each gait, when moving spontaneously in its paddock. They show that it avoided the expensive speeds near gait transitions.

Similar data have been obtained for humans. Graphs of metabolic rate against speed for walking and running cross at 2.2 m/s (Margaria 1976; Minetti et al., 1994). The speed at which adult humans change voluntarily from walking to running is about the same, 1.9–2.1 m/s (Thorstensson and Roberthson 1987; Minetti et al., 1994). Young adults in city streets walk on average at 1.5 m/s (Wirtz and Ries 1992), close to the speed that minimizes energy cost per unit distance.

Kangaroos similarly change from their slow shuffling gait to hopping at approximately the speed at which the faster gait becomes the more economical. A remarkable property of kangaroo hopping is that the rate of oxygen consumption remains almost constant throughout the range of hopping speeds (Dawson and Taylor 1973). This seems to be due to much more energy being saved by elastic storage in tendons at higher speeds, due to larger forces acting on the tendons (Biewener et al. 1998b). The metabolic cost of transport is high at low speeds, compared to quadrupedal mammals of similar mass, and similar to quadrupeds at high speeds (see the graph in Alexander [1982]).

Figure 7.11A shows three intersecting curves but, if we were not interested in differences between gaits, a single straight line would approximate the data reasonably well. Taylor et al. (1982) measured the metabolic rates of various mammals and birds running on treadmills, and analyzed them together with similar data for many other species, from earlier investigations. The species included in their analysis ranged from 20-g mice to 250-kg cattle, and from 18-g plovers (*Charadrius wilsonia*) to 100-kg ostriches. For each species they drew a straight-line graph of metabolic rate against speed, ignoring gait changes. From the results they calculated a general equation relating metabolic rate while running to speed and body mass, which can be applied to mammals or birds of any size. From their equation it is easy to derive an equation giving metabolic cost of transport (T_{metab}, in J/kg m) in terms of body mass (m kg) and speed (v m/s). The equation is

$$T_{metab} = 10.7m^{-0.32} + \frac{6.0m^{-0.30}}{v} \tag{7.2}$$

At high speeds, the metabolic cost of transport approaches its asymptotic value $10.7m^{-0.32}$. This minimum metabolic cost of transport is plotted against body mass in Fig. 7.12. The figure shows not only data for the mammals and birds studied by Taylor and his colleagues, but also data for reptiles, amphibians, and arthropods collected from other sources. Remarkably, all the metabolic data cluster around a single line. There are some quite large deviations from the line. The asymptotic metabolic cost of transport is 1.9 times as high for a penguin as for a turkey of similar mass, and three times as high for a lion cub as for a dog of similar mass. However, the one line expresses well the trend for arthropods as well as vertebrates.

The figure also shows the mechanical cost of transport (ignoring both internal kinetic energy and elastic savings) for a smaller range of species. This is always less than the metabolic cost, which is not surprising; muscles are not 100% efficient. What is surprising is that the efficiency (mechanical cost/metabolic cost) is much lower for small runners than for large ones. Efficiencies (calculated from the lines, not from individual points) are shown at the bottom of the graph. Runners of the size of ponies and ostriches (100 kg) have efficiencies of about 0.4, but a cockroach (*Periplaneta*, represented by the symbol *P*) has an efficiency of only 0.01.

These efficiencies have been calculated by dividing positive work by the metabolic energy cost calculated from oxygen consumption. Measurements of the oxygen consumption of people walking on sloping treadmills or pedaling bicycle ergometers have shown that, at the optimum rate of working, about 4 J of metabolic energy are needed to do 1 J of positive work (so the efficiency is 0.25). At equal negative rates of working, about 0.8 J metabolic energy is needed to do 1 J of negative work (an efficiency

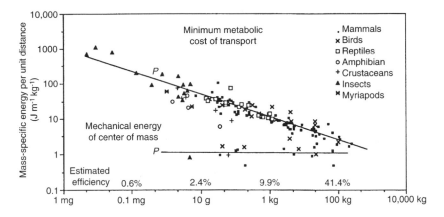

Fig. 7.12. Graphs of cost of transport against body mass for various legged animals. The sloping line shows the asymptotic value that the metabolic cost of transport approaches at high speeds (the first term on the right-hand side of Equation 7.2). The lower, horizontal line shows the mechanical cost of transport, calculated from the fluctuations of external kinetic energy and gravitational potential energy (ignoring both internal kinetic energy changes and savings due to elastic mechanisms). The symbols are as explained on the right, except *P*, the cockroach *Periplaneta*. From Full and Tu (1991).

of −1.25, negative because the work is negative). Thus, almost 5 J metabolic energy are needed to do 1 J negative work followed by 1 J positive work (Margaria 1976). The muscles have to do almost equal amounts of negative and positive work. Thus, we might expect the efficiencies shown by Fig. 7.12 to be 1/5 = 0.2.

The high efficiencies shown by the graph for large animals are easily explained. Some of the negative and positive work is done by tendons that stretch and recoil, reducing the work that has to be done by the muscles. We saw in Section 7.4 that in large mammals, such as kangaroos, horses, and humans, elastic mechanisms may reduce the work required of the muscles by 40–50%, so an efficiency of 0.4 calculated from Fig. 7.12 may be achieved by muscles doing 50–60% of the work with efficiencies of 0.20– 0.24, aided by tendon springs doing the rest. We also saw that elastic energy savings seem to be unimportant in small mammals, so we should not be surprised that small mammals do not show elevated efficiencies. However, this line of reasoning cannot explain why the efficiencies shown in Fig. 7.11 for small animals are far less than the values of around 0.2 that experiments on frog and mouse muscles lead us to expect (Woledge et al 1985; Barclay 1994; note that the efficiencies given by Barclay refer to the conversion of ATP energy to work, and should be divided by two to get efficiencies of conversion of food energy to work, as explained in Section 2.5).

Kram and Taylor (1990) proposed an ingenious theory to explain the paradox. They ignored the metabolic energy cost of doing work, and considered only the cost of exerting force. They explained their theory very concisely, and I have tried in this paragraph to fill in some of the steps in their argument. Equations 2.10 show that in isometric contraction (rate of shortening $v = 0$), the metabolic rate of a muscle is proportional to the force multiplied by v_{max}, the rate at which the muscle could shorten if the force was zero. Kram and Taylor argued that for animals of different sizes, the mean forces (averaged over a stride) that their feet exert on the ground must equal body weight, so the forces required of the muscles should be proportional to body mass. Also, the muscles that are active while a foot is on the ground must complete their contractions while it remains on the ground, in a time equal to (Step length)/Speed). Therefore, their unloaded shortening speeds v_{max} should be proportional to Speed/Step length). Metabolic rate, proportional to force multiplied by v_{max}, is thus expected to be proportional to (Mass × Speed)/(Step length). Metabolic cost of transport is Metabolic rate/(Mass × Speed), so should be inversely proportional to step length.

Kram and Taylor (1990) tested their theory by experiments on mammals ranging from 30-g kangaroo rats to 140-kg ponies, each species running at several speeds. They measured oxygen consumption and the time for which each foot remained on the ground, and calculated metabolic cost of transport and step length. They found that metabolic cost of transport was proportional to (body mass)$^{-0.30 \pm 0.05}$ (mean ± 95% confidence limits), in good agreement with the more extensive data shown in Fig. 7.12. Larger animals take longer steps. If animals of different sizes were geometrically similar to each other, and if they moved in dynamically similar fashion, step length would be proportional to (body mass)$^{0.33}$. Kram and Taylor's result was not significantly different from that prediction; they found that step length was proportional to (body mass)$^{0.25 \pm 0.09}$. The confidence limits are wider than might have been hoped, but the results are consistent with the theoretical prediction that metabolic cost of transport would be inversely proportional to step length.

In further experiments, Roberts et al. (1998b) and Roberts et al. (1998a) tested Kram and Taylor's theory on birds. Again, they found that metabolic cost of transport was inversely proportional to step length, but the constant of proportionality was 1.7 times higher than for mammals. They explained the difference by showing that turkeys have longer fascicles in their leg muscles than dogs of equal body mass. More energy is needed to develop a given force in a long fascicle than in a short one.

Alexander (1991a) challenged one of the assumptions of Kram and Taylor's theory, that v_{max} should be inversely proportional to the time for which each foot remains on the ground. I argued instead that v_{max} should be proportional to the maximum rate at which the muscle has to shorten in the course of a step, so that the muscles of animals of different sizes would work over the same range of values of v/v_{max}. The rate at which a muscle has to shorten is (Length change) / (Time available for shortening). Kram and Taylor took account of the time, but not of the length change. If this criticism is valid, their theory fails to explain why metabolic costs of transport are higher for small runners than for large ones.

The measurements of cost of transport reported so far were made while the animals were in a steady state, during a prolonged, aerobically powered run. Edwards and Gleeson (2001) measured the oxygen consumption of mice on a treadmill that was run intermittently, so that they were given periods of rest between short bursts of sprinting. The sprints were made as fast as the mice could manage, and during them the mice accumulated lactic acid by anaerobic metabolism. In a series of experiments, the mice made 1–13 bouts of sprinting, each 15 s long, evenly spaced over a period of 375 s. Edwards and Gleeson recorded their oxygen consumption throughout the period of intermittent activity, and continued to record it during the recovery period afterward, until the animals' oxygen consumption had returned to the resting level. They calculated the quantity of oxygen that was used in excess of the quantity that would have been used if the animals had been resting throughout. They found that the mice used more oxygen than they would have done while covering the same distance during a prolonged, uninterrupted run. However, the additional cost was incurred only once in a series of bouts of sprinting. The energy cost of a single 15-s sprint is many times higher than the cost of running the same distance in a steady state, but the cost of a long series of short sprints is little higher than that of continuous running. Mice and other small mammals habitually run in short bursts.

7.7. PREDICTION OF OPTIMAL GAITS

We have already seen that people walking or running at different speeds use different combinations of stride length, duty factor, and shape factor. Stride length increases as speed increases. Duty factor is high while walking, and low in running (Fig. 1.5A). Shape factor increases as walking speed increases, then drops sharply at the onset of running (Fig. 1.5B).

Alexander (1992a) sought to explain this. I presented a theory designed to predict mechanical costs of transport for the simple biped shown in Fig. 6.5G. This model has legs of appreciable mass, with telescopic actuators that do positive and negative work as required, and with springs in the feet. The masses of the legs were the same fraction of body mass as in humans, and the stiffness of the springs was chosen to imitate the elastic properties conferred on human legs by their tendons and ligaments. The forces that the model exerts on the ground always act along the length of the legs, but there are additional actuators at the hip joints to supply the torques needed to swing the legs forward and back. In computer simulations, I made this model walk and run at several different speeds. At each speed, I varied duty factor, shape factor, and stride length, and calculated the positive work that the actuators would have to do. I sought the combination of parameters that minimized the mechanical cost of transport for each speed. The optima that I found agreed reasonably well with the combinations of stride length, duty factor, and shape factor that people actually use, with one important exception: the predicted speed of the walk–run transition was much too high.

That model calculated the work required for locomotion, but this work is less directly important to us than the metabolic energy cost of locomotion. Minetti and Alexander (1997) developed the model to calculate the metabolic energy cost of locomotion, using the method explained in Section 2.5. Figure 7.13 shows results for four different speeds. At each speed, stride length was kept constant while duty factor and shape factor were varied. At a very low walking speed (A) the contours representing cost of transport show two minima. There is a local minimum (hollow star) at a duty factor of 0.2 and a shape factor of 0.1, representing a very slow jog. There is also a deeper global minimum (filled star) at a duty factor of 0.55 and a shape factor of 0.4, representing a walk. This is the gait that the model predicts to be most economical at very low speeds. At a moderate walking speed (B) there are still two minima, but the global minimum has moved to the highest possible shape factor. (Shape factors greater than 1.0 would imply negative vertical forces at midstep.) At 2.0 m/s (C) running has become the more economical gait, by a small margin; the global minimum is at a duty factor of 0.25 and a shape factor 0.1. This is approximately the speed at which subjects on treadmills change from walking to running. The position of the minimum changes only slightly with further increase of speed (D).

The model predicts that to minimize cost of transport, shape factor should be increased with increasing speed until, at about 2.0 m/s, it should be abruptly reduced. Also, duty factor should be reduced from high values below 2.0 m/s to low values at higher speeds. These predictions are in good qualitative agreement with the observed changes (Fig. 1.5). However, at each speed, the predicted duty factor is too low and the predicted shape factor too high.

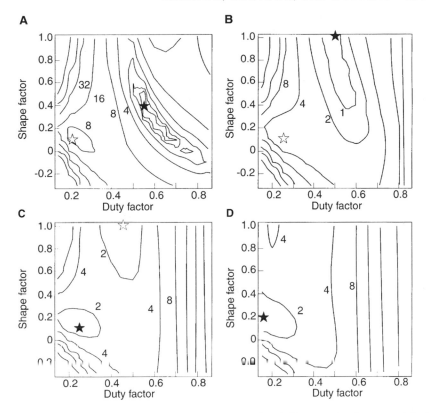

Fig. 7.13. Calculated metabolic energy costs for different gaits for a model of human walking and running. Each graph refers to a different speed: (A) 0.4 m/s; (B) 1.2 m/s; (C) 2.0 m/s and (D) 3.2 m/s. At each speed, duty factor (horizontal axis) and shape factor (vertical axis) were varied, and stride length was given a constant value, realistic for the speed. Metabolic cost of transport was calculated, and is represented by contours. It is expressed as joules per meter traveled, per newton of body weight. From Minetti and Alexander (1997).

In Fig. 7.13, stride length was given a constant value at each speed, approximately equal to the stride lengths that people are observed to use. In further calculations, we varied stride length and found that these stride lengths were very close to the optima predicted by the model. The optimum values of duty factor and shape factor predicted by the graphs are only slightly different from those found when stride length was also varied.

This model helps us to understand why people modify their gait as they do as they change speed. However, it is in some respects unsatisfactory. I am not troubled by the imperfect quantitative agreement of its predictions with observed gaits; it would not be reasonable to expect so simple a model to give perfect agreement. I am more concerned by the problem referred to in Section 3.6, that the physiological data on which Fig. 3.5 is based came from experiments in which muscles made single contractions at constant speed, and may not predict energy costs well for the repetitive movements of walking and running. Also, the model's leg ends at a point instead of a foot with a long sole like human feet. The human heel makes the initial impact with the ground in walking and (for most people) in running (Cavanagh and Lafortune 1980). The center of pressure travels forward along the sole of the foot, reaching the metatarsophalangeal joints (at the bases of the toes) by the time the foot leaves the ground. Throughout the step, the ground force remains in line with the hip joint in the model and (more or less) in real people. However, because the center of pressure moves along the sole of the foot, the range of angles traversed by the force in the course of a human step is much smaller than the range of angles traversed by the leg. The leg had to be made unrealistically long in the model to get the range of angles of the force right. This awkwardness could have been avoided by making the model more complex, but there are great advantages in keeping models simple, as explained at the end of Chapter 6.

7.8. SOFT GROUND, HILLS, AND LOADS

All our discussion of energy costs so far have assumed firm, smooth, level ground, but much of the ground on which animals travel is soft, uneven, and hilly.

The metabolic cost of transport is higher on soft than on firm ground. For example, Zamparo et al. (1992) found that the cost of human walking on soft sand was up to 2.5 times as high as on concrete, and Pandolf et al. (1976) found that walking in deep snow was up to five times as costly as on a treadmill. Different species may be affected to different extents by soft or uneven ground. For example, moving off a road onto tundra had less effect on the cost of transport for reindeer than for humans (White and Yousef 1978). The increased cost of transport on soft ground is presumably largely due to work done against the viscosity of the ground when the foot sinks in.

The shortest route between two points is a straight line, but it may be possible to save energy and/or time by making diversions round patches of soft or uneven ground. Alexander (2000a) illustrated this by considering the simple case of a route interrupted by a triangle of soft ground. Figure 7.14 shows the routes for which the energy cost is least from A_1 to F_1, from A_2 to F_2, and so on. In this example, the cost of transport is

Fig. 7.14. Examples of optimal routes past a triangular patch of soft ground, on which the cost of transport is 1.5 times as high as elsewhere. From Alexander (2000a).

1.5 times as high in the triangle of soft ground as elsewhere. I showed that to minimize energy costs, a walker should be prepared to diverge from the direct route by up to 18°. However, there would be no point in diverging so much in a journey from A_4 to F_4, because a smaller angular diversion is enough to avoid the soft ground altogether. If the angle at the apex of the triangle were larger, or if the soft ground were more costly to walk on, the maximum worthwhile angle of divergence would be greater than 18°.

An animal walking or running uphill does work against gravity. Accordingly, the metabolic cost of transport is higher than on level ground. This has been shown by measurements of oxygen consumption on sloping treadmills, both for vertebrates and for insects (Taylor et al., 1972; Full and Tullis 1990; Eaton et al., 1995b). The net efficiency of climbing is the work done against gravity divided by the additional metabolic energy consumption, over and above what would have been used at the same speed on level ground. The efficiency is high on shallow gradients, presumably because some of the work required for climbing is made available by reducing the negative work done in the first half of the step (Pugh 1971). We saw in Section 7.6 that the metabolic cost of level walking by humans is consistent with positive work being done with an efficiency of 0.25. The efficiency of walking up steep slopes is about the same (Margaria 1976). Smaller animals seem to do the work of climbing more efficiently than they do the work of level walking (Taylor et al., 1972). For example, cockroaches (*Periplaneta*) do the work of level walking with an efficiency of 0.012 (Full and Tu 1991), but climb vertical walls with an efficiency of 0.034 (Full and Tullis 1990). Consequently, the additional cost of running uphill is a smaller percentage of the cost on level ground for smaller animals on the same gradient.

Wickler et al. (2000) found that horses preferred to trot more slowly on an uphill gradient than on level ground. They measured the oxygen consumption of horses trotting on a treadmill, both with it level and with it sloping at a 10% uphill gradient. They found that the preferred speed was the speed at which the metabolic cost of transport was least, on the slope as well as on the level.

Because the net efficiency of climbing is higher on shallow gradients, it is more economical of energy to take a zigzag path up a steep hill than to walk straight up the slope. Minetti (1995) calculated that for humans, the direct route is the more economical on gradients up to 0.25. On steeper slopes the optimum path is a zigzag, with each straight section rising with a gradient of 0.25. Minetti examined contour maps showing footpaths in the Alps and Himalayas, and concluded that the paths conformed reasonably well to these predictions.

The metabolic energy cost of walking and running is less on downhill slopes than on level ground (Margaria 1976). However, it is greater on very steep downhill slopes than on shallower ones. On a shallow slope, the gravitational potential energy that is lost by descending can be used to do some of the positive work that would otherwise have to be done by muscles; but on a steep slope the muscles have to do additional negative work (Pugh 1971).

The energy cost of walking equal distances up a slope and then down a slope of the same gradient is always greater than the cost of walking the same total distance on level ground. For example, for humans on a gradient of 0.2, it is 2.5 times the cost of walking on the flat. Consequently, energy can be saved by taking a longer path round the shoulder of a hill, in preference to a shorter path that involves climbing to a greater height. Alexander (2000a) presented a simple model that predicted optimal paths around the shoulders of pyramidal hills. Old footpaths through mountain country seem generally neither to take the shortest route nor to follow the contours, but to compromise between them, as the model predicts.

Carrying a load generally increases the metabolic energy cost of locomotion. The simple models presented in Chapter 6 suggest that a load of x% of body mass should increase it by x%. This is approximately the case for soldiers carrying loads in backpacks but not for African women carrying loads on their heads (Maloiy et al., 1986). Thin African women carry loads up to 20% of body mass without energy cost, but fat women are less economical; the mass of body fat seems to count as part of the free 20% (Jones et al., 1987). Heglund et al. (1995) found that the work required for walking with modest head loads was no more than for unloaded walking, because changes of gravitational potential energy were more precisely matched to kinetic energy changes. The question of why they are more precisely matched remains unanswered. Measurements on mammals and ants carrying loads have generally shown that a load of x% of body mass increases the metabolic cost of walking by about x%. However, in experiments by Kram (1996), rhinoceros beetles (*Xylorctes*) with attached loads of up to 30 times body mass increased their metabolic rates by much smaller factors.

7.9. STABILITY

In this section, we shift our focus from energy costs to stability. A body is in equilibrium if the forces on it are balanced, in which case it is either stationary or moving with constant velocity and angular velocity. The equilibrium is stable if the body returns automatically to its initial condition after any small displacement. For example, a cone resting on its base is stable, but a cone balanced on its point is not.

For a standing animal to be stable, a vertical line through its center of mass must pass through the polygon of which the corners are its feet. For example, if three feet are on the ground, the center of mass must be over the triangle of support that they define. If the feet are small, stability requires at least three to be on the ground, but bipeds such as ourselves can be stable standing on two feet or even on one because each foot contacts the ground over a substantial area.

A small displacement will not topple a stable standing animal, but if the displacement is large enough to move the vertical through the center of mass out of the polygon of support, the animal will fall over. The larger the polygon of support, the larger the displacements that can be tolerated. The sprawling stances of reptiles, insects, spiders, and crustaceans, with the feet far out on either side of the body, give them high margins of stability. Alexander (1971) pointed out that wind or water currents tend to overturn animals standing on dry ground or in water. For geometrically similar animals of different sizes, these overturning forces are proportional to the surface area of the body, and so to length squared. However, body weight, which tends to prevent overturning, is proportional to length cubed. Thus, small animals are in more danger of being blown over than large ones. The sprawling stance of insects and spiders, with the legs well spread and the center of mass low, helps to stabilize them against wind loads. A water current exerts much larger forces on an animal than a wind of the same speed, which may explain why underwater walkers such as crabs spread their legs much more widely than terrestrial walkers of similar mass, such as mice. Martinez (2001) showed that crabs (*Grapsus*) walk under water with their legs more widely spread than on dry land, and that this reduces the danger of being overturned by water currents.

A statically stable gait is one in which the body is in stable equilibrium at all stages of the stride. A six-legged animal that moves its legs three at a time, as cockroaches do (Section 7.2), can maintain static stability. Ting et al. (1994) found that cockroach gaits are statically stable, except at the highest speeds. By moving legs 1 and 3 of one side with leg 2 of the other, cockroaches ensure that the center of mass is always over a triangle of support.

Statically stable gaits are possible also for four-legged animals, but in this case the legs must be moved one at a time, so that there are always three feet on the ground. In addition, the feet must be moved in an order that keeps the center of mass over the triangle of support. To have three feet on the ground at all times, a quadruped must walk with duty factors of at least 0.75. McGhee and Frank (1968) showed that for duty factors in the range 0.75 to 0.83, only one sequence of leg movements allows static stability. It is left fore, right hind, right fore, left hind, the sequence that mammals use when they walk (Section 7.2). However, quadrupeds seldom or never walk in statically stable fashion. Duty factors of 0.75 or more are used by turtles (Jayes and Alexander 1980), but very seldom by mammals (Hildebrand 1976). Even turtle walking is not statically stable because turtles have more than one foot off the ground at some stages of the stride.

For an animal walking or running at a steady speed, the forces on its body, averaged over a stride, must be balanced. However, there is no need for the animal to be in equilibrium throughout the stride. A galloping dog is obviously not in equilibrium at the stage of the stride at which all four feet are off the ground. The extent to which departures from equilibrium can be tolerated depends on the stride frequency and on the length of the legs (Alexander 1981). Consider an animal walking with stride frequency f, so that the duration of each stride is $1/f$. If the animal fell freely for this time with the gravitational acceleration g, it would fall a distance $g/2f^2$. If the length of its legs is h, the distance it can fall without hitting the ground is a little less than h. Thus, the dimensionless parameter $g/2f^2h$ can be used as a measure of the need for an animal to preserve equilibrium as it walks. It is 1 or less for a dog galloping, 5 for a dog walking very slowly, and about 200 for turtles such as *Geoemyda* walking at their normal, very slow speeds. Therefore, to walk effectively, turtles must keep themselves much closer to equilibrium than dogs.

Jayes and Alexander (1980) asked why, in that case, turtles do not use statically stable gaits? We pointed out that a quadruped performing a statically stable walk with a duty factor of 0.75 would have to exert forces on the ground as shown in Fig.7.15A. Notice that large, instantaneous changes of force are required whenever a foot is lifted or set down. The very slow muscles of turtles are incapable of abrupt changes of force, but their slowness enables them to maintain tension at low metabolic cost (Equations 2.10). If equilibrium is not maintained throughout the stride,

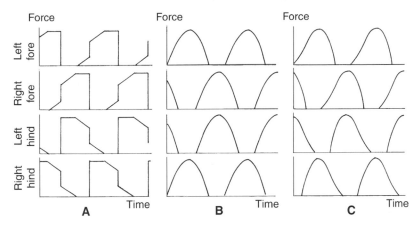

Fig. 7.15. Graphs of vertical force on the feet, against time, for three quadrupedal gaits. The duty factors of the feet are all 0.75. (A) The pattern required to maintain equilibrium throughout the stride; (B) The pattern that minimizes unwanted displacements if the force on each foot rises and falls like a half-cycle of a cosine curve ($q' = 0$ in Equation 7.3); and (C) The pattern that minimizes unwanted displacements if the forces rise and fall as described by Equation 7.3. From Alexander (1982).

unwanted displacements will occur: the animal will rise and fall, pitch and roll, as it walks.

Very slow muscles might be capable only of exerting forces that rose and fell like a half-cycle of a cosine curve, as in Fig. 7.9(iii). We showed by mathematical modeling that if this were the case, the gait that minimized unwanted displacements would be the one shown in Fig.7.15B, in which diagonally opposite feet move simultaneously. We went on to consider the possibility that if the muscles were a little faster, a foot that was on the ground from time $-\tau/2$ to time $+\tau/2$ might be able to exert vertical forces F_y according to

$$F_y = A \left[\cos \left(\frac{\pi t}{\tau} \right) + q' \sin \left(\frac{2\pi t}{\tau} \right) \right] \qquad (7.3)$$

where A is a constant and t is time. In this equation q' is a shape factor, like q in Equation 7.1; but whereas the effect of q was to make a graph of force against time bell-shaped or two-humped, the effect of q' is to make it asymmetrical. We showed that if all values of q' were possible, the gait that minimized unwanted displacements was the one shown in Fig. 7.15C. The forces rise and fall slightly asymmetrically, and there are times when only two feet are on the ground. The gaits and force patterns that turtles actually use are closely similar to Fig. 7.15C. It seems that turtle gaits are adapted to minimize unwanted displacements for an animal with very slow muscles.

We have been discussing static stability, and turn now to dynamic stability. In statically stable gaits, the animal is in equilibrium at all times. A gait that is not statically stable may nevertheless be dynamically stable in the sense that after a small disturbance the animal returns automatically to its original pattern of movement. Dynamic stability can be conferred by reflexes that stimulate the muscles to take appropriate correcting action after a disturbance, and until recently it was assumed that stability in walking depended entirely on reflex control. However, dynamic stability can also arise passively, due to the inherent mechanical properties of the system.

McGeer (1990a, 1993) investigated the stability of the walking biped shown in Fig. 6.5C. This is a mathematical model based on a traditional wooden toy that will walk down a downhill slope. The toy rocks from side to side as it walks, but the motion of this model is confined to two dimensions. McGeer studied its behavior as it walked downhill by mathematical analysis and by computer simulation. He also performed experiments with a physical model, a nonrocking version of the toy. He showed that once it had been set moving, its motion was dynamically stable. This suggests that the control of bipedal walking may be a much simpler problem than physiologists had previously thought. Kuo (1999) extended the analysis to a three-dimensional model that could rock from side to side. He showed that it was unstable in roll, implying that a walker would need reflex control to prevent it from falling over sideways.

Kubow and Full (1999) investigated the dynamic stability of a walking insect. They used a two-dimensional computer model, but whereas McGeer's (1990a) model of bipedal walking considered only the two dimensions of a vertical plane, they considered only movements in a horizontal plane. They set the model walking, using a realistic duty factor (0.6) and stride frequency (10 Hz). They made it move its legs in two groups of three, each group consisting of the fore and hind legs of one side and the middle leg of the other. They specified the position, relative to the body, at which each leg would be set down, and the magnitude and direction (relative to the body) of the forces it would exert while on the ground. They set the computer running and saw what happened, sometimes deliberately disturbing the gait. They found that if they altered the model's forward speed, it gradually returned, over many strides, to the speed at which it had previously been running steadily. It also recovered (in fewer strides) if it was given a transverse component of velocity, or an angular velocity. In these respects, the model was dynamically stable.

Figure 7.16 explains, as an example, the mechanism of stability against perturbations of angular velocity. In (A), a foot is exerting a force (indi-

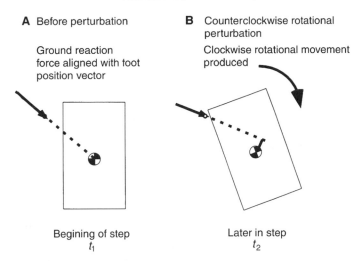

A Before perturbation

Ground reaction
force aligned with foot
position vector

B Counterclockwise rotational
perturbation

Clockwise rotational movement
produced

Begining of step
t_1

Later in step
t_2

Fig. 7.16. A diagram illustrating the mechanism by which a running cockroach recovers from a perturbation that affects its angular velocity. The animal's body, seen in plan view, is shown as a rectangle. Further explanation is given in the text. From Kubow and Full (1999).

cated by the arrow) that is in line with the animal's center of mass, and so has no tendency to cause any rotation. In (B), after a sudden anti-clockwise rotation, the foot is still at the same position on the ground, and the direction of the force *relative to the body* is unchanged, but the force now exerts a moment about the center of mass, tending to reverse the rotation.

7.10. MANEUVERABILITY

In this final section of this chapter we consider acceleration and the ability to turn at speed. The ability to decelerate rapidly in an emergency may be important to running animals, as well as the ability to accelerate, but I know no investigations of it.

Human sprinters leave the starting blocks with an acceleration of about 10 m/s² (Ballreich and Kuhlow 1986). Elliott et al. (1977) analyzed films of lions hunting prey (Fig. 1.1) and found that the mean initial acceleration was 9.5 m/s² for the lions and only 4.5–5.6 m/s² for the prey. The difference does not necessarily reflect a difference in athleticism; the prey may not have been fully prepared to run when the lions started.

The gravitational acceleration is about 10 m/s². That means that a forward acceleration of 10 m/s² requires a horizontal component of force equal to body weight, which is possible only if the coefficient of friction of the foot with the ground is at least 1.0. Nigg (1986) reports coefficients of friction of sport shoes with various surfaces, ranging from about 0.3 for a surface covered with loose granules to about 1.5 for artificial grass. I know no measurements of coefficients of friction of animal feet on natural ground surfaces. Acceleration on slippery surfaces seems to be limited by the coefficient of friction, but on others it may be limited by the ability of animals to develop the required force.

An animal running at speed v along a circular arc of radius r has a transverse acceleration v^2/r (Section 1.3). This requires a transverse force on the ground. Photographs of horses cornering sharply at speed in barrel races show them leaning over at angles of about 45° to the vertical. This indicates that the transverse component of force is about equal to the animal's weight, and that the coefficient of friction with the ground is 1.0 or more.

Greene (1985) measured the maximum speeds at which human athletes could run in circles of different radii. As expected, he found that maximum speeds were lower when the radius was smaller. Greene suggested that cornering speed on any radius might be limited by the ability of the body to generate the required forces. He presented a mathematical theory based on the plausible but uncertain assumptions that step length and stride duration were independent of the radius of the curve. He compared the observed relationship between speed and radius, with the relationship predicted by the theory, and found good agreement for large radii, but not for smaller ones. It seems possible that cornering speed on curves of smaller radius may be limited by the coefficient of friction of the feet with the ground (Alexander 1982).

When running around obstacles, the shortest and most direct route is not necessarily the fastest. It may be faster to take a slightly longer route with larger radii of curvature. Alexander (in press) used a very simple mathematical model to illustrate this point.

This long chapter has reviewed walking and running by tetrapods and arthropods. I have shown how they use different gaits at different speeds, and have argued that gaits are generally adapted to minimize energy costs. I have discussed both the mechanical work required for running and the metabolic energy cost, and have shown how tendon springs save energy. We have considered the stability and maneuverability of legged animals. I would like to finish by pointing to several topics that seem to me to be priorities for research. First, for reasons that I explained in Section 7.6, I am not satisfied with the attempts that have been made to explain why the muscles of small runners seem to work less efficiently than those of large ones. Secondly, Minetti and Alexander's (1997) theory of bipedal gaits needs refining; we need better means of predicting the metabolic costs of muscle activity, and a more satisfactory way of taking account of the movement of the center of pressure along the sole of the foot. Thirdly, I would like to see an equivalent theory for the trot/gallop transition. Finally, we still know very little about the maneuverability of running animals and the factors on which it depends.

Chapter Eight

....

Climbing and Jumping

*I*T SEEMS convenient to discuss jumping and climbing in the same chapter, because some animals jump to travel in and between trees. Lemurs and other prosimian primates leap between branches; monkeys and squirrels jump to cross gaps between trees; and gibbons swing from branch to branch. Many other jumping animals do not climb. Frogs jump to travel over the ground, locusts jump to get clear of the ground at the start of a flight, fleas jump to get onto a host, and all these animals jump to escape from danger. There are also many climbing animals that do not jump, for example, geckos that run up and down the trunks of trees and aphids that walk on the stems and leaves of plants, adhering (by different mechanisms) even to smooth surfaces. All these styles of locomotion are discussed in this chapter.

8.1. STANDING JUMPS

This section is about jumps in which the animal is initially stationary, not running as in a human long jump. We have already calculated the mechanical cost of transport for an animal that travels like a frog, by a series of standing jumps (Section 6.4). In that discussion we used the symbol v for the horizontal component of the velocity. Here we use the same symbol for the resultant velocity.

Consider an animal of mass m that takes off for a jump with velocity v, at an angle α to the horizontal, on level ground. At this stage we will ignore both air resistance and the mass of the legs. The work W that the animal's muscles must do equals the kinetic energy given to the body:

$$W = \frac{mv^2}{2} \tag{8.1}$$

The vertical component of the velocity is $v \sin \alpha$ and the horizontal component is $v \cos \alpha$. We can think of the kinetic energy as the sum of two parts: one, $m(v \sin \alpha)^2/2$, associated with the vertical component of velocity, and the other, $m(v \cos \alpha)^2/2$, associated with the horizontal component. As the animal rises to the highest point in its jump, the vertical component of velocity falls to zero, and the associated kinetic energy is converted to gravitational potential energy. The height h of the jump is given by

$$\frac{m(v \sin \alpha)^2}{2} = mgh \tag{8.2}$$

$$h = \frac{(v \sin \alpha)^2}{2g}$$

where g is the gravitational acceleration. For a given takeoff velocity, the height will be greatest (equal to $v^2/2g$) if the angle α is 90°. From this and Equation 8.1, the greatest height that can be jumped by doing work W is W/mg.

The time t that the animal takes to fall from the highest point of the jump to the ground can be calculated from

$$h = \frac{gt^2}{2}$$

From this and Equation 8.2

$$t = \frac{v \sin \alpha}{g} \tag{8.3}$$

The animal is off the ground for time $2t$ (time to rise plus time to fall). The length λ of the jump is the distance it travels in this time, moving with its horizontal component of velocity $v \cos \alpha$:

$$\lambda = 2 \left(\frac{v \sin \alpha}{g} \right) v \cos \alpha = \frac{v^2 \sin 2\alpha}{g} \tag{8.4}$$

This distance is greatest (equal to v^2/g) if the angle α is 45°. From this and Equation 8.1, the greatest distance that can be jumped by doing work W is $2W/mg$. Notice that both the height of a vertical jump and the length of a jump at 45° are proportional to work per unit body mass.

The most remarkable standing jump ever recorded seems be the one made by a bushbaby (*Galago senegalensis*) that jumped from the floor to the top of a 2.26-m door (Hall-Craggs 1965). In the calculations above, the height h is the height by which the animal's center of mass rises in the jump. This was probably less than 2.26 m for the bushbaby, because it started with its center of mass a little above the ground, and may have used its forepaws to pull itself up the last few centimeters. I will estimate $h = 2$ m. Muscle makes up 36% of the body mass of a bushbaby (Grand 1977). If all of this muscle contributed to the work of the jump, its work output would be 54 J/kg muscle. This is comfortably less than the 70 J/kg that vertebrate striated muscle can be expected to do in a single slow contraction (Section 2.1). However, the contraction has to be performed fast, as we shall see.

Frogs are much less remarkable jumpers, but more research has been done relating their jumping ability to the properties of their muscles. A small tree frog (*Osteopilus*) studied by Peplowski and Marsh (1997) made standing long jumps of up to 1.44 meters. The hind limb muscles that power the jump amount to 14% of body mass. Hence, the work requirement can be estimated as 50 J/kg muscle. Lutz and Rome (1994) studied one of the principal leg extensor muscles in a different species of frog. They showed that it shortened during takeoff by 24% of its length, and that the isometric stress that it could exert, averaged over the working range of lengths, was about 210 kN/m². Hence, the work it could do in a slow contraction over this range was $0.24 \times 210,000$ J/m³, or about 50 J/kg. This is the same as the work requirement of the tree frog, but it should be remembered that the tree frog was a different species, which jumped further.

Bushbabies, frogs, and many other animals power their jumps by rapidly extending their legs. We will work out how fast the extension has to be. The distance over which the animal accelerates to its takeoff speed is limited by the length of the legs; let it be s. The animal accelerates from rest to velocity v, so its mean speed over this distance (assuming constant acceleration) is $v/2$ and the time available for the legs to extend is $2s/v$. The height h attained in a vertical jump is $v^2/2g$, and the distance λ covered in a jump at 45° is v^2/g. Hence, the time t_{acc} in which the animal must accelerate to takeoff speed is

$$t_{acc} = s \sqrt{\frac{2}{gh}} \quad \text{for a high jump} \qquad (8.5)$$

$$t_{acc} = \frac{2s}{\sqrt{g\lambda}} \quad \text{for a long jump} \qquad (8.6)$$

The bushbaby that jumped to a height of 2 m accelerated over a distance of 0.16 m (Hall-Craggs 1965), so the acceleration time calculated from Equation 8.5 was 0.05 s. The actual acceleration time was probably a little longer, because the acceleration was not constant. Aerts (1998) recorded an acceleration time of 0.10 s for a bushbaby making a less high jump from a force plate. The frog that jumped a distance of 1.4 m (Peplowski and Marsh 1997) accelerated over a distance of 0.11 m, so Equation 8.6 gives an acceleration time of 0.06 s.

Smaller jumping animals have smaller acceleration distances, and so have to extend their legs in even shorter times to reach the same takeoff speed. The rabbit flea *Spilopsyllus*, which is about 1.5 mm long, accelerates to its takeoff velocity of about 1 m/s over a distance of 0.5 mm (Bennet-Clark and Lucey 1967). This implies an acceleration time of about 1 ms. Some small insects make wing beats in times as short as this, exploiting the oscillatory properties of fibrillar flight muscle (Section 2.6), but no known muscle can complete an isolated contraction in so short a time.

Bennet-Clark and Lucey (1967) argued that fleas and other jumping insects must use catapult mechanisms. Their muscles contract slowly, storing up elastic strain energy in elastic structures while the legs remain locked in a strongly flexed position. Then the leg is unlocked, allowing rapid elastic recoil. Thus, the rate at which the leg extends is not limited by the properties of its muscles. Bennet-Clark and Lucey (1967) showed that the catapults in fleas were blocks of resilin, a rubberlike protein, at the bases of the hind legs. They confirmed that these blocks were large enough to store the energy of the jump, and described the locking and unlocking mechanism. Bennet-Clark (1975) described the quite different catapult mechanism used by locusts, in which the springs are the apodemes of the knee extensor muscles, and cuticular structures at the knee joint. He measured the elastic properties of these structures and the force that the knee extensor muscles can exert on them, and confirmed that they are capable of storing the energy of the jump. Catapults have also been described in flea beetles, which jump by extending their hind legs (Brackenbury and Wang 1995), and in click beetles, which jump by a jackknife action of the back (Evans, 1972).

Aerts (1998) argued that the jumping ability of bushbabies must depend on catapultlike mechanisms, and Peplowski and Marsh (1997) argued the same for frogs. I will focus on the frogs because the argument for them is based on measurements of the physiological properties of their leg muscles. In the 1.44-m jump described above, the muscles apparently contacted in 0.06 s, doing work amounting to 50 J/kg. This implies a power output of $50/0.06 \approx 800$ W/kg muscle. Physiological measurements were made on the sartorius muscle, which was assumed to be typical of the leg muscles. Like other muscles (Fig. 2.3C) it gave maximum power output at a moderate rate of shortening. This maximum power output, measured at the temperature at which the animal jumped, was only 240 W/kg, less than one-third of the power required for the jump. It seems clear that an elastic mechanism must be involved. There is no apparent mechanism for locking the joints of the leg while the muscles build up tension and store elastic strain energy in the springs, as in fleas and locusts. However, it will become apparent in the next section that the enhancement of jumping ability that can be obtained by elastic recoil does not depend on the existence of a lockable catapult. The elastic structures that seem most likely to be important are the tendons of the leg muscles.

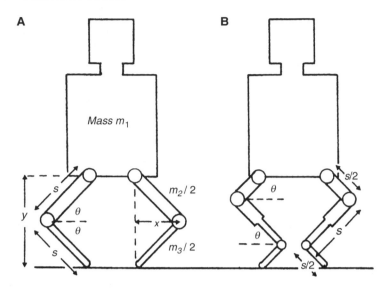

Fig. 8.1. Models used in computer simulations of standing jumps. These models jump vertically, by extending their legs. (A) the original model with two segments in each leg, and (B) a version with three segments. From Alexander (1995b).

8.2. Leg Design and Jumping Technique

Alexander (1995b) tried to establish design principles for jumping animals by analyzing the behavior of the simple mathematical model shown in Fig. 8.1A. This model is designed to be so general as to be applicable to jumpers as different in size and structure as fleas and humans. It consists of five rigid segments: a trunk, two thighs, and two lower legs, all of which have appropriate masses. It jumps vertically, powered by extensor muscles of the knee joints. The forces that these muscles can exert depend on the rate at which they are contracting, as illustrated in Fig. 2.3A. Their tendons have linear elastic properties.

Figure 8.2 shows simulations of three styles of jump. Part A represents a squat jump, which is a jump that starts with the knees maximally bent and the muscles inactive. At zero time the muscles are activated and start to build up force. Initially the force is low because the muscles are shortening quite rapidly, stretching their tendons. Soon it is large enough to overcome the weight of the body and start extending the knees. Thereafter, the knee angle rises at a progressively increasing rate. The force in the muscles increases but never reaches the isometric force (1.0 on the muscle force scale) because they continue to shorten. As knee extension continues, the rate at which the muscles have to shorten to keep the feet on the

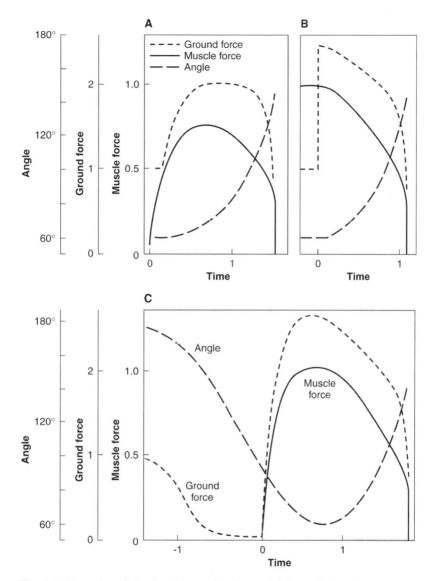

Fig. 8.2. Examples of simulated jumps by the model shown in Fig. 8.1A: (A) squat jump; (B) catapult jump; and (C) countermovement jump. The force exerted on the ground (expressed as a multiple of body weight), the force exerted by the knee extensor muscles (as a fraction of the isometric force) and the angle of the knee (2θ, Fig. 8.1A) are plotted against a dimensionless time parameter $t\sqrt{g/s}$, where t is time since the muscles were activated, s is leg segment length (Fig. 8.1A), and g is the gravitational acceleration. From Alexander (1995b).

ground increases, the force exerted by the muscles falls again, and the stretched tendons start to recoil elastically. Their recoil enables the muscles to extend the knees faster than they could do if the tendons were inextensible. Eventually, the muscles are no longer able to exert enough force to maintain the angular acceleration of the leg segments, which is needed to maintain contact with the ground. The feet leave the ground, and the animal continues to rise. The height of the jump is calculated.

Figure 8.2B represents a catapult jump, such as the jumps of fleas and locusts. The knees are locked at the same initial angle as in A. While they are locked, the muscles contract, stretching their tendons and storing elastic strain energy in them. Because this contraction can be completed slowly, the muscles develop their isometric force. At this stage the knees are unlocked and start extending. The muscles shorten at an increasing rate, so muscle force falls and the tendons recoil elastically. The jump is higher than a squat jump by the same animal, because the muscles exert larger peak forces and so store more energy in the stretched tendons.

Figure 8.2C represents a third style of jump. The jumper starts with the knees straight and makes a countermovement, bending them immediately before extending them. Human athletes start standing jumps with a countermovement, and tests with male volleyball players showed that this enabled them to jump 60 mm higher than in squat jumps (Komi and Bosco 1978). The simulation starts with the knees almost straight and the muscles inactive. The body falls, and the muscles are not activated until the appropriate moment to ensure that the fall is halted and the body starts to rise when the knee angle is 60° (which was the initial angle in the other simulations). During the fall, the force in the muscles may rise above their isometric force (active muscles exert greater than isometric forces when being stretched [Fig. 2.3A]). Muscle force may still be greater than isometric for a short while after the knees start to extend, because although the total length of muscle plus tendon is by then falling, the muscle may still be being stretched while the tendon starts its elastic recoil. Thus, knee extension starts with large muscle forces and a large store of elastic strain energy.

This may not be the full explanation of the advantage of countermovement jumping. Bobbert et al. (1996) pointed out that the countermovement allows time for the muscle to become fully activated before knee extension starts. The model did not show this advantage, because the muscle was assumed to be activated instantaneously. Another effect that may contribute to the advantage of countermovement jumping, which was ignored in the model, is stretch activation; the force that a muscle can exert remains elevated for a short time after a stretch (Edman et al., 1978).

The cross-sectional areas of muscles of geometrically similar animals would be proportional to (body mass)$^{2/3}$. Consequently, small animals can exert forces that are larger multiples of body weight than similar large animals can. For example, peak ground forces in standing jumps are generally 2–3 times body weight for jumping humans, up to 15 times body weight for bushbabies (Aerts 1998) and up to 135 times body weight for fleas (Bennet-Clark and Lucey 1967). Alexander (1995b) ran simulations for animals of three sizes, roughly corresponding to a human, a bushbaby, and a jumping insect. I found that for the human-sized model, catapult jumps and countermovement jumps were about equally high, and both were higher than squat jumps. For the bushbaby-sized model, catapult jumping was best, followed by countermovement jumping and then squat jumping. For the insect-sized model, countermovement jumping was little better than squat jumping, and catapult jumping was very much better. Humans and bushbabies lack the knee-lock mechanism that would be needed for catapult jumping, and use countermovements. Jumping insects do have catapult mechanisms, as we have seen.

In further calculations with the model, I varied the unloaded shortening speeds (v_{max}) of the muscles and the elastic compliance of the tendons. When the compliance was zero, the three jumping techniques gave jumps of equal height. Jump height increased with increasing muscle speed and tendon compliance for all three techniques. In further simulations, I kept the mass and properties of the muscles constant, while varying other quantities. I showed that increasing leg mass reduces the height of the jump, especially if the additional mass is located in the distal segment. We saw in Section 6.4 that heavy feet increase the work required for a jump of given length. Increasing leg length without changing leg mass increased the height of the jump, but the effect was small for catapult jumps if tendon compliance was high. Increasing the number of joints in the legs without altering their mass or overall length (Fig. 8.1B), and without changing the total mass of muscle increased the height of the jump.

The legs of many jumping animals are adapted in the ways that these simulations suggest. Jumping vertebrates such as bushbabies and frogs generally have longer legs than related animals of similar mass (Emerson 1985). Locusts have notably long hind legs but flea beetles (*Phyllotreta*), which also jump well by a catapult mechanism, do not. The elongated tarsal bones of bushbabies and frogs effectively add an additional segment to the legs. The movable iliosacral joints of frogs add both to the effective length of the legs and to the number of joints.

8.3. Size and Jumping

If jumping animals of different sizes were geometrically similar to each other, they would have muscle masses proportional to (body mass)$^{1.00}$ and leg lengths proportional to (body mass)$^{0.33}$. Frogs of a wide range of sizes deviate only a little from these proportionalities. Muscle mass has been found to be proportional (in different investigations) to (body mass)$^{1.03}$, (body mass)$^{1.08}$, and (body mass)$^{1.12}$; thigh length to (body mass)$^{0.29}$; and lower leg length to (body mass)$^{0.31}$ (Marsh 1994). Juvenile locusts are less close to geometric similarity, with leg lengths proportional to (body mass)$^{0.38}$ (Katz and Gosline 1993). Because the work that muscles can do in a single contraction is proportional to their mass, animals with equal proportions of jumping muscle in their bodies can be expected to take off at equal speeds and jump equal heights or distances (Hill 1950). Katz and Gosline (1993) found that takeoff speeds of locusts of different sizes were proportional to (body mass)$^{0.05}$, close to Hill's prediction. Similarly, Wilson et al. (2000) found that takeoff speeds and jump distances of marsh frogs (*Limnodynastes*) were independent of body mass. However, Marsh (1994) reported that jump lengths for frogs of different species were proportional to (body mass)$^{0.20}$.

The larger the leg muscles, the higher and further an animal can jump. On the other hand, muscles use metabolic energy even when inactive. The optimum size for leg muscles may depend on the balance between the benefit of stronger jumping and the metabolic cost of larger muscles. Alexander (2000b) presented a mathematical argument along these lines that predicted that locusts of different sizes should be geometrically similar, with equal proportions of leg muscle, and should be able to jump equal distances. The discussion of locusts was simple, because the jump seems to be powered entirely by elastic recoil of the catapult mechanism, but a discussion of frogs had to take account of the shortening speed (v_{max}) of the muscles, and so was more complex. It predicted that bigger frogs should have relatively larger leg muscles and jump further, which is consistent with Marsh's (1994) data but not with the results of Wilson et al. (2000).

Bennet-Clark (1977) and Bennet-Clark and Alder (1979) pointed out that air resistance would have a large effect on the jumping performance of small animals that took off at high speeds. Consider an animal of mass m and frontal area A (Fig. 3.2C) rising through the air in a vertical jump at velocity v. It is decelerated by drag as well as by gravity. The drag is $\frac{1}{2}\rho_{air}Av^2 C_D$, where ρ_{air} is the density of the air and C_D is the drag coefficient based on frontal area (Equation 3.8):

$$\frac{dv}{dt} = -g - \frac{\rho_{air}Av^2 C_D}{2m} \qquad (8.7)$$

where g is the gravitational acceleration. The height of a vertical jump with takeoff velocity $v_{takeoff}$ can be calculated by solving Equation 8.7:

$$h = Q \log_e \left(\frac{v_{take\ off}^{\ 2}}{2gQ} + 1 \right), \quad \text{where} \quad Q = \frac{m}{\rho_{air}AC_D} \qquad (8.8)$$

Table 8.1.

Jump heights for different takeoff speeds calculated for a spherical animal in a vacuum and (for two different sizes of sphere) in air at 1 atm. pressure

Takeoff speed, m/s	Height of jump in a vacuum, m	Height of jump in air, m	
		Radius 100 mm	Radius 1 mm
1	0.051	0.050	0.048
3	0.45	0.45	0.33
6	1.84	1.77	0.80

For simplicity, imagine the animal as a sphere of diameter d and density ρ_{body}. Its mass is $\pi \rho_{body} d^3/6$ and its frontal area is $\pi d^2/4$. Table 8.1 shows jump heights calculated for an animal of 100-mm diameter (mass 0.5 kg, a little heavier than a bushbaby) and for one of 1-mm diameter (mass 0.5 mg, in the size range of fleas). In a vacuum, with the same takeoff speed they would jump to the same height. In air they would both jump less high, but the flea-sized animal would be affected much more than the bushbaby-sized one, especially at high takeoff speeds.

It can be estimated from data in Bennet-Clark and Lucey (1967) that the combined mass of the muscles that power the flea's jump is about 3% of body mass, much less than the 36% in bushbabies. A flea with as large a proportion of muscle in its body as a bushbaby might be able to take off with as high a velocity as the bushbaby. However, it would jump much less high than a bushbaby. If jumping muscle mass is a compromise between the benefits of a high jump and the metabolic costs of big muscles, the benefit of the bigger muscles might not be matched by the cost. It may not be worthwhile for small insects to evolve large jumping muscles.

8.4. JUMPING FROM BRANCHES

Bushbabies and lemurs travel through forests by leaping from branch to branch, a style of locomotion that Napier and Walker (1967) described as vertical clinging and leaping. It is a very effective means of travel in forests in which a large proportion of branches are vertical or steeply sloping, as in the forest in Madagascar where Warren and Crompton (1997) studied the lemurs *Lepilemur* and *Avahi*. In forests with more horizontal branches it may be better to travel as squirrels and many monkeys do, running along branches and jumping only to cross gaps between one branch or tree and another (see, for example, Cannon and Leighton [1994] on *Macaca*).

There is an optimum angle of takeoff that minimizes the energy cost of jumping across a given gap. If the two branches are at the same height from the ground, this angle is 45° to the horizontal, as explained in Section 8.1. However, the animal will travel faster if it takes off at higher velocity, at a shallower angle. Crompton et al. (1993) filmed six prosimian species jumping between two perches at the same height, at different distances apart. They observed that *Galago moholi* always took off at about 45°, but that the other species used smaller takeoff angles except when the distance was near the limit of their jumping ability. If the jump is from a lower to a higher branch, the optimum angle is steeper than 45°, and if it is to a lower branch, the optimum angle is less steep. Warren and Crompton (1998) show that if the horizontal distance to be jumped is s and the increase of height is h, the angle of takeoff that enables the jump to be made at least energy cost is arctan $\{[h + (h^2 + s^2)^{0.5}]/s\}$.

The force exerted by an animal jumping from a branch deflects the branch, tending to set it vibrating (Alexander 1991b). The energy given to the branch would be returned to the animal by the branch's elastic recoil, if the time taken to extend the legs were half the period of vibration of the branch. However, the periods of vibration of branches are generally too long for this to be the case. For example, McMahon and Kronauer (1976) found periods of free vibration of 1–2 s for branches 10 m long (a length likely to be used by monkeys and squirrels). Demes et al. (1995) observed lemurs jumping from branches in their forest habitat in Madagascar. They found that the animal had generally lost contact with the branch before it recoiled. The branch was left vibrating, and the energy of the vibration was lost to the animal.

If the period of vibration is much longer than the duration of takeoff, the energy loss can be estimated by the principle of conservation of momentum, ignoring the elastic properties of the branch. If an animal of mass m projects itself with velocity v from a branch of effective mass m_{branch}, the branch is given velocity $-mv/m_{\text{branch}}$, and kinetic energy $m^2v^2/2m_{\text{branch}}$. The lighter the branch, the more energy will be lost.

Gibbons (*Hylobates*) and spider monkeys (*Ateles*) travel through trees by brachiation, swinging by their arms from one handhold to the next. If branches are close together, they may grasp each branch before letting go of the previous one. If the branches are far apart, gibbons must release one before the next is within reach, and fly briefly through the air. Gibbons can cross wider gaps in the canopy than macaques of similar mass (Cannon and Leighton 1994).

While holding a branch, a gibbon swings more or less like a simple pendulum of length L equal to the distance from the supporting branch to the animal's center of mass. Thus, brachiation can be represented as in Fig. 8.3, which shows brachiation (A) without and (B) with a flight phase. Bertram et al. (1999) used Fig. 8.3B as a model of gibbon locomotion. At the bottom of its swing the animal's velocity is v_b. As the swing continues the animal rises and slows down, until the flight phase is initiated when

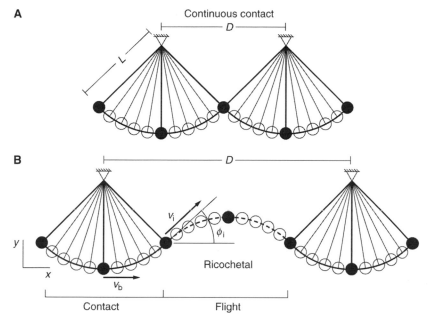

Fig. 8.3. Diagrams of brachiation by a gibbon. The animal is represented as a point mass with massless arms of length L. In (A) one hand does not let go until the other has grasped the next handhold. In (B) there is a flight phase in which neither hand contacts a branch. From Bertram et al. (1999).

the animal is traveling at velocity v_i at an angle ϕ_i to the horizontal. The sum E of the animal's kinetic energy and gravitational potential energy is constant, so

$$E = \frac{mv_b^2}{2} = \frac{mv_i^2}{2} + mgL\,(1 - \cos\phi_i) \qquad (8.9)$$

where g is the gravitational acceleration. Hence,

$$v_i^2 = v_b^2 - 2gL\,(1 - \cos\phi_i)$$

and the distance traveled in the flight phase (using Equation 8.4) is

$$D_{\text{flight}} = \frac{v_i^2 \sin 2\phi_i}{g} = \left[\frac{v_b^2}{g} - 2L\,(1 - \cos\phi_i)\right] \sin 2\phi_i \qquad (8.10)$$

The distance D from one handhold to the next (Fig.8.3) is this distance plus twice the distance traveled while swinging through angle ϕ_i

$$D = \left[\frac{v_b^2}{g} - 2L\,(1 - \cos\phi_i)\right] \sin 2\phi_i + 2L \sin\phi_i \qquad (8.11)$$

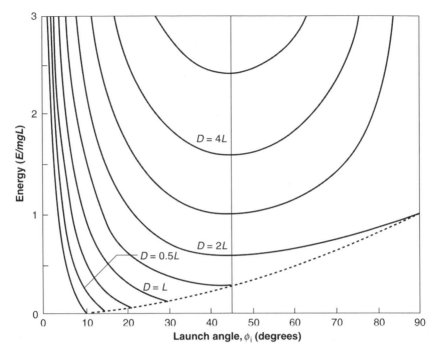

Fig. 8.4. Results from the simple model of brachiation (Fig. 8.3). The animal's total mechanical energy (expressed as E/mgL, Equation 8.9) is plotted against the launch angle ϕ_i for various support spacings D. Below the dotted line, the animal has too little energy to swing to the launch angle. From Bertram et al. (1999).

Figure 8.4 shows the energy E and the launch angle ϕ_i for various support spacings D. Below the dotted line, the animal has too little energy to swing to the launch angle. If the energy E is high, the animal has the option of using a low launch angle or a high one. It will travel faster if it uses the low angle.

Bertram et al. (1999) observed captive gibbons brachiating between equally spaced handholds. One of the handholds incorporated a force transducer (Chang et al., 1997) that revealed a discrepancy between the behavior of the model and of the real animal: the force on the handhold rose less abruptly at initial hand contact, and rose to higher peak values, than the model predicted. The animal behaved less like a simple pendulum than like one with a spring incorporated. The speeds at which the animals traveled for each handhold spacing were close to the minimum needed (according to the theory) to cross the gaps.

An ape may be able to make enough energy available to start brachiating by an initial downward swing. It can increase its energy while brachiating by bending and then extending the knees during the contact phase; the principle is the same as when children "pump" a swing. Fleagle (1974) showed that gibbons do this.

The theory outlined above assumes equally spaced supports, all at the same height above the ground. If its energy is high enough, the ape can travel between supports of variable height and spacing by choosing an appropriate launch angle for each gap (Preuschoft and Demes 1984). Also, the theory assumes rigid supports, but the branches from which wild gibbons swing are not rigid. Branches are left vibrating after a gibbon has passed. The energy of these vibrations is lost to the gibbon. The work that the ape must do to replace this energy probably makes the metabolic cost of transport much higher than for brachiation on rigid supports.

The model requires no work from the animal once the body has been given the kinetic energy needed to start brachiating. It might, therefore, be expected that the metabolic cost of transport on rigid supports would be low. Parsons and Taylor (1977) trained spider monkeys to brachiate from a continuous moving rope, designed on the same principle as a treadmill. They measured the monkeys' oxygen consumption and found, unexpectedly, that the metabolic energy cost of brachiating was 20–30% higher than the cost of running at the same speed.

8.5. Climbing Vertical Surfaces and Walking on the Ceiling

Squirrels, woodpeckers, and many other mammals and birds climb vertical tree trunks. Geckos and many insects climb vertical structures, such as tree trunks and other plant stems and walls, and can even run on inverted surfaces such as ceilings.

Figure 8.5A is a free-body diagram of a squirrel of mass m climbing a tree trunk. The following forces are acting on its body:

1. Its weight mg, where g is the gravitational acceleration.

2. An upward component of force mg, which may be divided in any convenient way between the fore and hind feet. Note that the line of action of this force is separated by a distance x from the line of action of force 1, so the two forces together exert a clockwise couple mgx on the animal.

3. A horizontal component of force mgx/y (where y is the distance defined by the diagram) acting toward the left of the diagram, on the fore feet. This force and force 4 (below) together exert an anticlockwise moment on the body, balancing the clockwise moment exerted by forces 1 and 2.

4. A horizontal component of force mgx/y acting toward the right of the diagram, on the hind feet.

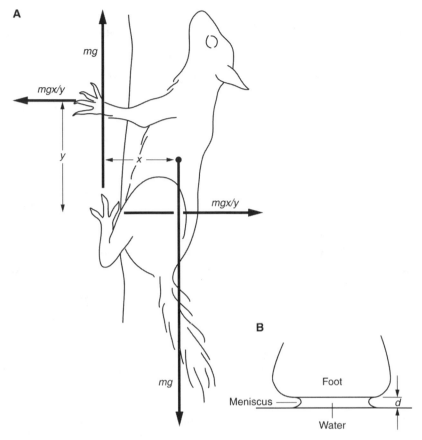

Fig. 8.5. (A) A free-body diagram of a squirrel climbing a tree trunk. (B) A diagram of an insect foot pad attached to a surface by capillary adhesion.

Notice that the animal cannot be in equilibrium unless its fore feet are pulling it toward the tree. The hind feet press on the tree, and friction between them and the bark may be enough to provide the upward force that must act on the paws. If the animal were descending the tree with its head pointing down, its hind feet instead of its fore feet would have to pull it toward the tree.

In the next few paragraphs, I will outline some principles that animals might exploit to attach themselves to vertical or inverted surfaces. After that I will examine the evidence that indicates which principle is used by a variety of climbing animals. We will find that different groups of animals adhere to vertical and inverted surfaces by different means.

If the animal's fore legs are long enough, it may reach round to the far side of a tree trunk and press on it. Alternatively, if it can reach only to the sides of the trunk, the resultant of the normal and frictional forces on its hand may be sufficient to hold it in place (Cartmill 1985).

If the animal has claws and the surface is soft enough to be penetrated by them, the animal may get the required grip by digging its claws in. Animals that can climb smooth, hard surfaces such as glass are unlikely to be dependent on claws.

The animal may attach itself to the surface it is climbing by suction. The force obtainable in this way is limited by atmospheric pressure (0.1 MPa at sea level). The obvious test to find out whether an animal is using suction is to reduce the air pressure; a sucker cannot work in a vacuum.

An animal might use electrostatic forces to attach itself to a surface. If it did, it would not be expected to be able to maintain its grip if its feet were wetted with an electrically conducting solution or if the air were ionized by means of an antistatic gun. This possibility has been considered for various insects, but no evidence has been found that any of them use it, and it seems unlikely that the feet would be well enough insulated from the substrate for sufficiently large potential differences to be maintained (Dixon et al., 1990).

An animal that could make sufficiently close contact with the surface it was climbing might be held in place by van der Waals forces (intermolecular attraction). The van der Waals force between a sphere of radius r and a flat surface separated by a distance d is

$$F_{\text{van der Waals}} = \frac{Hr}{6d^2} \tag{8.12}$$

(Tabor 1991). H is the Hamaker constant, which depends on the materials involved but is of the order of 10^{-19} J.

If there were a thin film of liquid between a foot and the surface it was clinging to, it might be held in place by Stefan adhesion. This is the effect that makes it very difficult to separate two sheets of wet glass. When two surfaces separated by a film of fluid are pulled apart, fluid has to flow through the gap between them. The viscosity of the fluid resists this flow. Large forces are required if the gap is initially very narrow and separation is fast. The force of Stefan adhesion between two disks of radius r, with a film of thickness d of a fluid of viscosity μ between them, is

$$F_{\text{Stefan}} = \left(\frac{1.5\pi\mu r^4}{d^3} \right) \frac{dd}{dt} \tag{8.13}$$

where dd/dt is the rate of separation (Denny 1993). The viscosity of water is 0.001 N s/m^2. Note that this mechanism cannot hold indefinitely; a small force that is maintained for long enough will eventually cause detachment.

Surfaces separated by a drop of liquid are also held together by capillary adhesion, if the liquid does not extend beyond their edges. This is due to surface tension in the meniscus at the edge of the drop (Fig. 8.4B), which because of its curvature reduces the pressure in the drop. If the thickness d of the liquid film is much less than the radius of the drop, the force of capillary adhesion is

$$F_{capillary} = \frac{(\cos\theta_1 + \cos\theta_2)\,A\gamma}{d} \tag{8.14}$$

where θ_1 and θ_2 are the contact angles of the liquid with the two surfaces, γ is its surface tension (0.073 N/m, in the case of pure water), and A is the area of the drop (Denny 1993). Obviously, this mechanism of adhesion cannot operate if all traces of liquid are removed from the surfaces.

Finally, surfaces may be held together by an adhesive.

Now we will consider some climbing animals, looking for evidence of how they attach themselves to the surfaces that they are climbing. Apes and monkeys (other than marmosets) have no claws, and depend on their long arms to grip tree trunks. Other mammals, and birds, depend on claws. When a mammal is descending a tree, the hind feet must be reversed to obtain the necessary grip. The ability to do this has been evolved by tree squirrels such as *Sciurus*, kinkajous (*Potos*), lemurs such as *Varecia*, and some other mammals (Jenkins and McClearn 1984; Meldrum et al., 1997). Domestic cats cannot reverse the hind feet and often have difficulty descending trees.

Geckos (Gekkonidae) have the soles of their feet covered by a carpetlike pile of fine setae (Russell 1975; Röll 1995). *Gekko gecko* has almost half a million setae 30–130 μm long, on each foot. Each seta, has between 100 and 1000 branches at its end, each ending in a tiny pad of 200–500 nm diameter, called a spatula. Irschick et al. (1996) showed that each foot of a *Gekko* of mass 43 g (weight 0.43 N) can exert an adhesive force of 20 N. That implies that each seta can exert 10 μN. Autumn et al. (2000) dissected off individual setae and glued their bases to entomological pins. They pressed the spatulae against a surface connected to an exceedingly sensitive force transducer, then measured the force needed to pull them free. The mean force required was 14 μN when the pull was perpendicular to the surface, and 190 μN (far more than the experiments with intact geckoes suggested) when the pull was parallel to the surface.

The forces are too large to be due to suction. Geckos adhere to dry surfaces and there is no evidence of any secretion that might provide capillary adhesion. It seems likely that they depend on van der Waals forces for adhesion. Autumn et al. (2000) used Equation 8.12 to make a very rough estimate of the van der Waals force on an attached seta. They supposed that each spatula might at some point be only 0.3 nm from the surface (the diameters of atoms are of the order of 0.3 nm [Tabor 1991]). They assumed that the surfaces of the spatulae could be regarded as segments of the surfaces of spheres of radius 2 μm. This gave them an estimated force of 0.4 μN per spatula, enough for a seta with 500 spatulae to exert the observed force of 190 μN. It must be emphasized that this is an exceedingly rough calculation, but it seems to show that van der Waals forces are of the right order of magnitude to explain gecko adhesion. Though the setae adhere strongly, they can be detached easily by peeling them like an adhesive plaster (Autumn et al., 2000).

Beetles have claws and also pads covered by a pile of setae, rather like the setae of geckos. Stork (1980) studied adhesion by the feet of *Chrysolina*, a beetle about 8 mm long. He used fine wires glued to the elytra to attach the beetles to a force transducer, and measured the forces they could exert parallel to the surface on various surfaces. They could exert forces of almost 40 times body weight on glass or an acrylic plastic, and rather more on cloth. Cutting off the claws with micro-scissors had no appreciable effect on the beetles' ability to cling to glass or acrylic plastic, but greatly reduced their ability to adhere to cloth. This indicated that attachment to cloth depended on the claws, but that the beetles adhered to glass and acrylic plastic by some other mechanism. Experiments in which the ambient air pressure was reduced had no appreciable effect on the beetles' adhesion to glass, showing that it did not depend on suction. An antistatic gun had no appreciable effect, indicating that electrostatic forces were not responsible. Beetles whose feet had been dried by walking over filter paper, and which were then tested in a very dry atmosphere, showed apparently undiminished adhesion to glass, suggesting (but not proving) that neither Stefan adhesion nor capillary adhesion was involved. Van der Waals forces seem the most likely means of adhesion to glass. The maximum forces that the beetles could exert on glass amounted to about 1 μN per seta. As there was only one spatula per seta, this is a little more than the force per spatula that geckos exert, but it is still low enough for it to be plausible to suggest that it is a van der Waals force.

The soles of the feet of blowflies (*Calliphora*) are also covered by a pile of setae. Walker et al. (1985) found that the forces needed to pull blowflies off glass average three times body weight for a pull perpendicular to the surface and nearly 30 times body weight for pulls parallel to the surface. There are about 3500 setae on a fore foot, each with a spatula of about 2 μm diameter at its end. If there are similar numbers of setae on the other feet, the detachment force for parallel pulls is about 1 μN per seta. This is low enough for van der Waals forces to offer a plausible explanation, but Walker and his colleagues found that flies leave oily footprints. They removed the oily secretion from the feet of flies by making them walk across filter paper moistened with hexane (which dissolves lipids), and found that this greatly reduced their ability to adhere to glass. They suggested that attachment was largely due to capillary adhesion. A calculation using Equation 8.14 shows that even with a liquid film 1 μm thick, capillary adhesion would be ample to explain the perpendicular force needed to pull blowflies off glass. However, it left the insects' resistance to pulls parallel to the glass unexplained. The most likely explanation for this resistance seems to be friction between the surfaces that are held firmly together by capillary adhesion (Gorb et al., 2001).

Aphids lack the setae found in beetles and flies, but can nevertheless walk upside down on clean glass surfaces. Dixon et al. (1990) studied *Aphis fabae*, which has a pair of claws on each foot and also a soft pad of about 17 μm radius. *Aphis* adhered to clean glass, with maximum forces of about 20 times body weight. Neither reduction of the ambient pressure nor use of an antistatic gun had a significant effect. Dixon and his colleagues considered the possibility that aphids might adhere by van der Waals forces. Instead of Equation 8.12, which predicts the force between a sphere and a flat surface, they used an equation that predicts the force between two parallel planes (Tabor 1991). They calculated that for van der Waals forces to provide such strong adhesion, the surfaces of the pulvilli would have to be within 8 nm (the diameter of a large protein molecule) of the glass surface. They thought it unlikely that such close contact could be achieved over so large an area. However, they found that aphids that had walked for 15 minutes on silica gel (a drying agent) were unable to adhere to glass for the next 30 min. The ability to adhere to glass could be restored very rapidly, by allowing an aphid that had walked on silica gel to walk on moist filter paper. This suggested that capillary adhesion was responsible for attachment to glass. Water could apparently serve as the liquid, but there was evidence that the feet could themselves secrete a liquid. Equation 8.14 showed that with a liquid film a few micrometers thick, capillary adhesion could explain the observed forces. A further calculation using Equation 8.13 shows that with the same film thickness, Stefan adhesion would be too weak to have any importance.

Though aphids with dry feet could not walk upside down on glass, they could walk on inverted plates of silica gel, presumably by means of their claws. They lost this ability if the claws were removed.

The toe pads of tree frogs adhere relatively weakly, apparently by capillary adhesion. Only a few species can support themselves from an inverted surface (Emerson and Diehl 1980).

The tube feet of starfishes attach to the substrate by means of an adhesive formed from the combined secretions of two types of glands (Flammang et al., 1998). This adhesive consists mainly of protein. A secretion from a third type of gland seems to release the adhesive, which is left behind as a "footprint" on the substrate when the starfish moves on.

This chapter has discussed jumping by animals ranging from fleas to humans, showing how different styles of jumping are best for animals of different sizes. It has described how gibbons swing through trees. It has also shown how some animals can climb vertical surfaces or even run across the ceiling, adhering by means that include claws, capillary adhesion, van der Waals forces, and adhesives.

As in previous chapters, many of the stories told in this one are incomplete. The dynamics of jumping is well understood, but we need experiments on jumping vertebrates to record the length changes of the tendons and muscle fascicles and to check our understanding of the muscle physiology involved. Research is needed on the mechanics and energetics of locomotion through trees, taking account of the flexibility and uneven spacing of branches. And I suspect that there is more to be learned about the roles of different physical mechanisms in the adhesion of insect and gecko feet.

Chapter Nine

Crawling and Burrowing

OR THE PURPOSES of this chapter, I define crawling as locomotion on land that depends principally on movements of the body rather than of limbs. Many crawling animals, such as earthworms and snakes, have no limbs. Others, such as caterpillars, have legs, but progress by bending and extending the body; the legs serve merely to anchor the anterior end of the body while posterior parts are drawn forward. An anomaly of this definition of crawling is that it excludes the crawling of human babies.

In many cases, animals crawl and burrow using similar movements. For example, earthworms crawl and burrow by peristalsis, and limbless lizards crawl and burrow by lateral undulation. It seems convenient to discuss crawling and burrowing in the same chapter.

9.1. WORMS

The animals discussed in this section are more or less cylindrical. The volume V of a circular cylinder of length l and radius r is

$$V = \pi r^2 l \qquad (9.1)$$

A closed cylinder filled with liquid must maintain constant volume even if its shape changes, so if its length increases, its radius and circumference must decrease, and vice versa. By rearranging Equation 9.1 and differentiating it, we get

$$\frac{dr}{dl} = \frac{-r}{2l} \qquad (9.2)$$

This tells us that for small strains (fractional changes of length), the strain in the circular muscles is minus one-half of the strain in the longitudinal muscles; 1% shortening of the circular muscles causes 2% lengthening of the longitudinal muscles. The incompressibility of the tissues and of the liquid in the body cavity makes the circular muscles antagonistic to the circular ones, just as the stiffness of a bony skeleton makes the flexor muscles in a vertebrate limb antagonistic to the extensor muscles. For this reason, the tissues and liquid are often said to function as a hydrostatic skeleton (Chapman 1958; Alexander 1995c).

A worm with circular and longitudinal muscles can bend its body by shortening the longitudinal muscles of one side while allowing those of the other side to be stretched. Earthworms have only longitudinal and circular muscles to change the shapes of their bodies, but leeches have, in addition, dorsoventral muscles that flatten the body, changing its cross section from a circle to an ellipse.

Leeches practice two-anchor crawling, using the principle explained in Section 6.1 (Gray 1968). Suckers at the two ends of the body attach alternately to the substrate. With the posterior sucker attached, the leech extends its body to push the anterior end forward. Then it attaches the anterior sucker and shortens the body to pull the posterior parts forward.

In contrast, earthworms crawl by peristalsis, sending waves of lengthening and shortening traveling backward along the body (Gray and Lissmann 1938). The principle was explained in Section 6.2. The anchorage required to prevent backward sliding is provided by chaetae, short bristles that protrude from every segment. Not only do the chaetae slope backward, making the coefficient of friction for backward sliding greater than for forward sliding, but they are protruded whenever the longitudinal muscles contract, which is when anchorage is required (Clark 1964). It is because of the chaetae that a worm feels rough when you run a finger anteriorly along its ventral surface, but smooth when you run the finger posteriorly.

Quillin (1999) filmed *Lumbricus terrestris* of a wide range of sizes crawling on moist fabric. Mean crawling speed was about 0.04 body lengths per second, and stride length was about 0.15 body lengths, for worms ranging from about 20 mg to 9 g body mass. Each segment was stationary for, on average, 38% of the time, and moving forward for 62% of the time. Thus, q (Equation 6.3) was 0.62, implying that the coefficient of friction for backward sliding was at least 1.6 times the coefficient for forward sliding. If a higher value of q were possible, the worm could crawl faster, as the following argument shows. As in Fig. 6.2A, consider a worm that lengthens and shortens each segment by Δl in the course of a cycle of its crawling movements. The duration of a cycle is τ, and one wavelength of the motion extends over n segments. The interval of time between the two positions shown in the Figure is τ/n. In this time, the anterior end of the worm advances a distance Δl at speed $n\,\Delta l/\tau$. It is advancing at this speed for a fraction q of the time and stationary for the rest, so the worm's mean speed v is

$$v = qn\,\frac{\Delta l}{\tau} \qquad\qquad (9.3)$$

and, other things being equal, an increase in q will result in an increase of speed.

We saw in Chapter 6 that the work that legless animals have to do when they crawl has two components. One of these is the work that has to be done to overcome friction with the ground. The other is the inertial work required to give parts of the body kinetic energy whenever they move forward. If these animals crawled fast, the inertial cost of transport would become very large. In Section 6.2, a simple model of earthwormlike crawling led to the conclusion that the inertial work would exceed the frictional work if the Froude number $v^2/\lambda g$ was greater than $2\mu_{forward} q^2$. Here v is speed, λ is stride length, g is the gravitational acceleration (9.8 m/s^2), $\mu_{forward}$ is the coefficient of friction for forward sliding of the body, and q has the same meaning as before. Consider a typical 1-g worm crawling at 0.004 m/s with a stride length of 0.015 m (Quillin 1999). Its Froude number is 1.1×10^{-4}. The coefficient of friction seems likely to be of the order of 0.1, and q is 0.62 (see above). Thus, the speed is much too low for the inertial cost of transport to be significant.

Septa cross the body cavity of earthworms, preventing the fluid in the body cavity from flowing between segments. Seymour (1969) recorded pressures in the body cavity by means of hollow needles connected by flexible tubing to pressure transducers. He showed substantial pressure differences between segments, confirming that the septa divided the body cavity effectively. Thus, shortening of the longitudinal muscles in each segment forces extension of the circular muscles of the same segment, and vice versa, independent of what is happening elsewhere in the worm. The pressure in each segment fluctuates as the worm crawls, with peak pressures of around 600 N/m^2 (Quillin 1998) occurring when the segment is long and thin. Seymour (1969) recorded pressures up to 7500 N/m^2 in a worm that was squirming violently.

Equation 9.2 implies that a small strain σ in the longitudinal muscle of a segment is accompanied by a strain $-\sigma/2$ in the circular muscle. Imagine a segment that changes slightly in length while the pressure inside it remains constant. The (positive) work done by the longitudinal muscle must be matched by the (negative) work done by the circular muscle. The work done by each is Force × Length change, which equals Stress × Cross-sectional area × Strain × Length, which equals the volume of the muscle multiplied by the stress multiplied by the strain. This implies that for the longitudinal and circular muscles to exert equal stresses, the volume of circular muscle should be twice the volume of longitudinal muscle. Chapman (1950) pointed this out, using a less concise argument. He went on to show that the volume of circular muscle is actually only 0.4 times the volume of the longitudinal muscle. Therefore, if the circular and longitudinal muscles can exert the same maximum stress (which is not necessarily the case), the longitudinal muscle is capable of withstanding five times as much pressure in the body cavity, as the circular muscle.

This would be puzzling if crawling was the worm's only mode of locomotion. However, worms also burrow, using movements like those of crawling. The swelling of shortened segments helps to anchor the worm and is probably also used to widen the burrow. The longitudinal muscles are needed to generate pressure in the body cavity, but while the body wall of a segment is pressed firmly against the walls of the burrow there is no need for stress in the circular muscles. The septa make it possible for high pressures to be confined to segments whose walls are pressed against the burrow wall, so there is no need for the circular muscles to be able to withstand high pressures. Seymour (1969) measured pressures up to 2500 N/m^2 in the body cavity of earthworms in the early stages of burrowing into loose earth, and it seems likely that much larger pressures are developed while burrowing in firm earth.

Nemertean worms are not segmented and have no partitions in the body. *Lineus* nevertheless crawls like an earthworm, by passing peristaltic waves posteriorly along its body (Gray 1968). The lugworm *Arenicola* has septa only near the anterior and posterior ends of the body, so fluid can be moved around within the body. Seymour (1971) has described how it burrows.

Polyphysia is a polychaete worm that crawls by passing peristaltic waves forward along the body, as shown in Fig. 6.2B (Elder 1973). This gait requires the segments to be long while they are stationary and short while they are moving. Burrowing requires the segments to be thick while they are stationary, to anchor them, and thin while they are advancing. Therefore, the segments must be fat while they are long and thin while they are short. Fluid must be free to move between the segments, so this style of locomotion is possible only for worms that, like *Polyphysia*, have no septa. Hunter and Elder (1989) recorded pressures in the body cavity of burrowing *Polyphysia*. They also measured the forces that the animal could exert when pushing with its anterior end or pulling with its tail. From these measurements they attempted to calculate the work of locomotion, but their calculation depended on the doubtful assumption that the forces measured when the animal was pushing or pulling on their instruments were the same as in unrestrained burrowing.

Nematode worms crawl like snakes (Section 9.4), by passing waves of bending backward along the body (Alexander 2001). The bends, however, are dorsoventral, not transverse as in snakes. Soil nematodes of the order of 1 mm long travel in this way in the spaces between soil particles (Wallace 1958). Other plant parasitic nematodes, such as the Chrysanthemum eelworm (*Aphelenchoides*), crawl by similar movements in the film of water covering wet leaves. The effectiveness of this technique of locomotion

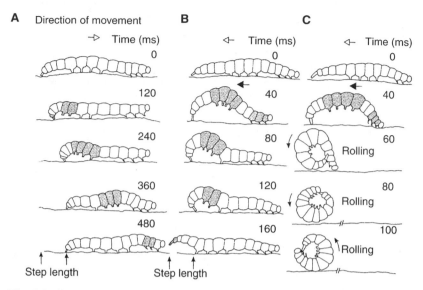

Fig. 9.1. Drawings from video films of *Pleurotya* caterpillars about 25 mm long (A) crawling forward at 10 mm/s, (B) crawling backward, and (C) rolling backward. Stippling indicates segments bearing appendages that are off the ground. From Brackenbury (1997).

depends on surface tension. When a worm is stationary in a water film, surface tension acts symmetrically on the two sides of its body. However, when it passes waves of bending down its body, its movements make surface tension act asymmetrically (Wallace 1959). Where a part of the body is moving transversely, a well-known property of liquid surfaces makes the surface of the film meet one side of the body at a steeper angle than the other. The equal surface tension forces on the two sides of the body act at different angles to the horizontal, so the horizontal components of the forces are different and the transverse movement is resisted. This effect gives the worm the purchase it needs to propel itself forward.

9.2. INSECT LARVAE

Figure 9.1 shows a caterpillar performing three different gaits. In (A) it is moving forward, using a gait that depends on the same principle as two-anchor crawling (Section 6.1), but involves three anchors: the true legs on body segments 1–3, the prolegs on segments 6–9, and the clasper on segment 13. In (B) it is retreating, using a similar gait in reverse. Finally, in (C) it is performing a remarkable backward roll that this species uses as an escape strategy. The momentum developed in the initial bending of the body is retained, and the animal rolls for up to five revolutions at a speed of around 0.4 m/s, many times faster than the forward walking speed of about 10 mm/s (Brackenbury 1997). *Ceratitis* is another dipteran whose larvae are capable of rapid escape movements, in this case by jumping (Maitland 1992). They jump distances up to 160 mm, and by making a series of jumps in rapid succession they can travel at 19 mm/s, whereas their normal crawling speed is only 2 mm/s.

Gypsy moth caterpillars (*Lymantria*) crawl much more slowly than adult insects of similar mass run, and take much shorter strides. Casey (1991) filmed them crawling on a treadmill and measured their rates of oxygen consumption. He recorded speeds up to 0.03 m/s and stride lengths up to 0.008 m, giving a maximum Froude number of 0.01. This is higher than for the earthworm (Section 9.1) but still too low to make the inertial cost of transport substantial. The metabolic cost of transport is nevertheless high, 4.5 times the value predicted for running arthropods of the same mass (Fig. 7.12).

Berrigan and Pepin (1995) studied crawling by the limbless larvae of four species of dipteran fly. Their gaits involved lengthening and shortening of the body in the manner of two-anchor crawling. Larvae of masses 0.5–220 mg crawled at speeds of 0.7–10 mm/s, with stride frequencies of 0.6–2.8 Hz. In contrast, adult insects of similar mass typically run at 70–100 mm/s, with stride frequencies near 10 Hz. Berrigan and Lighton (1993) measured the oxygen consumption of crawling dipteran larvae and calculated that the net metabolic cost of transport was ten times as high as predicted for running adult insects of the same mass.

9.3. MOLLUSCS

Gastropod molluscs crawl on a large, muscular foot at very low speeds; snails (*Helix*) crawl at about 2.5 mm/s and limpets (*Patella*) at about 1 mm/s. A specimen of *Haliotis* with a shell 99 mm long was induced to crawl at the exceptionally fast speed, for a gastropod, of 19 mm/s (Donovan and Carefoot 1997).

If a gastropod is viewed from below while crawling on glass, it can be seen that waves of muscular contraction travel along the foot (Lissmann 1945a). These waves may travel forward as in the garden snail *Helix*, or backward as in chitons. Muscular activity may be in phase across the whole width of the foot as in these examples. Alternatively, waves traveling along the left side of the foot may be half a cycle out of phase with waves on the right, as in *Haliotis* (forward waves) and *Patella* (backward waves [Trueman and Jones 1977]). If marks have previously been made on the sole of the foot, it can be seen through the glass that the waves involve local lengthening and shortening of the foot. The marks move forward relative to the glass at one stage of the crawling cycle, and remain stationary relative to the glass at another (Lissmann 1945a). It is clear that crawling depends on the same principle of peristalsis as earthworms and *Polyphysia* use (Section 9.1). If the waves travel forward, the shortened parts of the foot move forward while the extended parts remain stationary, as Fig. 6.2B suggests. If they travel backward, the reverse is the case, as in Fig. 6.2A.

Lissmann (1945b) believed that the parts of the foot that were to be moved were lifted clear of the substrate. Jones and Trueman (1970) showed that the muscle fibers in the foot of *Patella* (which uses backward-moving waves) seemed adapted for lifting the extended parts of the foot. Jones (1973) demonstrated a different arrangement of fibers in the foot of the slug *Agriolimax* (forward-moving waves) that seemed adapted for lifting contracted parts of the foot. However, Denny (1980a) pointed out that the thin layer of viscous mucus between the foot and the substrate would make it very difficult to lift the foot off the substrate (see the explanation of Stefan adhesion in Section 8.5).

The mucus of *Ariolimax* consists of 3–4% high molecular weight glycoprotein, together with water and salts (Denny 1980a). Whenever a part of the foot moves forward, the mucus under it is sheared as in Fig. 3.2B. When that part of the foot is halted, the shearing temporarily ceases. Denny (1980a) simulated this in an experiment with slug mucus in a cone and plate viscometer. This is a standard instrument for measuring viscosity (Fig. 9.2A). The material to be tested is sandwiched between a rotating cone and a stationary plate. The torque required to hold the plate stationary is measured, enabling the viscosity to be calculated. Denny put a sample of slug mucus into a viscometer. For one second he rotated the cone at constant speed, shearing the mucus; then he held the cone stationary for one second; then he rotated it again for a second, and so on (Fig. 9.2B).

To understand the result of Denny's experiment, we need to be clear about the difference between elasticity and viscosity. When an elastic solid is sheared, the stress in it rises, ideally in proportion to the strain. That implies that if strain continues at a constant rate (as in Denny's experiment), stress rises progressively. However, stress in a viscous liquid is proportional not to strain, but to strain rate. If the strain rate is constant, the stress in the liquid is also constant. In the experiment, each time the plate started rotating, the stress rose, initially at a more or less constant rate. The mucus was behaving like an elastic solid. However, there was a sudden change when the strain reached a value of about 5 (e.g., when the plate under a layer of mucus 1 mm thick had moved 5 mm). The stress fell to a lower level and remained constant at that level so long as shearing continued. The mucus was now behaving like a viscous liquid. When shearing

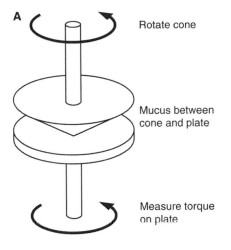

A — Rotate cone

Mucus between cone and plate

Measure torque on plate

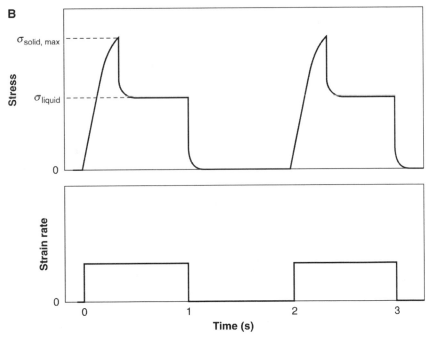

Fig. 9.2. (A) A cone and plate viscometer. (B) Results of Denny's (1980a) experiment with slug mucus in a viscometer. The upper graph shows the stress in the mucus resulting from the strain rates shown in the lower graph.

ceased, the stress fell to zero. Within one second the mucus had regained its solid properties, and the cycle could be repeated.

Let us consider how these properties limit the mollusc's speed. As in our discussion of earthworms (Section 9.1), let it crawl at speed v; let any point on the foot move for a fraction q of the time and be stationary for the rest; let the duration of a cycle of movement be τ; and let the strain involved in lengthening be ε. Let the wavelength of the pattern of lengthening and shortening on the foot be L. To get the speed, we need to modify Equation 9.3, the equation that we derived for worms. For the worm, L is equal to nl (the number of segments in a wavelength multiplied by the length of a segment), and ε is equal to $\Delta l/l$ (the fractional change of length of a segment). Thus, we can rewrite Equation 9.3 in a form applicable to unsegmented animals:

$$v = \frac{qL\varepsilon}{\tau} \tag{9.4}$$

An increase in q, L, or ε, or a decrease in τ, will enable the mollusc to move faster. Donovan and Carefoot (1997) found that as *Haliotis* increased speed, cycle duration τ fell and the distance traveled per cycle ($qL\varepsilon$) rose. The wavelength L in *Haliotis* is approximately equal to the length of the foot, so there is no scope for increasing it. The increase in $qL\varepsilon$ seems to have been at least largely due to an increase in q.

The forces on the stationary parts of the foot (area A_{stat}) must balance the forces on the moving parts (area A_{move}). Let the maximum stress that mucus can withstand in its "solid" state be $\sigma_{solid, max}$ and let the stress in the "liquid" mucus under the moving parts of the foot be σ_{liquid}. For the forces to balance,

$$A_{stat}\sigma_{solid, max} \geq A_{move}\sigma_{liquid} \tag{9.5}$$

If the lengthening and shortening of parts of the foot does not change their areas too much, A_{move}/A_{stat} is approximately equal to $q/(1-q)$. Thus, Equation 9.5 tells us that q cannot be greater than approximately $\sigma_{solid, max}/(\sigma_{liquid} + \sigma_{solid, max})$. This conclusion is less informative than we might wish, because both $\sigma_{solid, max}$ and σ_{liquid} depend on the strain rate (Denny 1980a). However, it does tell us that q is limited by the properties of the mucus. Also, the time for which a point on the foot is stationary, $(1-q)\tau$, must be large compared to the time required for the mucus to regain its "solid" properties. Thus, the properties of the mucus affect the range of feasible values of τ as well as of q. Denny (1984) made a much more rigorous analysis of the crawling of slugs and reached the same conclusion.

Denny (1980b) measured the oxygen consumption of crawling *Agriolimax* and calculated that the net metabolic cost of transport was 12 times as high as expected (Fig. 7.12) for a running mammal or arthropod of the

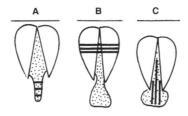

Fig. 9.3. Diagrams representing successive positions of a bivalve mollusc, during burrowing. Thick lines represent active muscles. From Alexander (1982).

same mass. Other gastropods that have been investigated have relatively lower costs of transport; the most economical is *Haliotis*, with a cost approximately equal to the predicted value for a mammal or arthropod (Donovan and Carefoot 1997).

In ideal Newtonian fluids subjected to shear, stress is proportional to strain rate. In mucus, stress is less dependent on strain rate; it is approximately proportional to (strain rate)$^{0.44}$ (Denny 1984). This still gives the animal scope for reducing the work required for crawling (and so presumably the cost of transport) by reducing the strain rate, which could be done by using a thicker layer of mucus. However, the thicker the mucus layer, the more glycoprotein will be left behind in the slime trail and lost to the animal. Denny (1980b) estimated that the energy cost to a slug of replacing the glycoprotein lost in the slime trail accounted for 35% of the net metabolic cost of transport.

Thus, gastropod crawling is slow and expensive of energy and materials. It has, however, the advantage that the animal adheres to the substrate. This advantage is exploited by shore-living gastropods, which are not easily dislodged from rocks by waves, and by slugs and land snails, which can climb the stems of plants to reach their food.

Many bivalve molluscs burrow in sand or mud, using a two-anchor mechanism (Fig. 9.3. [Trueman 1967]). Their two-valved shells can be closed by contraction of the adductor muscles, but spring open by elastic recoil of the hinge ligament when the adductors relax. Their muscular feet can be shortened by contraction of retractor muscles or extended by contraction of transverse muscles. There are blood-filled cavities around the viscera (the visceral hemocoel) and in the foot (the pedal hemocoel), connected through Keber's valve. With the valve closed, no blood can enter or leave the pedal hemocoel, so the volume of the foot remains constant; shortening of the retractor muscles inevitably stretches the transverse muscles, and vice versa. With the valve open, contraction of the adductor muscles can drive blood into the foot, and contraction of the muscles of the foot can drive blood back into the visceral hemocoel.

These movements are used in burrowing. With the adductor muscles relaxed, the elasticity of the hinge ligament presses the two shell valves firmly against the surrounding mud or sand. This anchors the shell while the foot is extended, pushing its tip into the sand (Fig. 9.3A). Then with Keber's valve open the adductor muscles contract, relaxing the pressure of the shell against the sand and forcing blood into the foot, whose tip swells (B). The swollen foot serves as an anchor while the retractor muscles contract, pulling the shell deeper into the mud or sand (C). Then the cycle is repeated. The mollusc moves down through the mud or sand with the foot leading, and with the shell and the foot serving as anchors alternately.

For most bivalves, burrowing is merely a means of burying themselves, not of traveling. However, surf clams (*Donax*) make use of burrowing in a remarkable mode of locomotion that enables them to travel rapidly up and down beaches (Ellers 1995). They emerge from the sand at the appropriate time for a wave to wash them up or down the beach, and burrow while the water is flowing in the opposite direction.

9.4. REPTILES

Snakes and limbless lizards most commonly travel by serpentine crawling. Waves that move backward along the snake's body remain stationary relative to the ground, making the snake move forward. We saw in Section 6.3 that friction makes serpentine crawling possible even on a smooth surface. However, when snakes crawl on smooth surfaces some backward slippage of the waves may occur (Gans 1984). On natural substrates, irregularities in the ground such as stones or tussocks of grass tend to prevent slippage of the waves. Walton et al. (1990) measured the oxygen consumption of black racer snakes (*Coluber constrictor*, mean mass 103 g) moving by serpentine crawling on a treadmill provided with projections designed to give the snake a purchase. The net metabolic cost of transport was 23 J/kg m, almost exactly the value predicted for running by mammals or birds of the same mass (Fig. 7.12).

Equation 6.6 gives the frictional part of the mechanical cost of transport for a very simple model of serpentine locomotion, in which the body moves along a zigzag path instead of a realistic smoothly curved one (Fig. 6.3B). The work done against friction with the ground contributes $\mu_{axial} g / \cos \phi$ to the cost of transport, where μ_{axial} is the coefficient of friction for sliding along the body axis, g is the gravitational acceleration, and ϕ is the angle shown in Fig. 6.3. The coefficient of friction μ_{axial} presumably varies between species and between substrates. Gans and Gasc (1990) obtained values of 0.29–0.39 for a limbless lizard (*Ophisaurus*) sliding

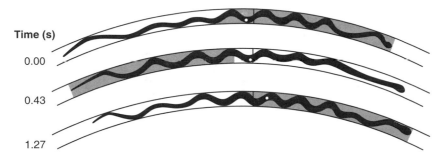

Fig. 9.4. Successive positions, drawn from video images, of a black racer snake (*Coluber*) concertina crawling in a narrow channel. From Jayne and Davis (1991).

forward on various substrates, and we will assume that 0.35 is typical also for snakes. In that case, with reasonably small angles φ we can expect the frictional part of the mechanical cost of transport to be around 4 J/kg m.

The inertial part of the mechanical cost of transport (Equation 6.7) is proportional to speed squared. It would be equal to the frictional part if the Froude number $v^2/\lambda g$ (where λ is the stride length) were equal to $\mu_{axial}/(\sin \phi \tan \phi)$. The stride length for the *Coluber* studied by Walton et al. (1990) was about 0.28 m at all speeds. Its maximum crawling speed, attained only in short bursts, was 1.5 m/s, giving a Froude number of 0.8. It is not clear what value we should give the angle φ, but if it were 36° (which does not seem unreasonable) and if μ_{axial} were 0.35, $\mu_{axial}/(\sin \phi \tan \phi)$ would be 0.8, equal to the Froude number. This crude estimate, based on the zigzag model of serpentine crawling, suggests that the inertial part of the cost of transport may be important for snakes traveling fast. However, the highest speed at which *Coluber* could sustain crawling by aerobic metabolism was only 0.14 m/s, so the measurements of its metabolic cost of transport were all made at 0.14 m/s or slower. At these speeds the Froude number would be very low, and the inertial part of the cost of transport would be negligible.

Thus, we may take our estimate of the frictional part, 4 J/kg m, as an estimate for the whole of the mechanical cost of transport, in the range of speeds for which the metabolic cost of 23 J/kg m was measured. This would imply an efficiency of 0.17, which lies within the range of efficiencies at which muscles commonly work (Section 2.5).

Snakes use a different gait, concertina locomotion, to crawl along narrow channels. This is a form of two-anchor crawling, in which the anterior and posterior parts of the body are anchored in turn. With the anterior parts bent to jam them against the sides of the channel (Fig. 9.4, 0.00 s) the posterior parts are drawn forward. Then with the posterior parts jammed against the sides of the channel, the anterior parts straighten and

Time (s)

0.00   0.60   1.16   1.64

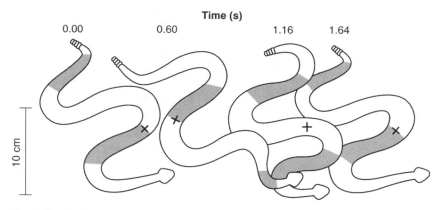

10 cm

Fig. 9.5. Outlines traced from video images of *Crotalus cerastes* sidewinding. The snake is about 0.5 m long, and is traveling at 0.17 m/s toward the right of the page. Only stippled parts of the body are resting on the ground. The cross is a paint mark. From Secor et al. (1992).

reach forward (Fig. 9.4, 0.43 s). Jayne and Davis (1991) filmed *Coluber* of mean mass 118 g (length 0.94 m) crawling in a treadmill formed as a circular channel, narrow enough to induce the snakes to use concertina crawling. The maximum burst speed that they observed was 0.21 m/s, in a channel 7 cm wide. Walton et al. (1990) measured the oxygen consumption of *Coluber* concertina crawling. They found a maximum aerobic speed of 0.06 m/s, and a net metabolic cost of transport of 170 J/kg m. Thus, concertina crawling is slower and much more expensive of energy than serpentine crawling. Its advantage is that it enables snakes to go where serpentine crawling would not be feasible. It even enables snakes to climb vertical tree trunks, using ridges on the bark to gain a purchase (photograph in Alexander [1992c]).

Sidewinding is another crawling technique used by snakes (Fig. 9.5). Waves of bending are passed posteriorly along the body as in serpentine crawling, but the body does not slide over the ground. Instead, each section of the body is lifted from one resting place to the next. In Fig. 9.5, stippled parts of the body are stationary, resting on the ground, and parts that are left white are moving and off the ground. This technique is effective on loose sand and on smooth surfaces, on which serpentine locomotion is difficult because they do not provide the snake with anything to push against. Many snakes sidewind on surfaces of this kind, but for a few, such as *Crotalus*, it is the normal mode of locomotion.

Sidewinding is not fast. The maximum speed attained in short bursts by *Crotalus* of mean mass 110 g was only 1.0 m/s (Secor et al., 1992), whereas the *Coluber* of similar mass discussed above attained 1.5 m/s in serpentine crawling. It is, however, remarkably economical. The net metabolic cost of transport for the *Crotalus* sidewinding was only 8 J/kg m, compared to 23 J/kg m for the *Coluber*'s serpentine crawling. The reason for its being so economical may be that because there is no sliding over the ground, no work has to be done against friction.

Rectilinear locomotion is yet another crawling technique, used in some circumstances by *Boa* and some other snakes (Lissmann 1950). No bending of the body is involved. Instead, rib movements are used to lift and move forward successive sections of the ventral body surface; the ribs are used almost like legs. Waves of rib movement travel posteriorly along the body.

Even fast snakes are relatively slow compared to legged lizards. *Coluber constrictor* has sufficient reputation for speed to be known as the black racer, but its maximum burst speed of 1.5 m/s (see above) is only one-quarter of the running speed of 6 m/s attained by the lizard *Cnemidophorus* (Bonine and Garland 1999).

Limbless lizards such as *Ophisaurus* crawl like snakes, using serpentine, concertina, or slide-pushing locomotion as appropriate (Gans and Gasc 1990). Amphisbaenians are reptiles related to lizards and snakes that have lost their limbs (apart from the forelimbs retained by one genus). They crawl by a technique that looks much like the crawling of earthworms but involves sliding the skin backward and forward over the underlying tissues (Gans 1968). They make systems of burrows in moist soil and apparently patrol them, searching for prey such as earthworms and termites. They dig with their heads, pressing the displaced soil into the roof of the tunnel.

Many lizards that live on sand dunes bury themselves in the loose sand, apparently to hide from predators. Arnold (1995) has described how they use leg movements or trunk and tail movements or both to do this. All of the lizards he describes retain legs of normal length, but other lizards that burrow in loose sand have rudimentary limbs or none. Once a lizard is buried, its limbs presumably only get in the way.

9.5. MAMMALS

Various small mammals make systems of burrows in soil. Moles (*Talpa*) use their shovel-like fore limbs to extend their burrows (Quilliam 1966), and mole rats such as *Cryptomys* dig with their teeth (Bennett 1991). Measurements of the oxygen consumption of various rodents have shown that the metabolic rate while burrowing is commonly around five times the resting rate (Ebensperger and Bozinovic 2000).

This chapter has discussed how worms, insect larvae, and snakes crawl, referring back to the simple models in Chapter 6. It has explained the role of mucus in snail crawling. And it has shown briefly how bivalve molluscs and mammals burrow. One of the topics on which I would particularly like to see further research is the energetics of burrowing in undisturbed natural substrates.

. . . .

Chapter Ten

Gliding and Soaring

A T THIS STAGE we turn from terrestrial locomotion to flight. Powered flight has been evolved only by insects, birds, bats, and (apparently) the extinct pterosaurs, and is the subject of later chapters. This
chapter is about gliding and soaring. Insects, birds, and bats glide, and so
also do various other animals, including flying fish and flying squirrels.

This chapter starts with explanations of some of the basic principles of
aerodynamics that are needed to understand flight. More detailed and
authoritative accounts can be found in Prandtl and Tietjens (1957) and
other textbooks of aerodynamics. Later sections of this chapter examine
the gliding performance of animals and show how some of them soar,
using natural air movements to keep them airborne.

10.1. DRAG

We have already encountered aerodynamic drag, which limits the jumping
performance of insects (Section 8.3). Drag is a force that resists the movement of bodies through fluids. It is due partly to the work that has to be
done against the viscosity of the fluid, as the body moves through it; and
partly to the work that is done giving kinetic energy to the fluid that is
left moving in the body's wake (Section 3.4). The relative importance of
viscous forces and inertial forces depends on the Reynolds number, which
takes account of the size of the body, the speed of movement, and the
properties of the fluid (Section 4.2). The wings and bodies of flying animals move through the air with Reynolds numbers high enough for inertial forces to be dominant. In this range of Reynolds numbers, the drag
F_{drag} on a body moving with velocity v through a fluid of density ρ can be
calculated using an equation that we have already met (as Equation 3.8):

$$F_{\text{drag}} = \tfrac{1}{2}\rho A v^2 C_{\text{drag}} \qquad (10.1)$$

where A is an area that can be defined in various ways, as explained in
Section 3.4, and C_{drag} is the corresponding drag coefficient. Both for aircraft and for flying animals, the largest aerodynamic forces act on the
wings, and the area generally used as A is the plan area of the wings,
stippled in Figure 10.1A. Notice that this area includes the strip of body

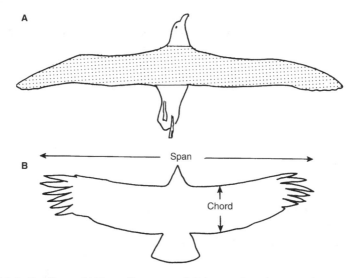

Fig. 10.1. Outlines of (A) an albatross and (B) a condor, showing the meanings of some terms. The stippled area in (A) is the plan area of the wings. From Alexander (1989b).

between the wing bases. The drag coefficients for wings depend on the Reynolds number and on how much lift they are providing, as we will see in Section 10.3.

For calculations of the drag on the fuselage of an aircraft or the body of an animal, the area A is sometimes taken to be the total surface area (the "wetted area") of the fuselage or body, and sometimes the frontal area (its area as seen in front view). Which is more convenient depends on how well streamlined the body is. The reason for this is that there are two kinds of drag. Pressure drag is due to aerodynamic forces acting at right angles to the surface of the body. For example, if a thin plate is moved through air with its plane at right angles to the direction of movement, it pushes on the air in front of it. The drag is all pressure drag, due to the pressure difference between the front and the back of the plate. However, if the plate is moved in its own plane, all the drag is friction drag due to the forces that act parallel to its surface as it drags the air along with it. The drag on bluff bodies (unstreamlined bodies such as spheres) is mainly pressure drag, but the drag on streamlined bodies is nearly all friction drag. Pressure drag depends on the shape of the body, but is affected more by frontal area than by total surface area, so it is generally convenient to use frontal area to calculate the drag on bluff bodies. Friction drag is proportional to total surface area, so it is often convenient to use total surface area to calculate drag on streamlined bodies.

The friction drag on a body of total surface area A_{total} moving with velocity v and Reynolds number $\Re$ through a fluid of density ρ is

$$\text{Friction drag} = \tfrac{1}{2}\rho A_{total}\, v^2\, (1.33\, \Re^{-0.5}) \tag{10.2}$$

at fairly low Reynolds numbers and

$$\text{Friction drag} = \tfrac{1}{2}\rho A_{total}\, v^2\, (0.074\, \Re^{-0.2}) \tag{10.3}$$

at higher Reynolds numbers (Prandtl and Tietjens 1957). The change occurs because, at lower Reynolds numbers, flow in the boundary layer close to the surface of the body is laminar; that means that the flow is smooth, always parallel to the body's surface. At higher Reynolds numbers, flow in the boundary layer is turbulent. The change from laminar to turbulent flow generally occurs at a Reynolds number between about 2×10^5 for bluff bodies and 2×10^6 for well-streamlined bodies. The length used to calculate the Reynolds number is the overall length of the body, in the direction of movement.

The body of a fruit fly 3 mm long flying at 1 m/s has a Reynolds number of 200, and the reasonably well-streamlined body of an albatross 1 m long flying at 15 m/s has a Reynolds number of 10^6 (see Section 4.2). Thus, we can generally assume that the boundary layers on the bodies of flying animals are laminar.

Drag on wingless bird bodies, frozen to make them rigid, has been measured in wind tunnels. Other measurements have been made on casts of bird bodies. These experiments have been reviewed by Hedenström and Leichti (2001), who also calculated body drag from radar observations of the speeds of birds in steep dives. These methods have given very variable drag coefficients, many of them in the range 0.2–0.4 (based on frontal area). These values are high compared to smooth, well-streamlined solids. Measurements by Maybury and Rayner (2001) show that they would be even higher if birds did not have tails. They found (unexpectedly) that drag on frozen wingless starling (*Sturnus*) carcasses was increased by removing the tail. Measurements of drag on insect bodies with the wings removed have also given rather high drag coefficients (for example, Wakeling and Ellington 1997).

10.2. Lift

Drag acts directly backward along the direction of motion through the air. The aerodynamic force on a symmetrical body moving along its axis of symmetry is entirely drag. However, if the body is asymmetrical, or moves at an angle to its axis of symmetry, a component of aerodynamic force may act at right angles to the direction of motion. This force, called lift, can be very much larger than the drag. It keeps aircraft and flying animals airborne. Aerofoils are structures designed to give high lift and low drag, such as the wings of aircraft and birds. The lift on an aerofoil of plan area A_{plan} moving at speed v through air of density ρ is

$$F_{lift} = \tfrac{1}{2}\rho A_{plan} v^2 C_{lift} \qquad (10.4)$$

where C_{lift} is the lift coefficient. Notice that this equation has the same form as the equation for drag (Equation 10.1).

Lift can be explained in two ways, which are really just two different ways of expressing the same thing. Here is the first. Figure 10.2A represents an aerofoil seen in section. It will be convenient to write as if the aerofoil is stationary and the air moving, but the effect would be no different if the reverse were the case; what matters is the velocity of the air, relative to the aerofoil. The aerofoil is shown tilted at an angle of attack α to the direction at which the air is approaching. This tilting and its asymmetrical shape (with the upper surface more curved than the lower), have the effect of deflecting the air downward (Fig. 10.2C). The air is being given downward momentum, so the aerofoil must be exerting a downward force on the air, and the air must be exerting an upward force on the aerofoil.

The span of an aerofoil is the distance from wing tip to wing tip (Fig. 10.1B). We can make reasonably accurate calculations about lift if we assume that the aerofoil deflects only the air that passes through the circle, of which the wingspan s is the diameter (Fig. 10.2B). The area of this circle is $\pi s^2/4$, so if air of density ρ passes through it with velocity v the mass of air deflected in unit time is $\pi \rho v s^2/4$. Let this air be deflected through an angle ψ, so that it is given downward velocity $v \tan \psi$. Momentum is mass multiplied by velocity, so rate of change of momentum is mass multiplied by acceleration, which by Newton's second law of motion equals force. The rate at which the air is being given downward momentum (the mass deflected in unit time multiplied by the downward velocity) is $(\pi \rho v^2 s^2/4)\tan \psi$. This is equal to the lift

$$F_{lift} = \frac{\pi \rho v^2 s^2}{4} \tan \psi \qquad (10.5)$$

We will need this equation when we calculate the extra drag that is an inevitable consequence of lift generation.

The second way of explaining lift depends on Bernoulli's principle, one of the fundamental principles of aerodynamics. Bernoulli's principle applies to steady flow along a streamline. Flow round a body is described as steady if the velocity at any particular point (relative to the body) is always the same, even though particles of fluid may speed up and slow down as they travel. A streamline is a curve whose direction is everywhere the same

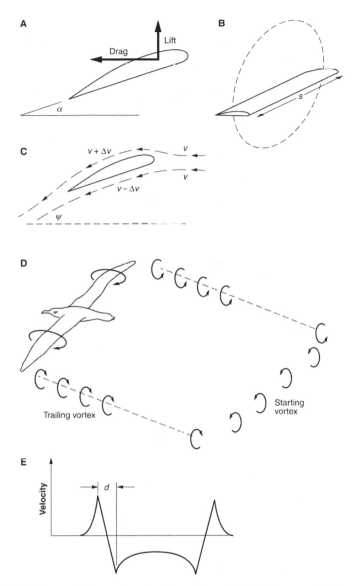

Fig. 10.2. Diagrams illustrating the discussion of lift. (A) An aerofoil in section, indicating the angle of attack α. (B) An aerofoil with the circle through which the air deflected by it is assumed to pass, in the derivation of Equation 10.5. (C) Air approaching an aerofoil with velocity v travels faster over the upper surface and less fast over the lower one, and is deflected through an angle ψ. (D) The circulation around a gliding bird's wings and the vortices in its wake. (E) The vertical component of the velocity of the air in a transverse section through its wake. d is the diameter of the core of one of the trailing vortices.

as the direction of flow of the fluid. In steady flow, the particles of fluid travel along the streamlines. The broken lines with arrows on them in Fig. 10.2C are streamlines.

Bernoulli's principle applies only if viscous forces are small enough for work done against viscosity to be negligible. A moving particle of fluid has kinetic energy, gravitational potential energy, and energy due to its pressure. These energies may change as it flows along a streamline, but if the flow is steady and there are no viscous losses the sum of these energies remains constant. A particle of fluid of volume δV and mass $\rho \, \delta V$ traveling with velocity v has kinetic energy $\frac{1}{2}\rho v^2 \delta V$. If it is at height h and the gravitational acceleration is g, its gravitational potential energy is $\rho g h \, \delta V$. If its pressure is p, its pressure energy is $p \, \delta V$. Thus, Bernoulli's principle tells us that as it travels along a streamline, its energy per unit volume is

$$\tfrac{1}{2}\rho v^2 \; + \; \rho g h \; + \; p \; = \; \text{constant} \tag{10.6}$$

Any changes of gravitational potential energy that occur as air flows past an aerofoil are generally negligible. Thus, Equation 10.6 tells us that along the length of a streamline, the pressure will be less where flow is faster and more where it is slower. Lift acts when air is flowing faster over one surface of an aerofoil than over the other, giving rise to a pressure difference.

Figure 10.2C shows two streamlines, one passing over the aerofoil and one under it. Before they reach the aerofoil, the velocity in both is v and the pressure is the same in both. As they pass the aerofoil, particles on the upper streamline accelerate a little, to $v + \Delta v$, and particles on the lower streamline slow down to $v - \Delta v$. Thus, the pressure is a little less above the aerofoil than below it. Equation 10.6 tells us that the difference of pressure is $\frac{1}{2}\rho[(v + \Delta v)^2 - (v - \Delta v)^2]$, which equals $2\rho v \Delta v$. If the plan area of the wings is A_{plan}, the lift is

$$F_{\text{lift}} = 2\rho A_{\text{plan}} v \Delta v = 2\rho s c v \Delta v \tag{10.7}$$

where s is the span of the aerofoil and c is its mean chord (Fig. 10.1B). The changes of velocity $\pm \Delta v$ are due to the shape of the wing section and to its angle of attack.

There is yet another way of thinking about lift that is sometimes useful. Figure 10.2C shows air velocities relative to the aerofoil, but Fig. 10.2D shows velocities relative to the air that has not yet reached the aerofoil. In this frame of reference, air is flowing backward over the upper surface of the aerofoil and forward over the lower surface; in effect, it is circulating around the aerofoil. Aerodynamicists define a quantity that they call circulation, which is represented by the symbol Γ. Circulation is very simple to calculate if it is possible to draw a closed loop in such a way that the velocity along the circumference of the loop is everywhere the same; in this case, the circulation is the circumference multiplied by the velocity. If we draw a loop that encloses the aerofoil tightly, its circumference is $2c$ (twice the chord, Fig. 10.1B), the velocity is Δv, and the circulation is $2c\,\Delta v$. Thus, Equation 10.7 can be written

$$F_{\text{lift}} = \rho s v \Gamma \qquad (10.8)$$

Figure 10.2D shows airflow in the wake behind a gliding animal or aircraft, as well as round the wing. There is a cylinder of rotating air forming a trailing vortex behind each wing tip. In some circumstances, water vapor condenses out in the trailing vortices behind aircraft, making them visible as vapor trails. If the air had no viscosity, the trailing vortices would persist all along the animal's path, connecting with the starting vortex that is formed when the aerofoil starts producing lift. Thus, the bound vortex (the circulation round the wing), the trailing vortices, and the starting vortex would form a rectangle enclosing the downward-moving air behind the wing. What actually happens, however, is that viscosity damps out the vortices some distance behind the wings. Spedding (1987a) made the trailing vortices behind a kestrel (*Falco tinnunculus*) visible by having it glide through a cloud of soap bubbles filled with a helium/air mixture (Section 5.4).

Aerodynamic theory predicts that the distance between the center lines of the left and right trailing vortices will be 0.79 of the wingspan. It also predicts that the diameters of the cores of the trailing vortices will be 0.17 of the distance between the vortices, or 0.13 of the span (Fig. 10.2E shows how this diameter is defined). These predictions depend on the plausible assumption that the induced drag factor, which will be explained in Section 10.3, is 1.00. Spedding (1987a) measured the vortex spacing and core diameter in the wake of the gliding kestrel and found good agreement with theory.

The trailing vortices are formed because the pressure under the aerofoil is greater than the pressure above. Consequently, air flows round the wing tips, from below to above. This in turn results in sideways airflow all along the wing; the air under the wing is given a component of velocity out toward the wing tip, and the air over the wing is given an inward component of velocity. The layers of air emerging behind the wing, flowing in slightly different directions, roll up to form the trailing vortices.

10.3. DRAG ON AEROFOILS

In Fig. 10.2C, lift is obtained by deflecting the air through an angle ψ, giving it downward velocity $v \tan \psi$. We have seen that the mass of air given this velocity in unit time is $\pi \rho v s^2 / 4$, so the rate at which kinetic energy is given to the air is $\frac{1}{2} (\pi \rho v s^2 / 4)(v \tan \psi)^2 = (\pi \rho v^3 s^2 / 8) \tan^2 \psi$. The lift is $(\pi\rho v^2 s^2/4)\tan\psi$ (Equation 10.5), so the rate at which kinetic energy is given to the air is $2F_{\text{lift}}^2/\pi\rho vs^2$. To get the lift, work must be done at this rate. This work is done against a component of drag, called induced drag, which acts only when lift is being generated. Work is force times distance, so rate of working is force times velocity, and the induced drag can be calculated by dividing the rate of working by the velocity:

$$\text{Induced drag} \approx \frac{2F_{\text{lift}}^2}{\pi\rho v^2 s^2}$$

That relationship is only approximate because we assumed unrealistically that all the air passing through the circle shown in Fig. 10.2B suffers the same change of velocity. The induced drag will never be lower than the equation shows, but is likely to be a little higher. To allow for this, we will insert an induced drag factor k_{induced}, which is expected to be a little greater than 1:

$$\text{Induced drag} = \frac{2k_{\text{induced}}F_{\text{lift}}^2}{\pi\rho v^2 s^2} \tag{10.9}$$

The aspect ratio $\mathcal{A}$ of a wing is the span divided by the mean chord (a high aspect ratio indicates long, narrow wings). The area of the wings is the span multiplied by the mean chord, so $s^2 = \mathcal{A}A_{\text{plan}}$. Hence,

$$\text{Induced drag} = \frac{2k_{\text{induced}}\,F_{\text{lift}}^2}{\pi\rho v^2\mathcal{A}A_{\text{plan}}} \tag{10.10}$$

Equation 10.2 told us that if flow is laminar, the friction drag coefficient based on total area is $1.33\mathcal{R}^{-0.5}$. The total area includes the under surface of the wing as well as the upper surface, so is twice the plan area. Thus, the friction drag coefficient based on plan area is about $2.7\mathcal{R}^{-0.5}$. The Reynolds number $\mathcal{R}$ should in this case be calculated using the mean wing chord as the length. We calculated in Section 10.1 that the Reynolds number for the body of an albatross 1 m long flying at 15 m/s is 10^6. The same albatross would have a wing chord of about 0.3 m, and the Reynolds number for the wing would be 3×10^5.

The total drag on a wing minus the induced drag is called the profile drag. It is a little more than the friction drag, because a little pressure drag acts on a wing, even when it is generating no lift. We will allow for this by writing the profile drag coefficient as $2.7k_{\text{profile}}\mathcal{R}^{-0.5}$, where k_{profile} is a profile drag factor. Thus, the total drag on a wing is

$$F_{\text{drag}} = \tfrac{1}{2}\rho A_{\text{plan}}v^2\,(2.7k_{\text{profile}}\mathcal{R}^{-0.5}) + \frac{2k_{\text{induced}}F_{\text{lift}}^2}{\pi\rho v^2\mathcal{A}A_{\text{plan}}} \tag{10.11}$$

We can substitute for F_{lift} using Equation 10.4:

$$F_{\text{drag}} = \tfrac{1}{2}\rho\, A_{\text{plan}} v^2 \left(2.7 k_{\text{profile}} \mathcal{R}^{-0.5} + \frac{k_{\text{induced}} C_{\text{lift}}^2}{\pi \mathscr{A}} \right) \qquad (10.12)$$

and the overall drag coefficient is

$$C_{\text{drag}} = 2.7 k_{\text{profile}} \mathcal{R}^{-0.5} + \frac{k_{\text{induced}} C_{\text{lift}}^2}{\pi \mathscr{A}} \qquad (10.13)$$

Figure 10.3A shows how lift and drag coefficients should be related, according to Equation 10.11, for the same aerofoil at three different Reynolds numbers. The induced drag factor and the profile drag factor have both been assumed equal to 1, so this graph represents the performance of an ideal aerofoil. The graph illustrates what is obvious from the equation, that the drag coefficient corresponding to any given lift coefficient is expected to decrease as Reynolds number increases. Figure 10.3B shows the results of similar calculations for aerofoils of different aspect ratios, all at the same Reynolds number. Induced drag is less for higher aspect ratio wings.

Figure 10.3C shows measurements of lift and drag for a dragonfly wing, mounted at various angles of attack in a wind tunnel. For angles of attack up to about 30°, an increase in the angle of attack results in increased lift, and drag increases with increasing lift more or less as Fig. 10.3B would lead us to expect. (The measurements were made on a single wing, so the length of the wing, rather than the wingspan of the intact insect, should be used to calculate aspect ratio. Calculated in this way, the aspect ratio was 2.3.) If the angle of attack of this particular wing is increased beyond 30°, the lift coefficient falls due to the phenomenon known as stalling. At low angles of attack the air flows parallel to the surfaces of the wing, but at high angles of attack the main flow separates from the upper surface of the wing, and large eddies form. The flow is deflected downward less effectively, so lift is reduced; and the kinetic energy given to the air in the eddies results in increased drag. All aerofoils stall at high angles of attack, commonly at angles of about 20°.

The wings of aircraft have streamlined sections, as shown in Fig. 10.4A. The inner parts of bird wings, which contain bones and muscles, have sections more or less like this (Fig. 10.4E), but the outer parts consist solely of feathers (Fig. 10.4F) and are in effect cambered plates (Fig. 10.4B). The wing membranes of bats also form cambered plates when stretched in flight. Insect wings are thin membranes stretched between thicker veins. Pleating enables these very light wings to be made stiff enough for flight (Fig. 10.4G). The wings of locusts and some other insects fold like fans when not in use. Wootton (1995) and Herbert et al. (2000) have explained how, when they are expanded, tension in the mem-

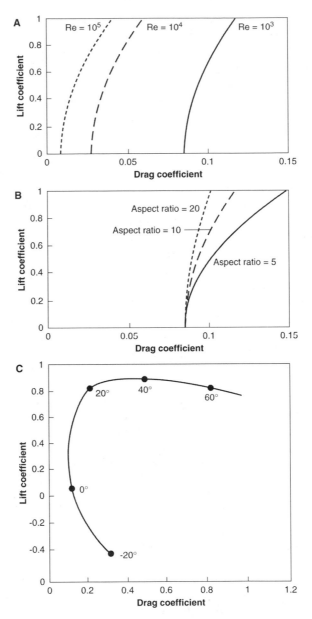

Fig. 10.3. Graphs of lift coefficient against drag coefficient (A) predicted by Equation 10.11 for an aspect ratio of 10 and Reynolds numbers of 10^3, 10^4, and 10^5; (B) predicted by the same equation for a Reynolds number of 10^3 and aspect ratios of 5, 10, and 20; and (C) measured for a fore wing of a dragonfly (*Calopteryx*) at a Reynolds number of 1480. The angle of attack is shown for some of the points on the graph. It is assumed in (A) and (B) that $k_{profile}$ and $k_{induced}$ are both equal to 1. The data for (C) are from Wakeling and Ellington (1997).

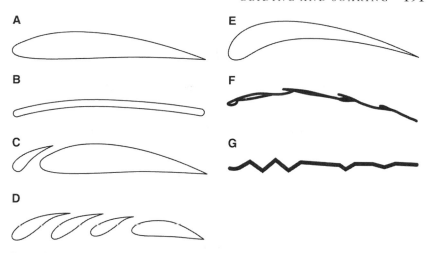

Fig. 10.4. Sections of wings: (A) a typical streamlined section, as used on aircraft; (B) a thin, cambered plate; (C) a slotted wing; (D) a multislotted wing; (E) and (F) inner and outer parts, respectively, of a wing of a pigeon (*Columba*, Nachtigall and Wieser [1966]); and (G) a dragonfly wing (Kesel 2000).

brane makes the veins bend, cambering the wing. Similarly, tension in the fabric of an umbrella bends the ribs when the umbrella is opened.

Good wings will give an aircraft or an animal the lift that is needed for flight with as little drag as possible. For some situations, for example for very slow flight, it is important to have the highest possible maximum lift coefficient. For others, for example, as we will see, for gliding at the shallowest possible angle, the priority is to have the highest possible ratio of lift to drag. Schmitz (1960) investigated the relative merits of streamlined wings and thin plates for model aircraft. He found that suitably cambered plates always perform better than similar, flat plates. At a Reynolds number of 168,000, a streamlined aerofoil performed a little better than a cambered plate; it gave both a higher maximum lift coefficient and a higher maximum ratio of lift to drag. At a Reynolds number of 42,000, however, the cambered plate performed much better in most respects than the streamlined aerofoil. This range of Reynolds numbers, at which the advantage shifts from cambered plates to streamlined aerofoils, is approximately the range in which the wings of medium-sized birds work. The wings of pigeons (*Columba livia*, mass about 0.4 kg) work at Reynolds numbers of around 40,000 when the bird is hovering, and 120,000 when it is gliding (Pennycuick 1967, 1968a). Insects and small birds work in the range in which cambered plates are better.

It might be thought that the pleating of insect wings might interfere with airflow over them. However, Fig. 10.3 shows that a dragonfly wing performs quite creditably, in comparison with an ideal aerofoil at the same Reynolds number. Kesel (2000) has shown how vortices that develop in the pleats give the wing an effectively streamlined profile, and excellent aerodynamic properties.

The maximum lift coefficient, obtained as the wing approaches the stalling angle, is generally no more than 1.5 at high Reynolds numbers and less at low ones. Pennycuick (1971c) calculated a maximum lift coefficient of 1.5 from his observations of a fruit bat (*Rousettus*) gliding in a wind tunnel at a Reynolds number of 30,000. Figure 10.3C shows a maximum lift coefficient of 0.9 for a dragonfly wing at a Reynolds number of 1480, and Vogel (1967) measured a maximum of 0.85 for a model *Drosophila* wing at a Reynolds number of 200. However, a small aerofoil set in front of the main one to form a leading-edge slot (Fig. 10.4C) makes lift coefficients up to 2 possible, and multislotted wings (Fig. 10.4D) can give even higher lift coefficients. The alula is a tuft of feathers on the second digit of bird hands (the first digit is missing). It normally lies flat on the wing surface, but can be raised to form a leading-edge slot. Nachtigall and Kempf (1971) experimented with bird wings in a wind tunnel, measuring lift with the alula raised or sewn down. They found that a raised alula increased the maximum obtainable lift, usually by about 10%. It is not surprising that the effect is small, as the alula is much shorter than the wing. Many photographs of birds landing show raised alulae, but they often also show small feathers on the upper surface of the wing fluttering, showing that the wing has stalled (see, for example, McGahan [1973]); the eddies that form when the wing stalls disturb the feathers).

Also in slow flight, the primary feathers may separate at the wing tips, as in Fig. 10.1B. This has the effect of making the outer part of the wing multislotted and might be expected to make high lift coefficients possible. Tucker and Heine (1990) calculated the lift coefficients of Harris' hawk (*Parabuteo unicinctus*) gliding at various speeds in a wind tunnel. At the lowest speed at which the bird would glide, the primaries were well separated and the lift coefficient was 1.6, which is only a little higher than would be expected of unslotted wings. Another property of wings with slots at the wing tip is that they suffer less induced drag than unslotted wings of the same aspect ratio. Tucker (1995) demonstrated this effect in experiments with Harris' hawk.

10.4. GLIDING PERFORMANCE

When an aircraft or animal is gliding at constant velocity, the forces on it must be in equilibrium. Figure 10.5 illustrates this. The forces in question are the lift F_{lift}, the drag F_{drag}, and the bird's weight mg. The bird is gliding

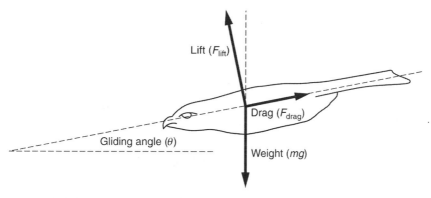

Fig. 10.5. A free-body diagram of a bird gliding.

at an angle θ to the horizontal. Forces along the direction of motion must balance, so

$$F_{drag} = mg \sin \theta \qquad (10.14)$$

Also, forces at right angles to the direction of motion must balance,

$$F_{lift} = mg \cos \theta \approx mg \qquad (10.15)$$

(If the gliding angle θ is small, as it will generally be, cos θ ≈ 1.) These equations imply

$$\tan \theta = \frac{F_{drag}}{F_{lift}} \qquad (10.16)$$

showing that, as already stated, a small gliding angle requires a large ratio of lift to drag. Finally, for equilibrium the resultant of lift and drag must be in line with the bird's weight, which implies that the center of pressure, where the lift and drag can be considered to act, must be vertically over the center of mass. The bird can adjust its gliding angle by moving its wings forward or back; the further back the wings are held, the steeper the glide.

The drag on an aerofoil, at the angle of attack at which it generates no lift is the profile drag, which is represented by the first term on the right-hand side of Equation 10.11. For a complete bird, the drag at zero lift includes also the drag on the body and can be expressed as $\frac{1}{2}\rho A_{plan} v^2 C_{zero\ lift}$, where $C_{zero\ lift}$ is the drag coefficient when the wing has the angle of attack at which it generates no lift. Using this and Equations 10.10 and 10.15, Equation 10.14 can be written

$$mg \sin \theta = \frac{1}{2}\rho A_{plan} v^2 C_{zero\ lift} + \frac{2 k_{induced} m^2 g^2}{\pi \rho v^2 \mathcal{A} \mathbf{A}_{plan}} \qquad (10.17)$$

$$\sin \theta = \frac{\rho \, v^2 C_{\text{zero lift}}}{2 \, N} + \frac{2 \, k_{\text{induced}} \, N}{\pi \rho \, v^2 \mathcal{A}} \qquad (10.18)$$

In this equation, N is the weight divided by the wing area, mg/A_{plan}. It is known as the wing loading.

The rate at which a gliding aircraft or animal loses height is $v \sin \theta$:

$$v \sin \theta = \frac{\rho \, v^3 C_{\text{zero lift}}}{2 \, N} + \frac{2 \, k_{\text{induced}} \, N}{\pi \rho \, v \mathcal{A}} \qquad (10.19)$$

As speed v increases, the first term on the right-hand side of this equation increases and the second term decreases. Height is lost rapidly at low speeds and at high ones, and there is an intermediate speed at which the rate of loss of height is least. This speed is known rather confusingly as the minimum sink speed; the minimum *sink* speed is the (forward) speed at which the (downward) *sinking* speed is least. It can be found by differentiating Equation 10.19 and finding the speed at which $d(v \sin \theta)/dv$ is zero:

$$\text{Minimum sink speed} = \left(\frac{4 \, k_{\text{induced}} \, N^2}{3\pi\rho^2 \mathcal{A} \mathbf{C}_{\text{zero lift}}} \right)^{1/4} \qquad (10.20)$$

The minimum sink speed is the speed at which an animal should glide if its aim is to remain airborne for as long as possible.

Alternatively, an animal might seek to glide as far as possible for given height loss, maximizing its range. To do that, it must minimize the gliding angle θ, which is, of course, the same thing as minimizing $\sin \theta$. By differentiating Equation 10.18 and finding the speed at which $d(\sin \theta)/dv$ is zero, we can find the speed at which the gliding angle is least.

$$\text{Maximum range speed} = \left(\frac{4 \, k_{\text{induced}} \, N^2}{\pi\rho^2 \mathcal{A} \mathbf{C}_{\text{zero lift}}} \right)^{1/4} \qquad (10.21)$$

Notice that this speed is $3^{1/4} = 1.3$ times the minimum sink speed. The minimum gliding angle can be obtained by inserting this speed in Equation 10.18.

$$\text{Minimum gliding angle} = \left(\frac{2 \, k_{\text{induced}} \, C_{\text{zero lift}}}{\pi \mathcal{A}} \right)^{1/2} \qquad (10.22)$$

The speed given by Equation 10.21 is the maximum range speed only in still air. The speed v that has appeared in our equations is the speed relative to the air. If the air is moving relative to the ground, the animal's speed relative to the ground is the vector sum of its speed relative to the air (the air speed) and the wind speed. In a wind, the animal will maximize the distance traveled for given loss of height, not by minimizing $\sin \theta$ (= Sink speed/Air speed), but by minimizing Sink speed /Ground speed. The air speed that gives maximum range is higher than the speed given by Equation 10.21 when the animal is gliding against the wind, and lower when there is a following wind (Pennycuick 1978). Liechti et al. (1994) have discussed the effect of sidewinds on optimal speed.

Equation 10.19 has been used to calculate the rates of loss of height shown in Fig. 10.6. When sinking speed ($v \sin \theta$) is plotted against air speed (v) in this way, the maximum range speed is the speed at which a tangent from the origin touches the curve. Figure 10.6B shows that increasing the wing loading increases both the minimum sink speed and the maximum range speed. However, the minimum gliding angle is unchanged. Increasing the aspect ratio decreases the minimum sink and maximum range speeds, and also reduces the minimum gliding angle.

Pennycuick (1968a) and later researchers have investigated the gliding performance of various animals by training them to glide in sloping wind tunnels. An animal gliding so as to remain stationary in a well-designed wind tunnel, in which air is blown at speed v at an upward angle θ to the horizontal, is acted on by the same forces as if it were gliding in still air at the same speed and at a downward angle θ. The width of the tunnel must be much greater than the animal's wing span (Section 5.2), so this experiment is practicable only for fairly small animals. The gliding performance of some larger birds has been investigated by direct observation from the ground, making allowance for the wind (Pennycuick 1982) or by gliding with them in their natural habitats, observing their movements relative to the observer's glider (Pennycuick 1971a). Wakeling and Ellington (1997) observed the gliding performance of dragonflies in a large greenhouse with the doors and vents closed to reduce drafts.

Figure 10.7 shows some results of observations of gliding performance. The sinking speeds shown are the minimum observed sinking speed for each airspeed. Animals can make themselves lose height more rapidly by changes of posture that increase drag. For example, birds increase drag by lowering their feet (Pennycuick 1971b; McGahan 1973). The graphs in Fig. 10.7 have the same general shape as the theoretical graphs in Fig. 10.6. However, we should not expect them to match the theoretical curves exactly, because we ignored some important points in deriving Equation 10.19.

First, we ignored the fact that an animal will stall, if it attempts to glide too slowly. For this reason, the experimental graphs do not extend to very low speeds. Secondly, we assumed that the gliding angle was small. When an animal is gliding at a shallow angle, as illustrated in Fig. 10.5A, its weight is balanced almost entirely by lift, with a very small contribution from drag. At steeper angles, drag is more important, and it dominates in very steep descents, which may be better described as parachuting than as gliding. Thirdly, we assumed that the zero-lift drag coefficient was con-

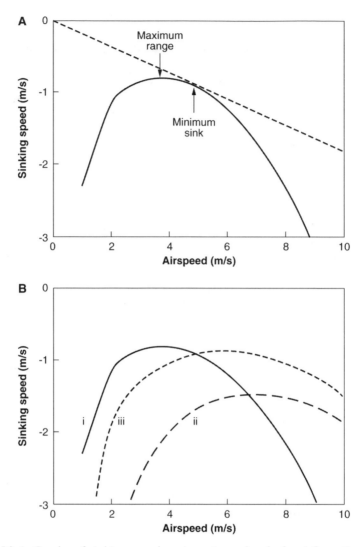

Fig. 10.6. Graphs of sinking speed against airspeed, calculated from Equation 10.19. (A) The minimum sink speed and the maximum range speed. (B) The effects of changing wing loading N and aspect ratio A, as follows:

	i	ii	iii
Wing loading, N/m²	30	100	100
Aspect ratio	7	7	14

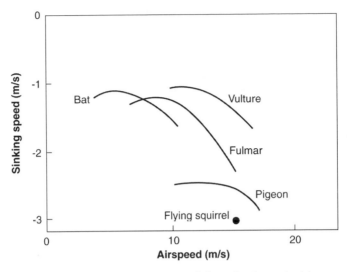

Fig. 10.7. Measured gliding performance of the animals marked by asterisks in Table 10.1. Sinking speed is plotted against airspeed, as in Fig. 10.6. Data are from Pennycuick (1960, 1968a, 1971c), Parrott (1970), and Scholey (1986).

stant, although we know from Equations 10.2 and 10.3 that the friction drag coefficient (a major component of the zero-lift drag coefficient) falls as Reynolds number increases. Finally, we assumed that an animal's wing area was the same at all speeds. However, birds reduce their wing areas and so increase wing loading when gliding fast. For example, the wing area of a pigeon was about 600 cm² when it was gliding at its minimum speed of 9 m/s, but when gliding at 20 m/s the bird partly folded its wings, reducing their area to 400 m/s (Pennycuick 1968a). At low speeds birds may spread their tails (Tucker 1992), and it has generally been assumed that the spread tail contributes to lift and so reduces the effective wing loading. However, Maybury et al. (2001) measured lift on a wingless starling (*Sturnus*) body in a wind tunnel, and found that very little additional lift could be obtained by spreading the tail. Bats alter their wing area less than birds, presumably because folding a bat's wing slackens the wing membrane and may allow it to bulge upward in an aerodynamically unsatisfactory way (Pennycuick 1971c).

There is an optimum wing loading $N_{\text{max range}}$ that minimizes the gliding angle θ, for any given air speed v. This wing loading can be found by differentiating Equation 10.18 with respect to N, and finding the value of N that makes $d \sin \theta / dN$ equal zero:

$$N_{\text{max range}} = \rho v^2 \sqrt{\frac{\pi C_{\text{zero drag}} \mathcal{A}}{4 k_{\text{induced}}}} \qquad (10.23)$$

This equation is less useful than one might hope, for predicting how wing area should be adjusted to suit changes of speed, because folding the wing alters the aspect ratio and probably also the drag coefficient. Rosén and Hedenström (2001) formulated a similar theory, but assumed that mean chord (rather than aspect ratio) was constant. Their theory predicted correctly that span and wing area should be reduced at high speeds, but failed to explain why a jackdaw (*Corvus monedula*) gliding in a wind tunnel partially folded its wings even at the lowest speeds at which it would glide. Both their theory and Equation 10.23 predict "optimum" wing areas that cannot be attained (because they exceed the area of the fully spread wings) at very low speeds. One would therefore expect that there would be a range of low speeds at which the wings would be kept fully extended.

Large birds generally glide fast because heavier birds do not have proportionately larger wing areas. Figure 4.2 shows that the wing areas of birds other than hummingbirds tend to be proportional to (body mass)$^{0.72}$, so bigger birds have larger wing loadings and (from Equations 10.20 and 10.21) higher optimum speeds. Because large gliding birds are both larger and faster than small ones, they glide at higher Reynolds numbers and so tend to have lower zero-lift drag coefficients. Equation 10.21 shows that a smaller zero-lift drag coefficient makes shallower gliding angles possible. Consequently, larger animals tend to have higher optimum gliding speeds than smaller ones, and shallower minimum gliding angles.

Table 10.1 shows wing loading and aspect ratio for the gliding animals included in Fig. 10.7, and a few others. The differences that it shows, between different species, help us to account for the observed differences in performance. The low wing loading of the dog-faced bat explains why its minimum sink speed is much lower than that of the black vulture. The "wings" of the flying squirrel are merely flaps of skin stretched between the fore and hind legs. This animal's low aspect ratio explains why its gliding performance is so poor compared to that of birds. The poor performance of the pigeon, however, may be an artifact of the experimental method. Pennycuick (1968a) induced it to glide in the wind tunnel by dispensing food from a device that the bird could reach only by flying in the required position. It is possible that the posture required for feeding was not the best possible for gliding.

So far, we have been considering animals gliding at constant speed and inevitably losing height. Now we will consider what may happen, if a gliding animal allows its speed to change. Let its speed at the present time be v and its acceleration dv/dt. Let it glide at an angle θ to the horizontal (a negative angle θ indicates an upward glide). Its mass is m, the gravitational acceleration is g, and the drag at airspeed v is $F_{drag}(v)$. The rate at which work is being done against drag is $vF_{drag}(v)$. The rate at which the animal is losing gravitational potential energy is $mgv \sin \theta$. The rate at which it is

Table 10.1.
Wing loading and aspect ratio of some gliding animals

	Mass, kg	Wing loading, N/m²	Aspect ratio
Wandering albatross, *Diomedea exulans*	8.7	140	15.0
White-chinned petrel, *Procellaria aequinoctealis*	1.4	80	11.6
*Fulmar, *Fulmarus glacialis*	0.82	65	10.3
Wilson's storm petrel, *Oceanites oceanicus*	0.038	19	8.0
Magnificent frigate bird, *Fregata magnificens*	1.5	37	12.8
Rüppell's griffon vulture, *Gyps rueppellii*	7.6	90	7.0
*Black vulture, *Coragyps atratus*	1.8	55	5.8
Kestrel, *Falco tinnunculus*	0.2	31	7.9
Marabou stork, *Leptoptilos crumeniferus*	7.1	74	7.4
White pelican, *Pelecanus onocrotalus*	8.5	84	8.5
*Feral pigeon, *Columba livia*	0.35	55	7.2
*Giant red flying squirrel, *Petaurista petaurista*	1.3	118	1.5
*Dog-faced bat, *Rousettus aegyptiacus*	0.12	21	5.4

Note. Asterisks mark the species included in Fig. 10.7. Data are from Pennycuick (1960, 1968a, 1971a, 1972, 1982, and 1983), Videler and Groenewold (1991), and Scholey (1986).

gaining kinetic energy is $d(\frac{1}{2}mv^2)/dt = mv\,dv/dt$. By the principle of conservation of energy

$$\Gamma_{\mathrm{drag}}(v) - mg \sin\theta + m\,\frac{dv}{dt} - 0 \qquad (10.24)$$

When $dv/dt = 0$, this equation reduces to Equation 10.14. However, if v is allowed to change, the equation shows that θ can be negative if dv/dt is negative (the gliding animal may rise while slowing down). Alternatively, dv/dt can be positive if θ is sufficiently positive (the animal can accelerate by gliding steeply downward). Both these strategies are used, for example, by the giant red flying squirrel, *Petaurista petaurista* (Scholey 1986). This rodent travels through forests in Borneo by climbing the trunk of a tree, gliding to a lower point on the trunk of a nearby tree, climbing up that trunk, and so on. At the start of a glide it dives steeply, gaining speed and kinetic energy. It continues less steeply, at a gliding angle of about 12° and at a speed of about 15 m/s. Then, as it approaches the tree on which it will land, it veers upward and loses speed, so that it is traveling quite slowly when it lands. If all the gravitational potential energy lost in the initial dive were converted to kinetic energy, a fall of 11 m would be needed to accelerate the squirrel to 15 m/s. In fact, some of the energy must be lost as work done against drag. Also, the height lost in the initial, steep part of the glide averaged only 7.5 m. To reach the speed recorded by Scholey (1986), the squirrels must have continued to accelerate in the shallower part of the glide.

Flying animals can dive at high speeds by folding their wings. Tucker et al. (1998) recorded speeds up to 58 m/s, in field observations of a trained gyrfalcon (*Falco rusticolus*). Tucker (2000) has pointed out that, due to the structure of their eyes, falcons and other raptors must turn their heads about 40° to one side to see prey directly ahead of them with maximum acuity. Wind tunnel measurements on models indicated that turning the head in this way would increase drag on the bird by about 50%, and reduce the speed of its dive. This may explain why dives generally follow a spiral course, enabling the falcon to see the prey with maximum acuity without turning its head.

10.5. STABILITY

Stability is an important design criterion for aircraft, but the stability of flying animals has been studied regrettably little. For that reason, this section must be brief. In it, I refer to some mechanisms that may be important for the stability of gliding animals, but do not explain them. Explanations will be found in textbooks on the aerodynamics of man-made aircraft.

Equation 10.14 shows that the sine of the gliding angle of an animal gliding at constant velocity is proportional to the drag. Consequently, the drag is least at the airspeed that gives the lowest gliding angle, the maximum range speed. If an animal is gliding more slowly than this, and some disturbance increases its speed slightly, the drag on the animal will fall and it will glide faster still. Similarly, if a disturbance reduces its speed, the drag will increase and it will glide yet more slowly. Thus, gliding is unstable at speeds below the maximum range speed.

The stability of a gliding animal is not merely a matter of maintaining constant speed. Disturbances may make the animal pitch (tilt nose up or nose down), yaw (turn to left or right), or roll about its long axis. Conventional aircraft are stabilized in pitch by their tail planes, and Thomas (1996) has argued that the tails of birds may serve the same function.

However, if the tail can generate little lift (Maybury et al, 2001) its stabilizing effect will be weak. Bats have no tail plane, and Pennycuick (1971c) has suggested that they may be stabilized in pitch by having their wings swept back and slightly twisted. Conventional aircraft are stabilized in yaw by a vertical tail fin, but no flying animal has any equivalent structure. Aircraft can be stabilized in roll by dihedral, that is by having wings that slope slightly upward from base to tip, so that the two wings form a shallow V. Birds, however, generally glide with little or no dihedral.

Polypedates is a tree frog that lives in South East Asian rain forests. It has large, webbed hands and feet that it uses as aerofoils, to glide down from trees. McCay (2001) observed it gliding in a tilted wind tunnel. He also investigated the stability of gliding by measuring the torque on casts of the frog held in various attitudes in a wind tunnel. He found that if the model was tilted nose up or nose down, an aerodynamic torque acted on it, tending to restore it to a normal gliding position. Thus, the model was stable in pitch. It was also stable in roll. However, if it was turned to left or right, the torque tended to deflect it more; it was unstable in yaw.

10.6. SOARING

An animal gliding in still air can keep itself airborne only for a limited time, but if the air is moving the animal may be able to keep itself airborne indefinitely, without beating its wings. The use of air movements to sustain gliding flight is called soaring. This section explains several soaring techniques and describes how animals use them.

Very few measurements have been made of the metabolic rates of gliding animals, but they are enough to show that soaring can be a very economical means of traveling. Baudinette and Schmidt-Nielsen (1974) measured the oxygen consumption of two herring gulls (*Larus argentatus*) gliding in a wind tunnel and found that it was only about twice the resting metabolic rate. In contrast, Tucker (1972) found that in flapping flight laughing gulls (*Larus atricilla*) metabolize at 7 times the resting rate. Bevan et al. (1995) used data loggers to record the heartbeat frequencies of free-ranging albatrosses (*Diomedea*) in the Antarctic (see Section 5.7). From their records they calculated metabolic rates (Section 5.3), and were able to show that the metabolic rate while soaring was little more than twice what it was when the bird was sitting on its nest. Adams et al. (1986) obtained a similar result by the doubly labeled water method (Section 5.3). By recording heartbeat frequencies, Weimerskirch et al. (2000) showed that albatrosses have higher metabolic rates when soaring into the wind than when the wind is behind them. One possible explanation is that they may have made more wing movements when flying into the wind. Weimerskirch et al. (2000) found that albatrosses flying long distances adapt their paths to the prevailing winds, so as to avoid flying against head winds.

Animals use at least four techniques of soaring: slope soaring, thermal soaring, wind-gradient soaring, and sea-anchor soaring. The first two of these are the most widely used and have been most studied, so the following account is principally about them.

Slope soaring is possible wherever the wind is deflected upward, for example, by a hillside, a building, or a wave. If the animal can glide in such a way that its rate of sink, relative to the air, is no greater than the rate at which the air is rising, relative to the ground, it can remain airborne indefinitely without having to flap its wings. For example, Videler and Groenewold (1991) observed kestrels (*Falco tinnunculus*) slope soaring over sea dikes in the Netherlands. Sea winds blowing at speeds around 9 m/s were deflected upward by the dikes, at angles of about 7° to the horizontal. Thus, the air was rising at about $9 \sin 7° = 1.1$ m/s. By gliding directly into the wind, matching their airspeed and angle of descent to the wind, the kestrels remained stationary over the dike. In experiments in a wind tunnel, Tucker and Parrott (1970) found that a different species of *Falco* had a minimum sinking speed when gliding of 0.9 m/s, attained at an airspeed of 9 m/s. The kestrels observed by Videler and Groenewold (1991) were hunting, scanning the ground below for voles or other prey. In other locations, where there is nothing to deflect the wind upward, hunting kestrels remain stationary by wind hovering, that is, by flapping flight into the wind, again matching their speed to the wind.

It is possible to travel by slope soaring, as well as to remain stationary. Figure 10.8 shows a seabird slope soaring along a wave. The plan view shows that to travel parallel to the crest of the wave, and so remain in the rising air, the bird must glide obliquely into the wind, at an airspeed greater than the wind speed. Its speed relative to the ground is the vector sum of its airspeed and the wind speed. To remain airborne, it must be capable of gliding at the required airspeed, with a sinking speed no greater than the upward component of the velocity of the wind. Albatrosses (*Diomedea*, etc.) commonly travel by slope soaring along waves (Pennycuick 1982). To travel in directions that are not parallel to the waves, they take a zigzag path, alternately soaring along a wave and gliding at an angle to the waves. If they can gain height or speed while soaring along a wave, they may be able to reach the next wave without flapping their wings. Pennycuick (1982) found that the lengths of the paths traveled by slope-soaring albatrosses, measured along the zigzags, averaged 1.5 times the straight-line distance between the end points.

Thermal soaring does not depend on wind, but on convection currents. Solar radiation may heat the ground to well above air temperature. Air

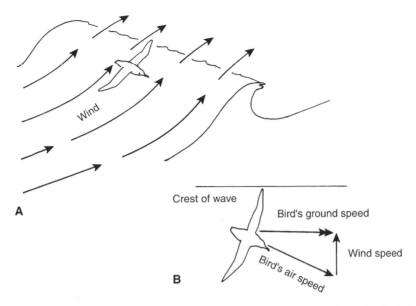

Fig. 10.8. Diagrams of a bird slope soaring along a wave. From Alexander (1989b).

close to the surface is heated by the ground, and so becomes less dense than the cooler air higher up. Consequently, the heated air rises in columns, like the convection currents in the liquid in a heated saucepan (Pennycuick 1972). These rising columns of air are called thermals. They are particularly likely to form over patches of ground that get hotter than the rest, for example, dark-colored rocks or slopes that face the sun. Where the ground is uniform, an array of more or less evenly spaced thermals will develop. The air cools as it rises, so water vapor may condense out, forming cumulus clouds. Glider pilots and presumably also birds find these clouds useful for locating thermals.

Birds can gain height by gliding in thermals, if the rate at which the air is rising is greater than the rate at which they are sinking relative to the air. In sunny regions, such as the East African plains, thermals commonly rise at speeds of 2–5 m/s (Pennycuick 1972). The minimum sinking speeds of many gliding birds are much less than this, around 1 m/s (Fig. 10.7). However, thermals are seldom if ever closely enough spaced for a bird to keep itself airborne by gliding straight through them, rising in each thermal and losing height between thermals. It is generally necessary to circle in the thermal to gain enough height to glide to the next thermal. Figure 10.9 represents a typical flight path. A bird gliding at speed v in circles of radius r has an acceleration v^2/r toward the center of the circles. If its mass is m, it needs a centripetal force mv^2/r to give it this accelera-

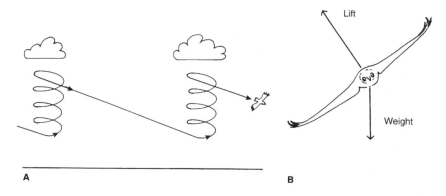

Fig. 10.9. Diagrams of a bird soaring in thermals. The downward slope of the glide is exaggerated. From Alexander (1989b).

tion. The bird develops this force by banking, as shown in Fig. 10.9B, so that the lift has a horizontal component mv^2/r as well as the vertical component mg (where g is the gravitational acceleration) needed to support the bird's weight. Thus, the lift required is $m\sqrt{g^2 + (v^2/r)^2}$. Additional lift implies additional induced drag, so the bird sinks faster relative to the air than it would if it were gliding along a straight path. To calculate the sinking speed ($v \sin \theta$), we must modify Equation 10.17, obtaining

$$mg \sin \theta = \frac{1}{2}\rho A_{\text{plan}} v^2 C_{\text{zero lift}} + \frac{2k_{\text{induced}} m^2 \left[g^2 + (v^2/r)^2 \right]}{\pi \rho v^2 \mathcal{A} A_{\text{plan}}} \quad (10.25)$$

$$v \sin \theta = \frac{\frac{1}{2}\rho v^3 C_{\text{zero lift}}}{N} + \frac{2k_{\text{induced}} N \left[1 + (v^2/rg)^2 \right]}{\pi \rho v \mathcal{A}}$$

This shows that the smaller the radius of circling, the greater the sinking speed must be for any given airspeed. Pennycuick (1971a) used this equation to calculate sinking speeds of a white-backed vulture (*Gyps africanus*) circling with different radii. His assumptions implied that, in a straight glide, the bird had a minimum sinking speed of 0.76 m/s. He calculated that when circling with radii of 30, 20, and 10 m, its minimum sinking speeds would be 0.8, 0.9, and 1.6 m/s, respectively.

Air rises faster in the core of a thermal than near the edge. A bird can benefit from faster rising air if it soars in smaller circles, but by reducing the radius of circling it increases the rate at which it sinks relative to the air. Its rate of gain of height is the rate at which the air is rising, minus the rate at which the bird is sinking relative to the air, and has a maximum value at an intermediate radius. Pennycuick (1971a) observed that *Gyps africanus* usually circles with radii between 15 and 25 m.

There is a minimum possible circling radius, which depends on the bird's lift coefficient. We have already seen that the horizontal component of lift, needed for circling, is mv^2/r. The resultant lift, taking account also of the vertical component, must be greater than that. Hence, using Equation 10.4,

$$\frac{mv^2}{r} < \tfrac{1}{2}\rho A_{plan} v^2 C_{lift} \qquad (10.26)$$

$$r > \frac{2m}{\rho A_{plan} C_{lift}} = \frac{2N}{\rho g C_{lift}}$$

The vulture in Pennycuick's (1971a) calculations had a wing loading N of 77 N/m^2. The density ρ of air is about 1.2 kg/m^3 and the gravitational acceleration g is about 10 m/s^2. The maximum lift coefficient of which the bird was capable was probably about 1.6. By putting these data into Equation 10.26 we find that the bird could not have circled, at any speed, with a radius less than 8 m. If its wing loading were lower or if it could develop higher lift coefficients, it could glide in smaller circles and use smaller thermals.

Conditions for thermal soaring are particularly good over the Serengeti plain in East Africa. The thermals there are not strong enough for soaring until about two hours after sunrise, but vultures spend much of the rest of the day soaring, looking out for dead mammals. They also travel long distances by soaring. Rüppell's griffon vulture (*Gyps rüppellii*) commutes daily between the cliffs where it nests and the herds of mammals, up to 140 km away, where it finds its food. Pennycuick (1972) used a motor glider to fly with them. On one occasion, he stayed with a *Gyps rüppellii* for 96 min, while it traveled 75 km on its way back to its nest. It traveled this distance entirely by soaring, circling in 5 thermals and passing through others without circling. Pennycuick (1972) also described the soaring behavior of storks, pelicans, and other birds. Some large fruit bats, including *Pteropus samoensis* (about 0.4 kg), soar in thermals (Norberg et al., 2000).

As a general rule, thermal soaring is confined to fairly large birds and bats, but Gibo and Pallett (1979) observed monarch butterflies soaring in thermals on their migrations between Canada and Mexico. Gliding tests with dead specimens showed that their minimum sinking speed was only 0.6 m/s, so they could gain height in weak thermals. However, their minimum gliding angle was 16°, so they could not glide far for a given loss of height.

Thermals generally occur only over land, but in the trade wind zone they occur also over the sea. The trade winds carry cool air from higher latitudes toward the equator. Cool air that is blown close over the surface of the warm tropical water is warmed and tends to rise. Pennycuick (1983) showed how frigate birds (*Fregata magnificens*) soar in the resulting thermals.

Now that we have discussed slope soaring and thermal soaring, we will consider briefly two other soaring techniques. Wind-gradient soaring (also called dynamic soaring) is a technique sometimes used by albatrosses (Wilson 1975). It depends on two principles. First, wind speed is generally lower close to the ground or the sea surface than it is higher up. Secondly, it is possible to gain height while gliding, at the expense of speed. Equation 10.24 shows that if drag is small compared to the other terms, then

$$\frac{dv}{dt} \approx -g \sin \theta$$

The rate of gain of height, dh/dt is $-v \sin \theta$. Thus,

$$\frac{dv}{dt} \approx - \frac{(g/v)dh}{dt} \qquad (10.27)$$

$$\frac{dv}{dh} \approx - \frac{g}{v}$$

This equation gives the rate of change of airspeed with height, if wind speed is independent of height. However, if the bird is gliding into a wind that is faster at higher levels, the increase in wind speed as the bird rises may be sufficient to keep the bird's airspeed v constant. The bird may be blown backward, but its airspeed will remain high enough to keep it airborne as it continues to rise. In wind-gradient soaring, birds glide downwind, losing height and gaining speed, then turn into the wind and rise, losing speed relative to the ground but maintaining their airspeed. Equation 10.27 indicates that for this soaring technique to be possible, the vertical gradient of wind speed must be at least g/v; for an albatross flying at 20 m/s, that would be 0.5 /s. The gradient of wind speed diminishes with height, so wind-gradient soaring is possible only over a limited range of heights. Pennycuick (1982) argued that it would seldom be as strong as 0.5 /s at heights exceeding 3 m. Albatrosses often rise considerably higher than this, to 15 or occasionally 20 m, before turning and gliding downwind (Pennycuick 1982). It seems likely that most of the energy for the climb comes from lost kinetic energy, rather than being extracted from the wind by the technique of wind-gradient soaring.

Sea anchor soaring is a technique used by storm petrels (*Oceanites*) to keep themselves airborne just above the surface of the sea, when they are searching for the small fish and squids on which they feed (Withers 1979; Sugimoto 1998). The birds look as if they are walking on the water with their wings spread. What they are actually doing is using their feet as paddles to prevent themselves from being blown backward at the speed of the wind. The bird's weight is balanced by the lift due to the wind blowing over the wings. The drag that the wind exerts on it is balanced by the drag (in the opposite direction) that the water exerts on its feet. The wings are close enough above the water for aerodynamic drag to be reduced by ground effect, that is, by interaction of the airflow with the surface of the water. They make small movements, quite unlike the movements of flapping flight, which may perhaps enhance lift (Withers 1979).

The optimum dimensions for wings depend not only on the mass of the animal, but also on its flying habits. Slope soarers that travel as albatrosses do, at high angles to the wind (Fig. 10.8), need to be able to glide considerably faster than the wind. To glide well at high speeds, they need high wing loading; Equations 10.20 and 10.21 show that the minimum sink speed and the maximum range speed are both proportional to the square root of wing loading. High wing loading is also an advantage for wind gradient soaring because higher speeds enable birds to soar in weaker vertical wind gradients, as the discussion after Equation 10.27 shows. In contrast, low wing loading is generally advantageous for thermal soarers because it enables them to soar in smaller thermals, as the discussion following Equation 10.26 shows. Some birds such as the Andean condor (*Vultur gryphus*) practice both slope soaring and thermal soaring (Pennycuick and Scholey 1984), but most soaring birds use one of these techniques much more than the other. Table 10.1 confirms that habitual slope soarers tend to have much larger wing loadings than thermal soarers of similar mass. An 8.7-kg albatross had a wing loading of 140 N/m², whereas three thermal soarers of only slightly lower mass (a Rüppell's griffon vulture, a stork, and a pelican) had wing loadings between 74 and 90 N/m². A 1.4-kg white-chinned petrel (a slope soarer) had a wing loading of 80 N/m², whereas a 1.8-kg black vulture (a thermal soarer over land) had a wing loading of only 55 N/m² and a 1.5-kg frigate bird (which soars in thermals over the sea) had a wing loading of 37 N/m². The table also illustrates the tendency for large birds to have higher wing loadings than smaller birds of similar habits. Figure 4.1B showed that the wing areas of birds tend to be proportional to (body mass)$^{0.72}$, which implies that wing loading (Body mass/Wing area) tends to be proportional to (body mass)$^{0.28}$.

A high aspect ratio is aerodynamically advantageous. Equation 10.19 shows that at any airspeed, other things being equal, an increase in aspect ratio reduces sinking speed. However, in Table 10.1, only the albatross, the petrel, and the fulmar (marine slope soarers) and the frigate bird (a marine thermal soarer) have aspect ratios greater than 10. Terrestrial thermal soarers such as vultures might soar better if they had higher aspect ratios, but if they combined low wing loading with high aspect ratio (as frigate birds do) their wings would be exceedingly long and possibly difficult to manage when taking off from the ground.

This chapter has introduced some of the basic principles of aerodynamics and applied them to the gliding performance of birds, bats, and flying squirrels. It has explained how albatrosses, vultures, and other birds can keep themselves airborne at low energy cost by slope soaring, thermal soaring, and other techniques, and it has shown that different wing designs are suited to different soaring techniques. It seems to me that the most promising topics for future research on gliding are stability and maneuvering, which have not so far been given the attention they deserve.

· ·

Chapter Eleven

Hovering

*H*OVERING MEANS flying so as to stay more or less stationary relative to the surrounding air. Kestrels (*Falco tinnunculus*, see Section 10.6) often keep themselves stationary *relative to the ground* by flying into the wind, matching their speed to the wind, but this is not hovering in the sense used in this chapter, because the bird is moving rapidly relative to the air.

Many insects hover: for example, Ellington (1984) studied hovering by flies, bees, moths, a beetle, and a lacewing. Hummingbirds (masses 2–20 g) hover in front of flowers, feeding on nectar, and can sustain hovering for many minutes (for example, up to 50 min in Lasiewski's [1963] experiments). Many other small birds and bats can hover, but only briefly; *Glossophaga soricina*, a 12-g nectar-feeding bat, hovered for no more than 4.5 s in the experiments of Winter et al. (1997). Birds up to about the size of pigeons (*Columba livia*, about 0.4 kg) can hover for even shorter times (Pennycuick 1968b) and larger birds cannot hover.

11.1. AIRFLOW AROUND HOVERING ANIMALS

Animals hover by beating their wings to drive air downward. The principle is the same as for hovering by helicopters, whose rotors similarly drive air downward. To keep an animal airborne, downward momentum must be given to the air at a rate equal to the animal's weight, as explained for gliding in Section 10.2.

Animals generally hover by beating their wings in a (more or less) horizontal plane. Each wing has length r and beats through an angle ϕ, so the air that the wings accelerate passes through the two sectors of circles shown in Fig. 11.1B. We will ignore the movement of the wings and think of the sectors as actuators that apply a sudden pressure change to the air that passes through them.

Initially, this air is stationary. Its density is ρ, and the total area of the two sectors is $r^2\phi$. The air passes through the sectors at velocity v_{ind}, and eventually reaches velocity v_{wake} in the wake far below the animal. Thus, the mass of air that is accelerated in unit time is $\rho r^2\phi v_{ind}$, and the rate at

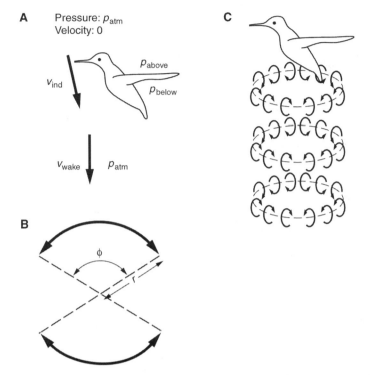

Fig. 11.1. Diagrams of the air movements produced by a hovering animal: (A) the air velocities and pressures referred to in the discussion of induced power; (B) a plan of the plane in which the wings beat; and (C) the vortex rings in the wake. Arrows represent air movement.

which momentum is given to this air is $\rho r^2 \phi v_{\text{ind}} v_{\text{wake}}$. This must equal the weight mg of the animal:

$$mg = \rho r^2 \phi v_{\text{ind}} v_{\text{wake}} \qquad (11.1)$$

A large part of the power needed for hovering is induced power, that is, power required to give kinetic energy to the accelerated air. The induced power P_{ind} equals the rate at which kinetic energy is given to the air.

$$P_{\text{ind}} = \tfrac{1}{2}\rho r^2 \phi v_{\text{ind}} v_{\text{wake}}^2 \qquad (11.2)$$

To calculate this power, we need values for the two velocities, v_{ind} and v_{wake}. We will obtain these by applying Bernouilli's principle (Equation 10.6). We can apply the principle to airflow above the wings and to flow in the wake below, but we cannot apply it to flow past the wings because the wings do work on the air; the principle assumes that the total energy of the air remains constant. Far above the wings, the air is stationary and has pressure p_{atm}, the ambient atmospheric pressure. Just above the plane of the wings it has velocity v_{ind} and pressure p_{above}. The change of height is too small for its gravitational potential energy to have changed appreciably, so Equation 10.6 tells us that

$$p_{atm} - p_{above} = \frac{1}{2}\rho\, v_{ind}^2 \tag{11.3}$$

The air immediately below the plane of the wings still has velocity v_{ind}, but its pressure is now p_{below}. Far below the wings, in the wake, the air has velocity v_{wake} and has returned to pressure p_{atm}. By applying Bernouilli's principle to this part of its path,

$$p_{below} - p_{atm} = \frac{1}{2}\rho\, (v_{wake}^2 - v_{ind}^2) \tag{11.4}$$

By adding these two equations together we get

$$p_{below} - p_{above} = \frac{1}{2}\rho\, v_{wake}^2 \tag{11.5}$$

The animal's weight is supported by this pressure difference acting across the sectors, whose total area (as we have seen) is $r^2\phi$:

$$mg = r^2\phi\, (p_{below} - p_{above}) = \frac{1}{2}\rho r^2\phi v_{wake}^2 \tag{11.6}$$

Comparison of Equation 11.6 with Equation 11.1 shows that $v_{wake} = 2v_{ind}$. By substituting this in Equations 11.1 and 11.2 we get

$$v_{ind} \sqrt{\frac{mg}{2\rho r^2\phi}} \tag{11.7}$$

and

$$P_{ind} - 2\rho r^2\phi v_{ind}^3 = \sqrt{\frac{m^3 g^3}{2\rho r^2\phi}} \tag{11.8}$$

This equation will help us to understand why large birds cannot hover. Geometrically similar birds of different sizes would have wing length r proportional to the cube root of body mass m. The equation shows that this would make the induced power proportional to $\sqrt{m^3/m^{2/3}} = m^{1.17}$; the power requirement per unit body mass would be greater for large birds than for small ones. However, large animals cannot produce as much power, per unit body mass, as smaller ones. Animals of different sizes tend to have similar proportions of muscle in their bodies (Section 4.3), and to be capable of doing equal amounts of work, per unit muscle mass, in a contraction (Section 2.1). However, larger animals move their limbs at lower frequencies than small ones, typically at frequencies proportional to (body mass)$^{0.25}$ (Section 4.1). Therefore, power output (work per cycle multiplied by frequency) is expected to be about proportional to (body mass)$^{0.75}$. Notice that the exponent (0.75) is much lower than the exponent (1.17) for the power required.

That argument tells us that if birds were geometrically similar to each other, there would be a critical size above which they could not produce enough power for hovering. Birds of different sizes are not, in fact, geometrically similar to each other. Hummingbirds have wingspans about proportional to $(\text{mass})^{0.53}$, which makes induced power proportional to $(\text{mass})^{0.97}$, and other birds have wingspans about proportional to $(\text{mass})^{0.39}$, which makes induced power proportional to $(\text{mass})^{1.11}$ (Rayner 1987). However, these exponents for power requirement (0.97 and 1.11) are still greater than the likely exponent for the available power (0.75). Thus, the conclusion stands, that large birds cannot generate enough power for hovering. This argument implies the assumption that induced power is a major part of the power required for hovering. Section 11.3 will confirm that this is the case.

The theory presented so far treats the plane of the wings simply as an actuator that increases the pressure of the air that passes through it. This implies that the wake will be a continuous column of downward-moving air. In fact, the air is driven by beating wings, each beat generating a puff of downward-moving air. This makes very little difference to the induced power (Rayner 1979). Each wing beat produces a starting vortex, a trailing vortex behind the wing tip, and a bound vortex round the wing, which separates from the wing at the end of the stroke (Section 10.3). In the wake below the animal, the vortices from the left and right wings combine to form vortex rings (Fig. 11.1C). This diagram is idealized. The closely packed vortex rings do not form a regular stack, but interfere with each other and coalesce into larger rings (Rayner 1979). I do not know any good illustrations of the wake of a hovering animal. However, Willmott et al. (1997) reproduce photographs of a tethered hawkmoth, *Manduca*, beating its wings in a wind tunnel in air that is flowing very slowly, at 0.4 m/s. Air movement in the wake is made visible by fine filaments of "smoke" (actually a suspension of oil droplets) introduced into the flow above the wings. These photographs show a vortex ring being formed by each stroke, but each ring becomes indistinct before the next stroke is complete.

The form of the wake, as a stack of vortex rings, suggests a hypothesis about the design of hovering animals. If evolution has optimized the designs of animals for hovering, we may expect similar hoverers of different sizes to have dynamically similar wakes. This implies, for example, that the distances between successive vortex rings will be proportional to wing length. Hovering is a cyclic motion, so one of the conditions for dynamic similarity is that hovering animals of different sizes will have equal Strouhal numbers (Section 4.2). Consider an animal of mass m and wing length r that beats its wings with frequency f and generates an induced velocity v_{ind}. The obvious relevant Strouhal number is rf/v_{ind}, so our hypothesis will be that this Strouhal number is the same for hoverers of different sizes. The induced velocity is proportional to $m^{0.5}/r$, so we expect $r^2fm^{-0.5}$ to be constant. For hummingbirds, r is proportional to $m^{0.53}$ and f to $m^{-0.60}$ (Rayner 1987), making $r^2fm^{-0.5}$ proportional to $m^{-0.04}$. For euglossine bees, r is proportional to $m^{0.42}$ and f to $m^{-0.35}$ (Casey et al., 1985), making $r^2fm^{-0.5}$ proportional to $m^{-0.01}$. In each case, $r^2fm^{-0.5}$ is more or less independent of body mass, as predicted.

11.2. LIFT GENERATION

Our discussion so far has shown that hovering animals must drive air downward, to counteract their weight and keep them airborne. Now we ask what are the aerodynamic effects that provide the required forces.

Our understanding of insect hovering depends largely on research by Charles Ellington and his colleagues at Cambridge University, on the hawkmoth *Manduca sexta*. This insect is convenient to work with because it is large (mass 1–2 g, wing length about 50 mm), and beats its wings at fairly low frequency (26 Hz). Also, it is not too difficult to keep and breed in the laboratory. Willmott and Ellington (1997b) took high speed video of *Manduca* hovering while taking sugar solution from an artificial flower. Figure 11.2A shows that it hovered with its body sloping at about 40° to the horizontal, with its wings beating in a plane that made an angle of 20° to the horizontal. Like most other insects, *Manduca* has fore and hind wings that beat together as a unit. The two wings of one side of the body are described as a wing couple.

The dorsal surface of the wing couple is uppermost in the forward stroke, but for the backward stroke the wings are turned upside down so that their ventral surfaces are uppermost. Willmott and Ellington (1997a) devised a method that enabled them to calculate from their video images the angles of attack of the wings. Figure 11.3 shows information obtained in this way both for hovering (0 m/s) and for forward flight. It shows the paths of a wing couple relative to the air, which is moving downward at the induced velocity. Notice that because the wings turn upside down for their forward stroke in hovering, their stiff anterior edges always lead while their flexible posterior edges trail behind (Fig. 11.3B). Also, the angles of attack ensure that both the forward and the backward stroke produce upward forces to support the body's weight. Obviously, the mean vertical force equals the weight of the body; otherwise, the insect could not hover. How large must the lift coefficients be?

This question can be answered by blade-element theory, as applied to the aerodynamics of helicopter rotors. The frequency of the wing beat cycle is f, so each forward or backward stroke of the wings is made in

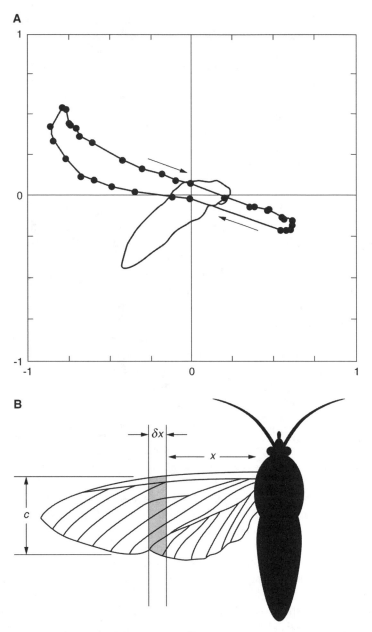

Fig. 11.2. (A) The path of the wing tip of a hovering hawkmoth, *Manduca sexta*, from Willmott and Ellington (1997b). (B) A wing couple of *Manduca* (i.e., the fore and hind wings of one side) showing a wing strip as used in blade-element theory.

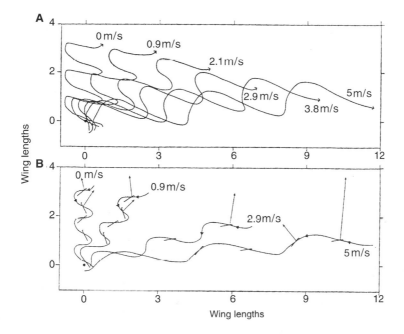

Fig. 11.3. Paths relative to the surrounding air of the wings of *Manduca* (seen in side view) flying at various speeds. (A) The path of the wing tip. (B) The path of a point halfway along the wing. Lines at the mid point of most strokes represent the wing chord, drawn to scale, with a ventral projection added at the leading edge to show which way up the wing is. Thin arrows show the estimated direction and relative magnitude of the aerodynamic force (the resultant of lift and drag). Arrowheads on the wing paths in (B) show the direction of movement of air relative to the wing. From Willmott and Ellington (1997c).

time $1/2f$. In this time the wing moves through an angle ϕ, with mean angular velocity $2f\phi$. The narrow strip of wing shown in Fig. 11.2B is at a distance x from the wing base, so its velocity relative to the insect's thorax is $2fx\phi$. However, its velocity relative to the air is the vector sum of this velocity and the induced velocity v_{ind}. The two components of velocity are approximately perpendicular to each other, so their vector sum is $\sqrt{(2fx\phi)^2 + v_{ind}^2}$. The area of the strip is $c\delta x$ and the density of air is ρ, so if the lift coefficient is C_{lift}, the lift on the strip of wing is $\frac{1}{2}\rho c\delta x C_{lift}[(2fx\phi)^2 + v_{ind}^2]$. Now imagine the entire wing couple divided into narrow strips. The total lift on all the strips of the wings of one side of the body must equal half body weight. Hence,

$$mg = \Sigma\{\rho c\,\delta x\,C_{lift}[(2fx\phi)^2 + v_{ind}^2]\} \tag{11.9}$$

where Σ indicates the sum of the values for all the strips from the wing base to the wing tip. If we use Equation 11.7 to get the induced velocity, and assume (because we know no better) that the lift coefficient is constant along the whole length of the wings, we can calculate the coefficient. Willmott and Ellington (1997c) performed a more sophisticated version of this calculation and obtained lift coefficients (for three individual moths) ranging from 1.3 to 1.8. They also measured lift coefficients for isolated *Manduca* wing couples held at various angles of attack in a wind tunnel. The highest lift coefficient that they could measure in the wind tunnel, in the appropriate range of Reynolds numbers, was only 0.7. Thus, the wings of hovering *Manduca* provide far more lift than conventional aerodynamic theory allows. Lift coefficients that are improbably high, according to conventional aerodynamics, have also been calculated for other insects, including bees (Dudley and Ellington 1990b) and flies (Ennos 1989), and for birds, including hummingbirds (Weis-Fogh 1973) and a flycatcher (*Ficedula* [Norberg 1975]).

Conventional aerodynamics fails to explain how animals hover. This seems to be because conventional aerodynamics deals with aerofoils such as wings and propeller blades, which move through the air at more or less steady speeds. In contrast, the wing movements of hovering animals are markedly unsteady. The wings beat backward and forward, repeatedly reversing their direction of movement. As a rough general rule, calculations using steady-state lift and drag coefficients work reasonably well only if the distance traveled by the wing relative to the air in each cycle of movement is at least 12 chord lengths. Figure 11.3B shows that *Manduca* wings move much less far than this in a hovering wing beat cycle.

The phenomenon called delayed stall plays a large part in making high lift coefficients possible for *Manduca* and other hovering animals. Figure 11.4A shows air flow over a streamlined aerofoil such as an aeroplane wing at a moderate angle of attack. The flow clings closely to the surfaces of the aerofoil. However, if the aerofoil is a thin plate with a sharp leading edge, or if the angle of attack is high, or both, the flow pattern shown in Fig. 11.4B develops. The flow over the upper surface separates from the leading edge and reattaches further back, trapping a leading-edge vortex in which air circulates as shown. Initially, after the aerofoil starts moving from rest, the vortex is small and there is little lift. As movement continues, the leading-edge vortex grows, the bound circulation strengthens, and lift increases. Eventually, if the angle of attack is high, the leading-edge vortex detaches and is lost in the wake, the wing stalls, and lift is lost. In the short time before this happens, however, the lift may be much higher than can be maintained in a steady state.

Ellington et al. (1996) used an enlarged model to simulate airflow over the wings of a hovering *Manduca*. Each wing couple was made of cloth stretched over an appropriately shaped framework of brass tubes. A motor with a complex gearbox made them beat in a cycle imitating the wing

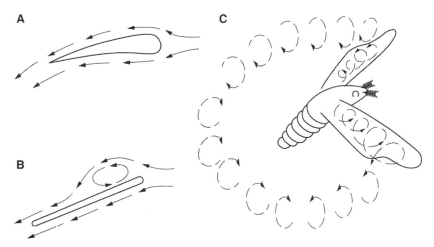

Fig. 11.4. (A) Flow around a streamlined aerofoil at a moderate angle of attack. (B) Flow around an aerofoil with a sharp leading edge, at a high angle of attack. (C) Flow around the wings of a hovering moth.

movements of the moth. Smoke released from holes in the tube that forms the leading edge made the air movements around the wings visible. The wings were ten times as long as in the moth but beat at one hundredth of the moth's frequency, so the speed of any point on a wing was one-tenth of the speed of the corresponding point on the wing of the real moth. With ten times the length of the real moth wing and one-tenth the speed, the Reynolds number is the same as for the moth (see Section 4.2). Also, with these ratios the Strouhal numbers are equal. Thus, flow around the model's wings was expected to be dynamically similar to flow around the wings of the real moth.

The experiments with the model showed a leading-edge vortex that developed during the forward stroke. At any stage of the stroke, the vortex was larger over the faster-moving distal parts of the wings than over the slower-moving proximal parts. Also, the air in the vortex had a component of velocity out toward the wing tips, due to centrifugal effects and to the lower pressure over the faster-moving wing tips. Thus, the air in the vortex flowed in conical spirals (Fig. 11.4C). At the wing tips the leading-edge vortex merged into the developing vortex ring.

Ellington et al. (1996) also used smoke to show airflow around the wings of real, tethered *Manduca*, beating their wings as if flying. The tether was attached to a balance that showed that the moth's flying movements were supporting at least 70% of its weight. For some of their observations, the airflow was so slow (0.4 m/s) that the moth was effectively hovering. Less detail could be seen than in the experiments with the model, but the pattern of airflow seemed (as expected) to be the same. Liu et al. (1998) obtained further confirmation of the pattern by computational fluid dynamics. Their computations showed that a leading-edge vortex is formed in the backward as well as the forward stroke, and predicted sufficient lift to support the insect's weight.

A leading-edge vortex on a wing moving steadily at a high angle of attack grows and eventually detaches, resulting in a stall. The flow toward the wing tip of the moth removes air from the vortex, restricting its rate of growth and so stabilizing it. This is probably more important in forward flight (in which the wings travel further in each wing beat cycle) than in hovering.

Hummingbirds hover like *Manduca* and some other insects, turning the wings upside down for their backward stroke and getting upward lift in both strokes. Other birds and bats generate upward lift only in the forward stroke, partially folding the wings in the backward stroke and holding them at angles that keep the aerodynamic forces small. Photographs of birds other than hummingbirds hovering show the primary feathers bent upward by lift during the forward stroke, but not in the return stroke (Norberg 1975). Dickinson and Götz (1996) found evidence that the fruit fly *Drosophila* also obtains lift from the downstroke only. They tethered *Drosophila* to a delicate force transducer with their feet unable to touch the ground; this stimulated them to beat their wings as if hovering. They used a smokelike suspension of *Lycopodium* (club moss) spores to make air movements visible, and found that the downstroke, but not the upstroke, produced a vortex ring. The output of the force transducer confirmed that only the downstroke produced lift.

Dickinson et al. (1999) used a model to learn more about the aerodynamics of *Drosophila* flight. They made a pair of wings, each 25 cm long, cut to the shape of *Drosophila* wings. Each wing was connected to three motors, which could be used to make it rotate about any axis through its base. The wings were made to beat in a bath of mineral oil, at speeds that made the Reynolds number the same as for the wings of a real fly flying in air. Force transducers at the bases of the wings measured the forces on the wings. A pattern of beating designed to imitate the wing beat cycle of *Drosophila* generated upward lift both in the downstroke and in the upstroke, in contrast to the observations described in the paragraph above.

The importance of these model experiments was that they demonstrated two other aerodynamic effects, as well as delayed stall, that may be important in insect flight. Figure 11.5A is a schematic graph of lift on the model against time. Notice that there is a peak of lift just after the beginning of each stroke and just after the end.

Figure 11.5B shows what seems to have happened at each stage of the cycle. It will be convenient to start at stage 2, in mid downstroke. There

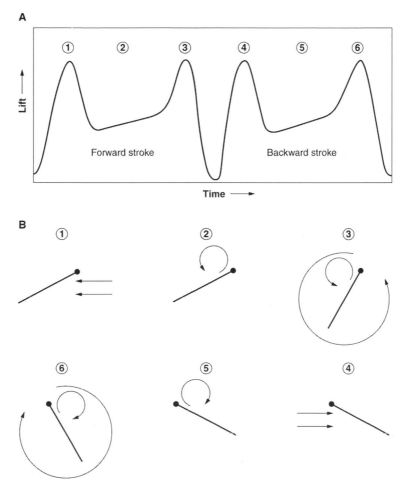

Fig. 11.5. (A) A schematic graph of lift against time, based on records of forces on a model simulating *Drosophila* flight (Dickinson et al., 1999). The numerals indicate the stages of the wing beat cycle illustrated below. (B) Diagrams showing a wing (seen in section) and the airflow near it, at successive stages of the cycle.

is a leading-edge vortex like the ones observed over *Manduca* wings, and the lift is attributable to delayed stall. At stage 3, near the end of the downstroke, the wing is turning over in preparation for the upstroke and has a very high angle of attack. Rotation of the wing enhances the circulation, increasing lift. At the same time, because the angle of attack is high, the drag becomes very large, decelerating the wing. The airflow immediately behind the wing has a large component toward the right, corresponding to the drag, as well as the downward component corresponding

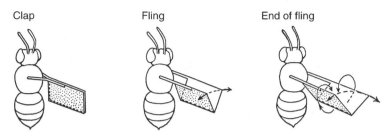

Fig. 11.6. Diagrams showing successive stages in the clap and fling of a hovering *Encarsia*. The wasp has a wing span of 1.5 mm and beats its wings at a frequency of 400 Hz. From Weis-Fogh (1973).

to the lift. (Remember that if the air is exerting a backward force on the wing, the wing must be exerting a forward force on the air.) The enhancement of circulation by rotation of the wing explains the peak in the lift. By stage 4 the wing has turned over completely and is moving toward the left of the diagram, at the beginning of the upstroke. It moves back into its own wake and meets the air that it set moving to the right at the end of its previous stroke. This generates another peak of lift. At stage 5, the situation is the same as at stage 2, and the cycle continues. Delayed stall is important throughout both strokes, but the lift it generates is supplemented by rotational circulation at stages 3 and 6, and by wake capture at stages 1 and 4. In some insects, such as hoverflies (Syrphidae), which beat their wings through smaller angles than most other insects, rotational circulation and wake capture may be more important for flight than delayed stall.

Birch and Dickinson (2001) used the *Drosophila* model for further experiments, focusing on the leading-edge vortex. Ellington et al. (1996) had found strong flow outward toward the wing tip in the leading-edge vortex of *Manduca* (Fig. 11.4c). In contrast, Birch and Dickinson (2001) found only very slow outward flow in the vortex on their *Drosophila* model. They used baffles to stop even this slow flow, and found that the vortex persisted. It seems that the aerodynamic mechanism that keeps the vortex in place in *Drosophila* is different from the one that operates in *Manduca*. This is not too surprising, as the Reynolds numbers are 10–20 times lower in the case of *Drosophila*.

There is another unsteady mechanism that seems to be important for hovering in some insects such as the tiny wasp *Encarsia*, for which a lift coefficient of 2.3 has been calculated. This mechanism is the clap and fling (Weis-Fogh 1973). *Encarsia* hovers with its body vertical and its wings beating in a horizontal plane (Fig. 11.6). At the end of the backward stroke, the wings clap together. Their leading edges separate before their trailing edges, drawing air around the leading edges into the growing space between the wings and establishing a strong circulation rapidly. Lighthill (1973) worked out the theory of the clap and fling, and Spedding and Maxworthy (1986) showed how effective it can be. The clap and fling, or the related clap and peel in which the wings peel apart instead of opening like the leaves of a book, is also used by butterflies (Brackenbury 1991) and some other insects. It is also used at takeoff by many birds and bats. In the case of pigeons (*Columba livia*) the wings can be heard clapping together.

Most insects have two pairs of wings, but move the two wings of each side of the body as a unit. However, dipteran flies have only one pair of wings, as the hind pair have become halteres. In beetles, the fore wings have become elytra, which are held spread but stationary while the insect flies by beating its much longer hind wings. At rest, the hind wings are folded under the protective elytra. Dragonflies and damselflies (Odonata) have two pairs of wings that beat out of phase with each other. Ellington (1984) filmed hovering by members of all the groups mentioned in this paragraph. The wings of some insects bend markedly in flapping flight (Wootton 1999).

11.3. POWER FOR HOVERING

In this section we will discuss estimates of the mechanical power required for hovering, and compare them to measurements of the rates of oxygen consumption of hovering animals.

We have already derived an equation for induced power (Equation 11.8). For a 1.7-g moth (*Manduca*) beating wings 50 mm long through an angle of 115° (2.0 rad) in air of density 1.2 kg/m^3, it gives a power of 11.5 W/kg body mass. A more sophisticated calculation by Willmott and Ellington (1997b) gives 14–16 W/kg. The same authors also estimated the power required to overcome profile drag, treating the wings as assemblies of strips moving at different speeds as in the blade element calculations of lift coefficients (Fig. 11.2B). These calculations are subject to considerable uncertainty, because the profile drag coefficients in the unsteady motion of hovering are not necessarily the steady-state values. Willmott and Ellington's (1997b) estimates for three individual moths, using three different sets of assumptions, ranged from 3 to 10 W/kg. Their estimates of the total mechanical power required for hovering ranged from 18 to 26 W/kg.

They also considered the energy needed to accelerate the wings for each stroke. If the moment of inertia of a wing couple about its base is I and it reaches a peak angular velocity ω_{max} in the course of a wing stroke, the kinetic energy that it gains and loses in the course of the stroke is $\frac{1}{2}I\omega_{max}^2$.

The moment of inertia used in this calculation should include the added mass of air that is accelerated and decelerated with the wings. Willmott and Ellington (1997b) measured the moments of inertia of *Manduca* wings, took account of added mass, and calculated that the inertial power required for hovering was 23–38 W/kg. This is the total positive work required per unit time to accelerate the wings.

The inertial power seems to be greater than the aerodynamic power, so there is scope for saving energy by elastic storage (Section 3.6). Are such savings made? In the discussion that follows I will use midrange estimates of 22 W/kg for aerodynamic power and 30 W/kg for the inertial power. If there is no elastic storage, the kinetic energy lost by the wings in the second half of each stroke can be used to do the aerodynamic work for that half of the stroke, so the total mechanical power requirement can be estimated as the inertial power plus half the aerodynamic power, or 41 W/kg. If there is perfect elastic storage, so that no inertial work has to be done by the muscles, the mechanical power requirement is 22 W/kg. Casey (1976) measured the oxygen consumption of hovering *Manduca* and calculated that the metabolic power consumption was 237 W/kg. Thus, we can estimate the efficiency of the muscles as $41/237 = 0.17$ if there is no elastic storage, and $22/237 = 0.09$ if there is perfect elastic storage. In work loop experiments Josephson and Stephenson (1991) measured efficiencies of 0.04 to 0.10 for locust flight muscle, which suggests that *Manduca* probably benefits from elastic mechanisms. Similar calculations can be done for the hummingbird *Amazilia*. Weis-Fogh (1972) estimated that its power requirements for hovering were 26 W/kg (aerodynamic) and 29 W/kg (inertial). Berger and Hart (1972) found that the metabolic power was 239 W/kg. Thus, the efficiencies can be estimated as $42/239 = 0.18$ for no elastic storage and $26/239 = 0.11$ for perfect elastic storage. Neither estimate seems impossible for vertebrate striated muscle (Section 2.5).

Dickinson and Lighton (1995) performed an ingenious experiment to try to discover whether elastic storage is important for the fruit fly *Drosophila*. Aerodynamic and inertial power requirements for hovering by this insect are estimated to be about 15 W/kg each. A fly glued to a support beat its wings as if flying. It was stimulated to attempt to turn alternately to left and right by a moving pattern of lights. In trying to turn, it varied the frequency f of its wing beats and the angle ϕ through which its wings moved. Its oxygen consumption was measured and found to fluctuate with the fluctuations of frequency and angle.

The method depends on the different effects that fluctuations of frequency and angle have, on the components of mechanical power. Induced power is proportional to $\phi^{-0.5}$ (Equation 11.8). Profile drag is proportional to the speed of the wings squared (ignoring the effect of Reynolds number, Equation 10.11), and profile power is profile drag multiplied by speed, and so is proportional to speed cubed or to $f^3\phi^3$. The inertial work required for each wing beat is proportional to angular velocity squared or to $f^2\phi^2$, and inertial power is inertial work times frequency so is proportional to $f^3\phi^2$. Thus, each of these three components of power is proportional to a different function of frequency and angle. Dickinson and Lighton (1995) analyzed the fluctuations of metabolic rate of the flies as wing beat frequency and angle varied. Their observations suggested that the muscles worked with an efficiency of about 0.11, and that modest energy savings were made by elastic storage.

The question remains, what potentially energy-saving springs have insects got? Weis-Fogh (1960) discovered various structures in insect thoraxes, made of a rubberlike protein that he called resilin. They may save energy by elastic storage, as also may the walls of the thorax. In advanced insects, such as flies, beetles, bugs, and bees, that have fibrillar flight muscles (Section 2.6), the muscles themselves may serve as springs (Josephson 1997a). A passive spring dissipates a little energy by hysteresis each time it is stretched and recoils, whereas these muscles do net work in each cycle; in effect, they are springs with negative hysteresis.

The possible energy-saving role of elastic mechanisms in bird flight will be discussed in Section 12.2.

This chapter has shown that only small animals can hover. Hummingbirds and moths generate lift both in the upstroke and in the downstroke, but other small birds and some flies seem to obtain lift only in the downstroke. Delayed stall and several unsteady aerodynamic mechanisms are important for hovering animals. It is not clear to what extent hovering animals save energy by elastic storage.

Insect flight has presented very difficult problems in previously unfamiliar branches of aerodynamics. The challenge has been met by research of outstanding quality, involving imaginative and technically difficult experiments. Great progress has been made in the past few years, but there is scope for more work, especially on the possibility of important differences between major groups of insects. Also, we still do not know how much energy is saved by elastic mechanisms in animal flight, or which structures are important as energy-saving springs.

••

Chapter Twelve

Powered Forward Flight

Insects, birds, and bats are the only modern animals capable of powered flight. All of them fly by flapping their wings.

12.1 AERODYNAMICS OF FLAPPING FLIGHT

Helicopters and animals hover by driving air vertically downward, as explained in Chapter 11. To fly forward, they must drive air downward and backward. The horizontal component of the momentum given to the air provides the thrust force that is needed to overcome drag.

To hover, animals beat their wings in a near-horizontal plane, with their bodies tilted at a steep angle to the horizontal (see, for example, Fig. 11.2). For forward flight, the wings beat in a more tilted plane. For example, the angle of the wing stroke plane of the moth *Manduca*, which is about 20° to the horizontal in hovering, increases to about 40° in forward flight at 2 m/s, and 50–60° at 5 m/s (Willmott and Ellington 1997b). Similarly in birds, the angle of the stroke plane to the horizontal increases with increasing speed (Tobalske and Dial 1996). Tilting of the wing stroke plane ensures that the induced velocity imparted to the air has the horizontal component needed to provide thrust. As the stroke plane becomes more vertical with increasing speed, the long axis of the body generally becomes more horizontal. This has two consequences. First, it keeps the angle of the stroke plane relative to the body more nearly the same at different speeds than if the body angle remained constant while the angle of the stroke plane to the horizontal changed. Secondly, it tends to align the body with the airflow, reducing drag.

Wing beat frequency remains more or less constant as speed increases in insects (Dudley and Ellington 1990a; Willmott and Ellington 1997b). In birds, it may be higher in hovering than in forward flight (Pennycuick 1968b), but is more or less independent of speed in forward flight (Tobalske and Dial 1996).

Figure 11.3 shows the movements of the wings of *Manduca* relative to the air flowing past them in forward flight, as well as in hovering. In hovering, the wing travels forward during the downstroke (the ventralward stroke), and backward during the upward (dorsalward) stroke. As speed increases, the wings still move forward and backward relative to the body but the path of the upstroke relative to the air becomes vertical (at 2.9 m/s) or inclined forward (at higher speeds). The angular speeds of wing movement relative to the body in the upstrokes and downstrokes are fairly nearly equal. However, as forward speed increases, the speed of the wing relative to the air becomes higher in the downstroke than in the upstroke, enabling the downstroke to produce the higher aerodynamic forces.

Figure 11.3 also shows that the wings turn upside down for the upstroke both in hovering and at low forward speeds. At 5 m/s, however, the path of the upstroke is forward and the dorsal surfaces of the wings of *Manduca* remain uppermost throughout. In contrast, due to its faster wing movements, the velocity of the upstroke of the bumblebee *Bombus* has a backward component even at the highest speed (4.5 m/s) observed by Dudley and Ellington (1990a).

In hovering, the angle of attack of the wing is positive (i.e., the air strikes its ventral surface) in the downstoke, and negative (the air strikes the dorsal surface) in the upstroke. Lift acts upward, supporting the weight of the body, in both strokes. When the upstroke is vertical, the lift is horizontal and can provide thrust but not weight support. At 5 m/s, the path of the upstroke slopes forward, and if the angle of attack were negative the lift would have a downward component. It appears from the observations of Willmott and Ellington (1997c) that at this speed the angle of attack remains positive throughout the wing beat cycle. This implies that the aerodynamic force on the wings in the upstroke may act upward and backward, contributing to weight support but giving negative thrust.

The forces on the wing can be inferred more reliably from observations of the airflow in the wake. Willmott et al. (1997) used smoke filaments to show the flow around and behind the wings of *Manduca* in simulated flight. The moths were tied to a thin rod, which held them in a wind tunnel that blew air past them. They flapped their wings as if flying. The rod was attached to a balance, which showed that aerodynamic forces were supporting only 70–96% of the moths' weight, so simulation of free flight was not perfect. No measurements were made to discover whether thrust was the same as in free flight.

Figure 12.1 shows the airflow that was observed when the wind speed was 1.8 m/s. The downstroke (Fig. 12.1A, C) produces a near-horizontal vortex ring, implying that air is driven downward, giving a near-vertical aerodynamic force. The strong leading-edge vortex is stabilized, as in hovering (Fig. 11.4C) by flow out toward the wing tips. The upstroke (Fig. 12.1B, D) produces a near-vertical vortex ring, implying that it provides thrust but little or no weight support. Grodnitsky (1999) and Dudley (2000) have reviewed observations of the wakes of other insects.

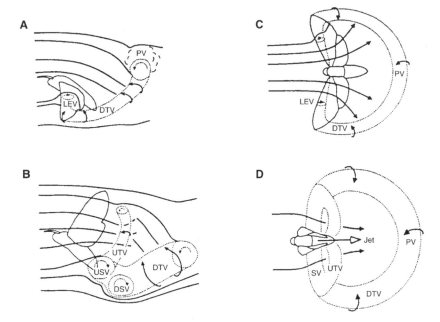

Fig. 12.1. Diagrams of airflow in the wake of *Manduca* in tethered flight in a wind tunnel, at an airspeed of 1.8 m/s: (A, C) the downstroke and (B, D) the upstroke. DSV, downstroke stopping vortex; DTV, downstroke trailing vortex; LEV, leading edge vortex; PV, SV, vorticity due to changes in angle of attack between strokes; USV, upstroke starting vortex; UTV, upstroke trailing vortex. From Willmott et al. (1997).

    Spedding et al. (1984) used helium-filled soap bubbles to discover the pattern of airflow in the wake of a flying pigeon (*Columba livia*; see Section 5.4). The bird flew slowly in the conditions of the experiment, at about 3 m/s. In each downstroke, the wings produced a downward puff of air with a vortex ring around it, as shown in the diagram of a bat in Fig. 12.2A. In the upstroke, the elbow and wrist were flexed and the wings generated no strong air movements. It appears that the downstroke was made with the wings at a positive angle of attack, producing lift; but that in the upstroke the wings had a near-zero angle of attack and provided little or no lift. The authors calculated from the observed air movements that the rate at which downward momentum was being imparted to the air was only 60% of what was needed to support the bird's weight. It was later shown that, with the laboratory arranged as it was, the bird was indeed not fully supporting its weight as it flew through the cloud of bubbles, but was accelerating downward (Rayner and Thomas 1991).

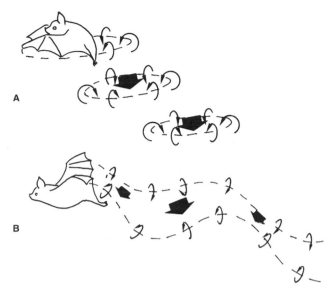

Fig. 12.2. Diagrams of airflow in the wake of a noctule bat (*Nyctalus noctula*) flying at (A) 3 m/s and (B) 7.5 m/s. Thick arrows represent airflow through the vortex rings or between the trailing vortices. This diagram, based on the experiments of Rayner et al. (1986) is reproduced from Alexander (1999).

Spedding (1987b) repeated the experiment with a kestrel (*Falco tinnunculus*), which flew much faster than the pigeon, at about 8 m/s. He observed an entirely different wake pattern, shown in the diagram of a bat in Fig. 12.2B. Instead of discrete vortex rings, continuous trailing vortices were formed. These resembled the trailing vortices behind a gliding bird (Fig. 10.2D), but undulated up and down following the paths of the beating wings. They also came closer together for the upstroke (when the wings were partially folded), and moved further apart for the downstroke (when the wings were extended). Air was driven downward, between the two vortices, throughout the wing beat cycle. The wings must be held at a positive angle of attack, so as to produce upward lift, throughout the cycle. As well as supporting the bird's weight, they must generate the thrust needed to overcome drag on the body, but upward lift in the upstroke must be accompanied by negative thrust, as already noted for *Manduca* flying at 5 m/s (Fig. 11.3B). To ensure that the mean thrust over the wing beat cycle is sufficient, the upstroke must generate less lift than the downstroke. This could be done in either of two ways, as Equation 10.8 shows. Lift in the upstroke could be reduced by reducing the circulation Γ (by reducing the angle of attack), or by reducing the wingspan s. Spedding

(1987b) measured the velocities of soap bubbles in the trailing vortices and calculated that the circulation was approximately the same in the up-stroke as in the downstroke. If the circulation had fluctuated, the wake would have had a ladderlike form, with starting and stopping vortices running across it between the two trailing vortices. It seems that the reduction of wingspan as the wings partially fold for the upstroke is responsible for the reduction of lift.

The difference between the vortex ring wake of the pigeon and the continuous vortex wake of the kestrel is not due to their being different species, but to their flying at different speeds in the conditions of the experiment. Later work has shown that other species produce a vortex ring wake at low speeds and a continuous vortex wake when they fly faster. The change of wake pattern with speed was first shown by Rayner et al. (1986) for the bat *Nyctalus noctula* (Fig. 12.2). The patterns of wing movement that produce the two types of wake seem to be distinct gaits, like walking and running (Section 7.2). So far as I know, no intermediate between them has ever been observed.

Unfortunately, we have very little information so far about the speeds at which flying animals change gaits. *Nyctalus* uses the vortex ring gait at 3 m/s and the continuous vortex gait at 7–9 m/s (Rayner et al., 1986). The wakes of pigeons flying fast have not been studied, but a change in the pattern of movement in the upstroke seems to show that pigeons change gaits at 12–14 m/s (Tobalske and Dial 1996). The wakes of different-sized animals could be dynamically similar only if they had equal Strouhal numbers (Section 4.2), so one possible prediction is that animals will change gaits at equal Strouhal numbers sf/v, where v is airspeed, s is wing span, and f is wing beat frequency. However, the data given above seem to show that pigeons change gait at a much higher Strouhal number than *Nyctalus*. The span and frequency are 0.34 m and 9 Hz for *Nyctalus* (Rayner 1986; Norberg and Rayner 1987), and 0.6 m and 6 Hz for pigeons (Tobalske and Dial 1996). Thus, sf is 3.1 m/s for *Nyctalus* and 3.6 m/s for pigeons, and at equal Strouhal numbers the birds would be flying only slightly faster than the bats.

12.2. POWER REQUIREMENTS FOR FLIGHT

Equations that we derived for the gliding performance of animals can be used also to predict power requirements for flapping flight. The energy required for gliding is supplied by the gravitational potential energy that the animal loses as it sinks through the air. The rate of loss of potential energy by an animal of mass m gliding with airspeed v at an angle θ to the horizontal is $mgv \sin \theta$, where g is the gravitational acceleration.

Hence, from Equation 10.19, the power P required for flight at speed v is expected to be

$$P = mg \left(\frac{\rho v^3 C_{\text{zero lift}}}{2N} + \frac{2 k_{\text{induced}} N}{\pi \rho v \mathscr{A}} \right) \qquad (12.1)$$

where ρ is the density of the air, $C_{\text{zero lift}}$ is the zero-lift drag coefficient, N is wing loading, k_{induced} is the induced drag factor, and $\mathscr{A}$ is the aspect ratio. Because this equation has been obtained by multiplying Equation 10.19 by mg, a graph of power against speed derived from it has the same shape as a graph of sinking speed against airspeed derived from Equation 10.18, but turned upside down (compare Fig. 12.3B with Fig. 10.6A). The power requirement is high at low speeds and at high speeds, and lower at intermediate speeds. It is least at the minimum power speed, which is predicted to be the same as the minimum sink speed for gliding. The work required per unit distance is least at the maximum range speed, which is predicted to be the same as the maximum range speed for gliding. As for gliding, the maximum range speed can be found by drawing a tangent to the curve through the origin (Fig. 12.3B).

Equation 12.1 predicts that infinite power will be required for flying at zero speed, but small birds can nevertheless hover. The discrepancy is due to the equation being based on the aerodynamics of fixed-wing aircraft. To predict power requirements for hovering, helicopter theory must be used (Section 11.1). For forward flapping flight, Equation 12.1 may be expected to underestimate power requirements, because the flapping movement of the wings increases their speed relative to the air, and so increases profile drag. However, this effect may not be very large. The wing tip of a pigeon moves up and down relative to the body at speeds of about 6 m/s as the wing flaps when the bird is flying at 10–20 m/s. Thus, the speed of this point relative to the air can be estimated as $\sqrt{10^2 + 6^2} = 11.7$ m/s when the bird is flying at 10 m/s and $\sqrt{20^2 + 6^2} = 20.9$ m/s when it is flying at 20 m/s. The speed increment due to flapping is less for more proximal parts of the wing and zero at the wing base.

Birds and bats that glide well have a minimum sinking speed in still air of around 1 m/s (Fig. 10.6B). At this sinking speed, they lose gravitational potential energy at a rate of about 10 W/kg body mass. This suggests that the mechanical power required for flapping flight should be around 10 W/kg.

Figure 12.3A shows metabolic rates of flying birds and bats, obtained by measuring rates of oxygen consumption of animals flying in wind tunnels (Section 5.3). Most of the rates lie between 50 and 120 W/kg, and so are consistent with our rough estimates based on the sinking speeds of gliding

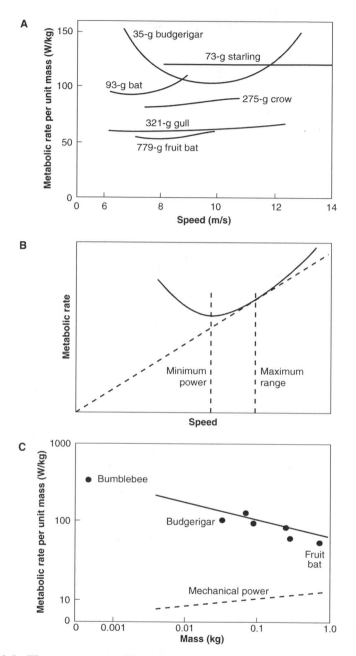

Fig. 12.3. The energy cost of level flapping flight. (A) Metabolic rates per unit body mass, plotted against speed. (B) A U-shaped graph of metabolic rate against speed, as calculated from Equation 12.1. (C) Minimum metabolic rates in flight plotted against body mass, for the same species as in (A) plus the bumblebee *Bombus*. From Alexander (1999).

birds if the muscles work with efficiencies in the range 0.08 to 0.2. How-ever, among the curves shown, only the one for the budgerigar is markedly U-shaped, as Equation 12.1 suggests it should be. The graph for the star-ling shows no dependence of metabolic rate on speed, but this may be an artifact; more recent measurements of metabolic rates of flying starlings show metabolic rate increasing with speed (Rayner 1999).

Figure 12.3A shows metabolic power, but Equation 12.1 predicts me-chanical power. Dial et al. (1997) measured the mechanical power output of the pectoralis muscle (the principal flight muscle) of magpies flying in a wind tunnel. To do this they needed to know the rates at which the muscle shortened and the forces it exerted throughout the wing beat cycle. They calculated the length changes of the muscle from the angle of the shoulder joint, as seen in films taken from above and from the side. (Biewener et al. [1998a] have since measured the length changes more directly by sonomicrography; see Section 5.6.) To measure muscle force, they took advantage of the fact that the pectoralis muscle inserts on a flange on the anterior face of the humerus. In a surgical operation prior to the experiment, they glued a strain gauge onto this flange. Forces ex-erted by the muscle bent the flange and elicited a signal from the strain gauge, which was connected to recording equipment through long, light electrical leads that trailed behind the flying bird. After the experiment, the bird was anesthetized and the strain gauge was calibrated by exerting known forces on the muscle origin.

Dial et al. (1997) used the forces and length changes that they measured in this way to draw work loops (Section 2.3). From the areas of the loops they calculated the work done in each wing beat cycle and multiplied this by the frequency to get mechanical power. They found that the power was 20 W/kg body mass in hovering, fell to 8 W/kg at a speed of 6 m/s, and rose again to 12 W/kg at 14 m/s. Thus, a graph of power against speed was U-shaped, as predicted by Equation 12.1. At any particular speed, the power was a little less than predicted from aerodynamic theory (Alexander 1997c), but it should be remembered that only work done by the pecto-ralis muscle was measured; the contributions of other muscles, notably the supracoracoideus, were ignored. The measured mechanical powers of 8–20 W/kg are consistent with the metabolic powers measured for other species of similar size if the muscles work with an efficiency of the order of 0.12 (Dial et al., 1997).

Magpies seem unable to fly faster than 14 m/s (Tobalske and Dial 1996), although the power requirement at this speed is much less than the power required for hovering. They can hover for 1–4 s, so why can they not fly briefly at speeds above 14 m/s? Rayner (1999) has suggested that speed may be limited, not by power requirements, but by the problem of producing enough thrust. Total drag (the sum of induced drag and profile drag) is least at the maximum range speed. As speed increases beyond this, the force required to support the weight of the bird remains constant but the thrust required to drive it forward through the air increases. However, as speed increases it becomes more difficult to exert high thrust, for the following reason. Consider a bird flying at speed v, beating its wings through an angle ϕ radians with frequency f. If the upstroke and downstroke each occupy half the cycle time, the bird advances a distance $v/2f$ during the downstroke. At the same time, a point on the wing at a distance r from the base moves ventrally through a distance of about $r\phi$. Thus, the path of this point through the air makes an angle of approximately $2fr\phi/v$ with the horizontal. Lift acts at right angles to this path, at an angle $2fr\phi/v$ to the vertical. As speed v increases, the lift becomes more vertical, reducing the ratio of thrust to weight. Birds flying fast partially overcome this problem by making the downstroke faster (so that it occupies less than half of the wing beat cycle), and by tilting the plane of the wing beat (so that the wing moves posteriorly, relative to the body, as it flaps down [Tobalske and Dial] 1996]). However, beyond a certain speed even these adjustments cannot produce enough thrust.

Another question that we have not yet answered is, why do birds and bats change from the vortex ring gait to the continuous vortex gait at high speeds? We saw in Section 7.6 that horses save energy by changing gaits; each gait is the most economical, in the range of speeds at which it is used. Can the same be true for flying animals?

It would be difficult if not impossible to train birds or bats to use each gait on command, even at speeds at which they would normally use the other. Consequently, the equivalent experiment to the one on horses illustrated in Fig. 7.11A has not been performed. We can only tackle the problem through aerodynamic theory.

Equation 12.1, which predicts power requirements for flight at different speeds, is based on the theory of fixed-wing aircraft. It gives reasonable approximations to the energy cost of animal flight, but cannot predict the different power requirements of the two gaits. More sophisticated theories, which take account of the momentum and kinetic energy of the vortices, have been formulated both for the vortex ring gait (Rayner 1979) and for the continuous vortex gait (Rayner 1993). Rayner (1999) used these theories to calculate the power requirements of both gaits for a pigeon over a range of speeds. Figure 12.4 shows the power requirements predicted in this way. They indicate that the vortex ring gait is the more economical at speeds up to 14 m/s and the continuous vortex gait is more economical at higher speeds. It appears from their wing movements that pigeons make the change of gait at 12–14 m/s (Tobalske and Dial 1996).

In addition to these gaits, there is a class of gaits called intermittent gaits, because the bird alternates periods of a few wing beat cycles in which

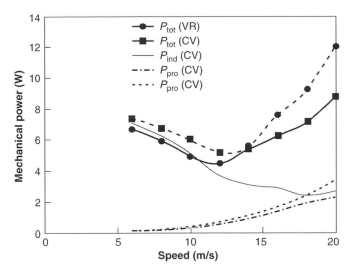

Fig. 12.4. Calculated mechanical power plotted against speed for a pigeon, for the vortex ring (VR) and continuous vortex (CV) gaits. P_{tot}, total power; P_{ind}, induced power; P_{pro}, profile power (required to overcome zero-lift drag on the wings); P_{par}, parasite power (required to overcome drag on the body excluding the wings). From Rayner (1999).

it flaps its wings with periods when it does not. Sparrows (*Passer domesticus*) and many other small birds commonly use bounding flight, in which the wings are folded while they are not beating. For example, zebra finches (*Taeniopygia guttata*) flying fast alternately flap their wings for about 0.15 s (four wing beat cycles) and fold them for about 0.10 s (Tobalske et al., 1999). Bounding flight is commonly used by small birds with low wing loading, low aspect ratio, and rounded wing tips (Rayner 1985b). The largest bird known to use it is the green woodpecker (*Picus viridis*, mass 0.2 kg).

The other common intermittent gait is undulating flight, also known as flap-gliding. In this gait, bouts of wing flapping alternate with bouts of gliding with the wings spread. This gait is used by many medium- and large-sized birds, including crows and gulls (Rayner 1985). Budgerigars (*Melanopsittacus undulatus*) and starlings (*Sturnus vulgaris*) use undulating flight at low speeds and bounding flight at high speeds (Tobalske and Dial 1994; Tobalske 1995).

Aerodynamic theory suggests that bounding flight should need less energy than continuous flapping flight at high speeds. When a bird is bounding, the wings are spread and generate lift only during the flapping phases. Consequently, in the flapping phases of bounding flight they must generate more lift than in continuous flapping, and suffer more induced drag. However, while folded they suffer no profile drag. As speed increases, induced drag becomes a smaller proportion of total drag and profile drag becomes a larger proportion. If the speed is high enough, the reduction in profile drag, due to bounding, more than compensates for the increased induced drag. Rayner (1985b) showed that this would be the case only at speeds above 1.2 times the maximum range speed, but that undulating flight could save energy at all speeds. However, zebra finches and some other birds bound at low speeds, and even when hovering (Tobalske et al., 1999). Rayner (1985b) suggested that bounding might enable the muscles to work more economically than in continuous flapping, whenever the power output required for flight was less than the maximum that the muscles could provide. Muscles work most efficiently at a particular rate of shortening (Fig. 2.3D). Intermittent flapping may enable the muscles to operate at their most efficient shortening rate, at speeds at which the power requirement is less than continuous flapping (using the same shortening rate) would provide. However, Tobalske et al. (1999) found that zebra finches do not keep the rate of shortening of the muscles constant, but reduce the angular velocity of the wing in the downstroke from 7 degrees per millisecond in hovering to about 5.5 degrees per millisecond in fast flight. This reduction in angular velocity results from a marked reduction of the amplitude of the wing beat.

So far, I have not mentioned inertial power requirements for forward flight. As we noted in Section 11.3, kinetic energy given to each wing in each stroke is $\frac{1}{2}I\omega_{max}^2$, where I is the moment of inertia of the wing about its base, and ω_{max} is the maximum angular velocity attained during the stroke. Birds have two wings, and there are two strokes (one up and one down) in each wing beat cycle. Thus, the inertial power is $2I\omega_{max}^2 f$, where f is the wing beat frequency. The wing beats through an angle ϕ in time $1/2f$, so $\omega_{max} = 2\phi f$ (if the stroke is made at constant angular velocity) and the inertial power is $8I\phi^2 f^3$. Alternatively, if the angular velocity varies sinusoidally, $\omega_{max} = \pi\phi f$ and the inertial power is $2\pi^2 I\phi^2 f^3$. Notice that in these expressions, the frequency f is cubed and the angle ϕ is squared, so small changes in these quantities may have a large effect on power.

Van den Berg and Rayner (1995) measured the moments of inertia of the wings of birds ranging from a 10-g tit (*Parus caeruleus*) to a 2-kg heron (*Ardea cinerea*) and calculated inertial power, assuming sinusoidal wing motion. In their paper, they count negative work done decelerating the wings as well as positive work done accelerating them, but I will give their result counting positive work only, as in the previous paragraph. They found that the inertial power for a bird of mass m kilograms was $3.9m^{0.80}$ watts (0.10 W for a 10-g bird and 6.8 W for a 2-kg one). They point out that these values may be a little too high as they did not allow for the wing being partially folded (reducing the moment of inertia) in the upstroke. Rayner (1995) calculated that the aerodynamic power for bird flight at the minimum power speed is $12m^{1.1}$ watts (0.076 W for a 10-g bird and 26 W for a 2-kg one). These data indicate that the work done accelerating the wings of a 2-kg bird in the first half of a wing stroke is less than the aerodynamic work to be done in the second half. Thus, the inertial work may be recovered at the end of the wing stroke, if the kinetic energy of the wings is used to do aerodynamic work. However, the kinetic energy given to the wings of a 10-g bird is greater than the aerodynamic work to be done in the second half of the stroke, so the inertial work cannot all be recovered as aerodynamic work. There is scope in small birds, but apparently not in large ones, for energy saving by elastic mechanisms, as explained in Section 3.6. It is possible that the tendons of the principal flight muscles may be compliant enough to have a significant role as energy-saving springs. Pennycuick and Lock (1976) suggested that feather elasticity may be important, but Alexander (1988) raised an objection.

Most measurements of the metabolic rates of insects in forward flight have used tethered insects, which may not have been beating their wings sufficiently vigorously to support their weight. Ellington et al. (1990) achieved the remarkable feat of measuring the oxygen consumption of individual free-flying bumblebees (*Bombus*). They were unable to fit the bees with masks, as has been done in most experiments with birds, and so flew them in a miniature wind tunnel that recirculated the air, so that its oxygen content gradually fell. They measured the falling oxygen consumption and calculated that the metabolic rate was about 350 W/kg throughout the range of speeds at which the bees would fly, 0–4 m/s.

Figure 12.3C shows minimum metabolic rates in flight of the birds and bats included in Fig. 12.3A, together with the bee. Metabolic rate per unit mass falls as body mass increases. The same graph shows mechanical power requirements, estimated by Rayner (1995). This shows mechanical power per unit mass increasing as mass increases. Together, the two lines on the graph show that smaller animals have less efficient flight muscles. Similarly, muscles used for running operate at lower efficiencies for smaller legged animals (Fig. 7.12).

The mechanical power required for flight, shown in Fig. 12.3C, is proportional to (body mass)$^{1.10}$, implying that larger flying animals need more power per unit mass than small ones. However, as we noted in Section 12.2, larger flying animals cannot produce as much power per unit body mass as small ones. If the required power per unit mass increases with body size, and the available power falls, there must be an upper limit to the size range of animals that can fly. The largest modern flying animal is possibly the Kori bustard (*Ardeotis kori*; I have weighed a 16-kg specimen). However, some extinct birds and pterosaurs were substantially larger (Alexander 1998a). That argument for a size limit for flying animals seems clear, but, surprisingly, a parallel argument based on metabolic rates is not. The metabolic power per unit mass required for flight (Fig. 12.3A) is proportional to (body mass)$^{0.83}$. It does not seem to be known how maximum metabolic rates of birds are related to body mass, but maximum aerobic metabolic rates for mammals are proportional to (body mass)$^{0.81}$ (Taylor et al., 1981), and it seems likely that the exponent for birds is about the same. The exponents 0.83 and 0.81 are not significantly different, so these metabolic data fail to explain the size limit for flying animals.

So far, we have considered only the cost of steady, level flight. Much higher power may be required for takeoff, making repeated short flights much more expensive of energy than a single flight over the same total distance. Nudds and Bryant (2000) trained zebra finches (*Taeniopygia guttata*) to fly repeatedly between two perches 5.5 m apart. Each perch was periodically drawn tight against the wall of the aviary, obliging the bird to leave it and fly to the other. The metabolic rates of the birds and of controls that perched undisturbed were measured by the doubly labeled water technique (Section 5.3). From the results, it was calculated that the metabolic rate during the short flights was 6.6 W. The expected metabolic rate for a bird of the same mass (13 g) in steady flight is only 1.6 W (Rayner 1995). Much of the cost of these short flights was probably due to the need to accelerate and climb at takeoff. The aviary was deliberately arranged to prevent the birds from gaining kinetic energy at takeoff by dropping from the perch and gliding, in the manner described for flying squirrels in Section 10.4.

12.3. OPTIMIZATION OF FLIGHT

This section asks how wings should be adapted to different lifestyles, how fast animals should fly, and whether they should flap their wings in bursts or all the time.

We saw in Section 10.6 that soaring birds seem generally to have wing designs well suited to their styles of soaring; slope soarers have higher wing loadings and aspect ratios than birds of the same mass that soar in thermals over land. Analysis of masses and wing dimensions of a much wider selection of bats (Norberg and Rayner 1987) and birds (Rayner 1987) has revealed other correlations of wing design with flying habits that seem adaptive. For example, bats and birds that take large prey tend to have relatively low wing loadings that enable them to carry the prey. Bats and birds that live in cluttered forest habitats tend to have short wings with low aspect ratios. Auks and other birds that use their wings both for swimming and for flight tend to have small wings, a compromise between the requirements of the two modes of locomotion.

The direct energy cost of a journey is least if the animal flies at its maximum range speed, but time spent on a journey is unavailable for other activities, so there may be an advantage in flying faster and arriving sooner. Let P be the metabolic rate when flying at speed v. The maximum range speed is the speed for which P/v is least. However, if the animal is able to feed and lay down energy reserves at a rate Π when it is not feeding, the true energy cost of traveling unit distance is not P/v, but $(P + \Pi)/v$. Thus, the most energetically advantageous speed may not be the maximum range speed, but the higher speed that minimizes $(P + \Pi)/v$. This is unlikely to be much higher than the maximum range speed because Π will generally be much smaller than P. Lindström (1991) showed that even in the most favorable circumstances, birds generally cannot lay down fat reserves at a rate faster than 2.5 times the standard metabolic rate; but metabolic rates in flight are generally about 10 times the standard metabolic rate (Alexander 1998b). Thus, we may expect to find that birds commonly fly a little faster than the maximum range speed.

Pennycuick (1997) has measured the speeds of various birds, most of them flying distances of several hundred meters between their nests or roosts and their feeding areas. He measured the direction and speed of the wind as near as possible to the birds, and subtracted the wind vector from the ground speed to obtain the airspeed. The mean measured airspeeds for different species ranged from 9 m/s for a little blue heron (*Egretta caerulea*, wing loading 25 N/m²) to 19 m/s for a common murre or guillemot (*Uria aalge*, wing loading 171 N/m²). He compared them to estimates of the minimum power speed and the maximum range speed, based on measurements of the masses and wing dimensions of the same species. Unfortunately, there is considerable uncertainty about the conclusions that should be drawn from the comparisons.

The problem concerns the profile drag. Equation 12.1 makes the profile power (mg times the first term in the brackets) proportional to (Speed)³. The zero-lift drag coefficient in this term refers to the wings and body together. Pennycuick's (1975) theory on which his estimates are based treats the profile power as the sum of two terms, one (power to overcome profile drag on the wings) independent of speed and the other (power to overcome drag on the body) proportional to (Speed)³. The assumption that profile drag on the wings is independent of speed is contentious, and it makes estimates of minimum power and maximum range speeds very sensitive to the drag coefficient for the body. Pennycuick (1997) made estimates for two different drag coefficients, a high one that he had previously used and a lower one based on recent wind tunnel measurements. The new estimates are not necessarily better than the old ones. Minimum sink speeds for gliding (which are expected to equal the minimum power speed for flapping flight) have been measured for two of the species in his paper (*Fulmarus glacialis* and *Coragyps atratus* [Pennycuick 1960; Parrott 1970]. In each case, the minimum sink speed is closer to Pennycuick's old estimate of the minimum power speed than to his new one.

Pennycuick (1997, 2001) found that the speeds observed in the field for most of the species lay between the old estimates of minimum power speed and maximum range speed. However, if the new estimates were used, almost half the species flew at speeds below the minimum power speed. These conclusions contrast with the argument above that birds should fly faster than the maximum range speed. Pennycuick (1997) points out that the predicted optima are rather flat, so little energy is lost by flying slightly below the optimum speed. On the other hand, a small increase of speed, in the neighborhood of the maximum range speed, requires a substantial increase in power and so in the mass of muscle required. He argues that this may make it advantageous to fly a little more slowly than the maximum range speed.

Migrating birds often have to cross deserts or oceans where they can find no food. For example, garden warblers (*Sylvia borin*) crossing the eastern part of the Sahara Desert have to fly 2200 km, apparently with no opportunity to feed on the way (Biebach 1998). Bristle-thighed curlews (*Numenius tahitiensis*) flying from the Hawaiian Islands to Alaska cross 4000 km of ocean with no opportunity to feed (Piersma 1998). These journeys are possible only because the birds start out with a very large store of fat in the body. The warblers start very fat with masses of around 24 g, and arrive two or three days later at the other side of the desert having lost a mean of 7.3 g, of which 5.1 g is estimated to have been fat and 2.2 g protein. The curlews start off with a mean body mass of 675 g, of which 289 g (43%) is fat. Fat has the advantage over other foods that its energy content per unit mass is higher.

A rough calculation will show why so much fat is needed. Birds of the size of garden warblers would have metabolic rates in flight of at least 100 W/kg (Fig. 12.3C). They might fly as fast as 14 m/s, in which case flying 2200 km would take about 160,000 s (44 h) and use 16 MJ/kg body mass. Metabolism of 1 kg fat yields 40 MJ, so this calculation suggests that 0.4 kg fat would be needed per kilogram of bird. The birds lose mass on the journey; their mean mass is about 20 g, so they might be expected to use 8 g fat. In fact, as we have seen, they only use 5 g. Possible reasons for the discrepancy are that there was a little protein metabolism (but only enough to supply about 2 MJ/kg); the birds may have been aided by wind; and some of the data used in this rough calculation may not be accurate.

Interestingly, much of the lost protein seems to be muscle (Biebach 1998). As the fat load falls, the bird becomes lighter and less muscle is needed to power flight. The bird can make itself lighter still by burning off the excess muscle. Gut protein is also metabolized, greatly reducing the mass of the gut, which will not be needed until the bird has crossed the desert or other unfavorable terrain and can feed again. The bird reduces the energy cost of flight, by reducing its mass as much as possible. Kvist et al. (1998) took advantage of natural fluctuations of the body mass of knots (*Calidris*) to measure the energy cost of flight of the same birds at different body masses. They flew the birds in a wind tunnel for periods of up to 10 h, and measured their metabolic rates by the doubly labeled water technique (Section 5.3). They confirmed that the metabolic rate was less at lower body masses, but found that the effect was less pronounced than aerodynamic theory predicted.

Geese, pelicans, and some other large birds habitually fly in V-formations (skeins), with the left wing tip of one bird close behind the right wing tip of another, or vice versa. This raises the possibility that energy may be saved by interference between neighboring birds' trailing vortices. Vortices behind left wing tips and behind right ones have circulations in opposite directions, so the bird that is behind in the skein will tend to cancel out the trailing vortex of the bird in front. Ideally, this canceling out might leave vortices only behind the outer wings of the birds at the extreme ends of the arms of the V. The skein of birds would resemble, aerodynamically, a single aerofoil of exceedingly high aspect ratio. A high aspect ratio reduces induced drag, saving energy. Weimerskirch et al. (2001) used data loggers (see Section 5.7) to record the heart beat frequencies of trained pelicans (*Pelecanus onocrotalus*) flying behind a boat or an ultralight aircraft. Their results indicated that the bird leading a skein had as high a metabolic rate as when flying alone, but that the other birds had lower metabolic rates.

In this chapter we have seen how the angles of the body and the wing stroke change as flying insects increase their speed. We have seen how many birds and bats change from a vortex ring gait at low speeds to a continuous vortex gait at higher speeds, and we have discussed the intermittent gaits of some birds. Our discussion of optimum flight speeds was hampered by uncertainty about maximum range speeds. Further research is needed to check, for other species, Rayner's (1999) explanation of gait change in pigeons (Fig. 12.4). We would like to know more about the circumstances in which birds use intermittent gaits, and the advantages (if any) that they gain by doing so. It would also be useful to have better information about maximum range speeds that might help us to understand the speeds at which birds actually fly. For this purpose, maximum range speeds should be determined by metabolic measurements.

....

Chapter Thirteen

Moving on the Surface of Water

W ATER STRIDERS (*Gerris* and *Halobates*), some other insects, and fisher spiders (*Dolomedes*) walk on the surface of water, supported by surface tension. Basilisk lizards (*Basiliscus*) run on water, relying on quite different forces. This chapter considers both these means of moving over the surface, and also discusses swimming on the surface as performed, for example, by ducks. More general aspects of swimming that apply to submerged swimmers as well as to those that swim at the surface are discussed in later chapters.

13.1. FISHER SPIDERS

Water striders and fisher spiders stand on water with the distal leg segments (the tarsus) resting on the surface (Fig. 13.1A). Surface tension acts horizontally in a level water surface, but the animal's weight depresses the surface so that it slopes where it meets the animal's foot (Fig. 13.1B). Consequently, the forces F_{st} exerted on the foot by surface tension have an upward component that supports the animal's weight. If the load on the foot is increased, more of the foot is submerged, and surface tension acts at a steeper angle to the horizontal (Fig. 13.1C). The contact angle (α, Fig. 13.1B) remains constant; for water on insect cuticle it is about 110° (Denny 1993).

A liquid of surface tension γ meeting a solid along a line of length l exerts a force γl on the solid. If it meets the solid at an angle β to the horizontal (Fig. 13.1B) the vertical component of the force is $\gamma l \sin \beta$. Thus, the maximum possible vertical force on a foot acts when $\beta = 90°$, and can be calculated by multiplying the surface tension of the liquid by the perimeter of the foot. The surface tension of water at 20°C is 0.073 N/m (Denny 1993), so a large fisher spider (*Dolomedes*) with a mass of 1 g (weight 0.01 N) could be supported by feet of total perimeter 0.01/0.073 = 0.14 m. The spider would have legs about 50 mm long, of which around 15 mm would rest on the surface (Suter et al., 1997). The total perimeter of contact (counting both sides of all eight legs) would be 15 × 2 × 8 mm or 0.24 m, which is more than sufficient to support the animal.

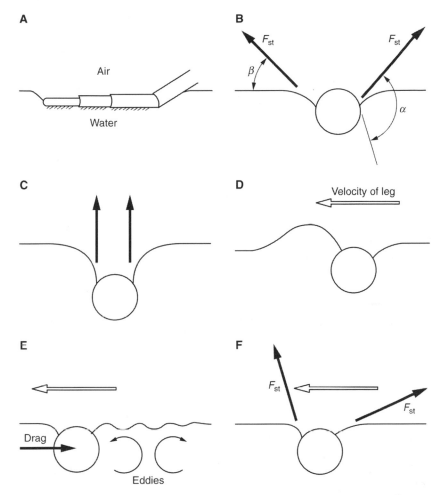

Fig. 13.1. Diagrams of a foot of a fisher spider on the surface of water. (A) shows the foot in posterior view, and the other diagrams show it in section. (B) and (C) show it pressing down on the water with a small force and a larger force, respectively. (D), (E), and (F) illustrate three possible effects that could enable a foot moving toward the left to exert propulsive forces: (D) shows a wave pushed by the moving foot; (E) shows water flowing round the foot and the dimple in the water, exerting drag; and (F) shows asymmetrical surface tension forces. Further explanation is given in the text.

The forces that surface tension exerts on geometrically similar animals are proportional to their lengths, but their weights are proportional to length cubed. Therefore, surface tension cannot support large animals. For example, a 1-kg animal (1000 times the weight of the spider that we have been considering) would need feet with a perimeter of 140 m.

To be able to move on the water surface, spiders must be capable of exerting horizontal as well as vertical forces on it. *Dolomedes* swim by backward sweeps of their second and third pairs of legs. The two pairs of legs sweep back in rapid succession, accelerating the animal, which then glides along passively for a while before making the next leg stroke. Suter and Wildman (1999) observed a spider accelerating at up to 2.3 m/s², implying forces up to 0.23 times its weight. The forces exerted by the legs that propel the animal must have been larger than this, because there was presumably some resistance to the movement of the other two pairs of legs, which slide passively forward over the water surface.

Suter et al. (1997) asked how the propulsive forces are generated. One possibility is that a wave builds up behind the backward-moving foot, which then pushes on the wave (Fig. 13.1D). Another is that the movement of the foot is resisted principally by hydrodynamic drag (Fig. 13.1E). A third is that the movement of the foot would make the dimple in the water surface asymmetrical, so that the surface tension forces on the two sides of the foot would act at different angles, giving a resultant with a horizontal component (Fig. 13.1F). Suter et al. (1997) performed experiments with an amputated spider leg to investigate these possibilities. They built a device that held the distal parts of the leg horizontal on the surface of water, as in swimming. The leg pressed down on the water with a force that could be varied. The water was made to flow smoothly past the leg at various speeds, and the horizontal component of the force that it exerted on the leg was measured. This experiment is referred to repeatedly in the paragraphs that follow.

First, we consider the possibility that the propulsive force is mainly due to a wave. We will have to consider the speeds of waves, because an animal cannot push on a wave that travels faster than it can move its foot. Waves on the surfaces of liquids travel at speeds that depend on their wavelengths (Denny 1993). For small waves, the principal force tending to flatten them is surface tension, and the shorter their wavelength the faster they move. For large waves, the principal flattening force is gravity, and the longer their wavelength the faster they move. The slowest waves have an intermediate wavelength, $2\pi\sqrt{\gamma/\rho g}$, where γ is the surface tension (0.073 N/m for water), ρ is the density (1000 kg/m³ for water), and g is the gravitational acceleration (9.8 m/s²). Their speed is $(4g\gamma/\rho)^{1/4}$. These formulas tell us that the slowest waves on water have a wavelength of 17 mm and travel at 0.23 m/s. A spider will not get a wave to push on unless it sweeps its feet back at at least 0.23 m/s. Suter et al. (1997) confirmed with their apparatus that the foot produced no observable wave unless it was moved at at least 0.21 m/s. When the living spiders swam, they swung their swimming legs back at angular velocities that made their distal ends move at speeds ranging from 0.16 to 0.62 m/s. More proximal parts of the foot traveled more slowly, so in some swimming sequences little or none of the foot was traveling fast enough to make waves. On these occasions, at least, the spiders cannot have been relying on wave forces for propulsion.

Next, we consider the possibility that drag is the important force. In the experiments with amputated legs, the horizontal component of force increased approximately in proportion to the square of the speed, as expected for drag, but this is not inconsistent with the other possibilities. At any speed and dimple depth, the measured horizontal force was fairly close to a theoretical estimate of the drag. The surface of the water was disturbed in the wake of the leg (Fig. 13.1E), showing that eddies were formed. Work must have been done to give kinetic energy to the disturbed water, so eddies imply drag.

Finally, there is the possibility that asymmetrical surface tension forces may be important. The closeness of the measured forces to the theoretical drag suggests that this effect contributed only a little, if anything, to the propulsive force. Also, 25% reduction of surface tension (by replacing the water with dilute ethanol) had little effect on the force. The conclusion of Suter et al. (1997) is that drag is certainly important, but that waves may be important when the legs move fast, and that surface tension may also contribute.

Suter and Wildman (1999) pointed out that *Dolomedes* has two gaits. At speeds up to 0.2 m/s (averaged over a complete cycle of acceleration and deceleration) it swims as described above, by oarlike movements of the second and third pairs of legs. The tarsi are kept approximately horizontal, and the spider is always in contact with the water surface. This gait is described as rowing. A different gait, described as galloping, is used at speeds from 0.3 m/s up to a maximum of 0.7 m/s. The animal advances in a series of leaps, losing contact with the surface of the water. Between leaps it strikes the water with the first three pairs of legs. Instead of laying the tarsi flat on the water surface as in the rowing gait (Fig. 13.2A) it moves them in a near-vertical plane. They hit the water so fast that an air-filled cavity opens up in the water anterior to the leg (Fig. 13.2B). We will see something similar happening with different animals in the next section of this chapter.

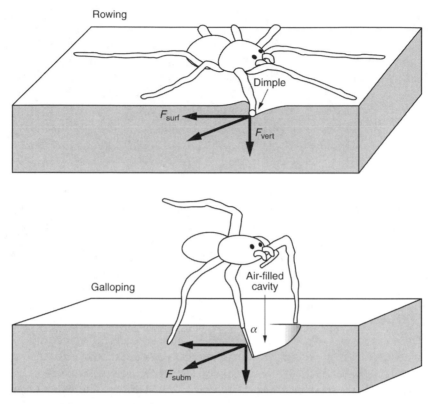

Fig. 13.2. Diagrams of a fisher spider performing (above) the rowing gait and (below) the galloping gait. Arrows represent the components of force exerted by the foot on the water. The angle α between the foot and the water surface is less than 30° when the foot hits the water, and 90–150° when it leaves it again. From Suter and Wildman (1999).

13.2. BASILISK LIZARDS

Basilisk lizards (*Basiliscus basiliscus*) with masses up to at least 130 g run short distances over the surface of water on their hind legs (Glasheen and McMahon 1996b). At 100 times the weight of the spiders discussed in the previous section, they are much too large to be supported by surface tension. Glasheen and McMahon (1996a) explained how they are supported. As in running on solid ground, the mean vertical force on the feet, over a complete stride, must equal body weight. It will be convenient to express this in another way. During one stride, gravity exerts on the animal an impulse (force multiplied by time) equal to $mg\tau$, where m is body mass, g is the gravitational acceleration, and τ is the stride duration. To support

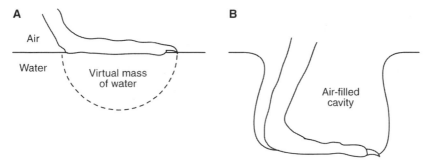

Fig. 13.3. Diagrams of the foot of a basilisk lizard (A) at the instant of hitting the water surface and (B) a little later.

the body, the feet must exert an equal impulse on the ground (or in the case of the basilisk on the water). In the discussion that follows we will consider a 90-g lizard (weight 0.9 N). Basilisks of this mass have a stride frequency of about 8 Hz, so the stride duration is 0.125 s and the impulse required in each stride is $0.9 \times 0.125 = 0.113$ N s, or 0.056 N s from each hind foot.

Three phenomena are important. First, the sole of the foot is slapped down on the surface of the water with high velocity, typically 3 m/s in the case of 90 g lizards (Glasheen and McMahon 1996a). This accelerates the water under the foot, giving the water momentum (mass multiplied by velocity). Newton's second law of motion tells us that force equals mass multiplied by acceleration. Impulse is force integrated over time, and momentum is mass multiplied by (acceleration integrated over time). From this it follows that the momentum given to the water equals the impulse applied to it. Different parts of the water are given different accelerations, but we can define a virtual mass of water as the mass that would have the same momentum as is actually given to the water if the whole of the virtual mass were accelerated to the velocity with which the foot hits the water (Fig. 13.3). Glasheen and McMahon (1996a) dropped a weighted model foot onto the surface of water and used a force transducer to measure the impulse. From the results of dropping the foot from various heights they calculated that the virtual mass accelerated by the foot of a 90-g lizard was 3.3 g. This mass, accelerated to 3 m/s, would have momentum equal to $0.0033 \times 3 = 0.010$ N/s, 18% of what is needed.

Secondly, after hitting the water the foot sinks down into it. The water does not immediately close over the foot; instead, the foot makes a temporary air-filled hole in the water (Fig. 13.3B). The hole is still intact when the foot is withdrawn, 0.06 s after hitting the surface. While the foot is at the bottom of the hole, its upper surface is exposed to atmospheric pressure but its under surface is exposed to the hydrostatic pressure at that depth. Thus, an upward force acts on the foot, equal to the hydrostatic pressure multiplied by the area. Glasheen and McMahon's (1996a) photographs of a 90-g basilisk show that the final depth of the hole was about 100 mm. The hydrostatic pressure at that depth is the depth (0.10 m), multiplied by the density of water (1000 kg/m^2), multiplied by the gravitational acceleration (9.8 m/s^2). It is about 1000 N/m^2. That is the final pressure when the hole is deepest; the mean pressure over the time while the foot is in the hole is half that, 500 N/m^2. The area of the foot was 6×10^{-4} m^2, so the mean hydrostatic force on it was $500 \times 6 \times 10^{-4} = 0.3$ N. If this force acted for the 0.06 s that the foot was in the water, its impulse was $0.3 \times 0.06 = 0.018$ N s, about 32% of what was needed.

Thirdly, drag acts on the foot as it is pushed down into the water. As already noted, the area of the foot was 6×10^{-4} m^2. It was pushed 0.1 m down into the water in 0.06 s, so its mean speed was 1.7 m/s. The density of water is 1000 kg/m^3, and the drag coefficient of the foot would probably be about 1.0 (similar to disks and cylinders). With these data, Equation 3.8 gives a drag of $0.5 \times 1000 \times 6 \times 10^{-4} \times 1.7^2 \times 1.0 = 0.9$ N. In 0.06 s, this would give an estimated impulse of 0.05 N s, 90% of what was needed.

Those rough estimates of the hydrostatic and drag forces are apparently too high. Glasheen and McMahon (1996a) pushed their model down into water at various speeds, measuring the force on it. After the initial slap on the surface, the measured force was the sum of the hydrostatic and drag forces. They found that it was only 68% of what my calculation gives. Thus, my estimate of the sum of the impulses from these two forces should be reduced to $0.68(32 + 90) = 83\%$ of the requirement. With the 18% from the slap impulse, that gives almost exactly the required total. However, Glasheen and McMahon (1996a) noted that individual steps differ from each other; in some the foot is slapped down harder on the water, or driven deeper, than in others. They calculated from the results of their model experiments that the impulses delivered in different steps ranged from about 70% to about 130% of the required mean.

13.3. SURFACE SWIMMERS

Ducks, penguins, water snakes, and many other animals often swim at the surface of water, with parts of their bodies exposed to the air. Ducks swim by paddling with their feet, penguins by using their wings as hydrofoils, and water snakes by undulating their bodies. These methods of propulsion are discussed in the chapters that follow. In this chapter we are concerned only with the resistance they have to overcome when swimming at the surface, due to the waves that their motion generates.

Moving ships make waves that spread out on either side in the shape of a V. So also do surface-swimming animals, as can be seen by watching ducks on a smooth pond. Waves formed at the bow and stern travel at the speed of the boat or animal, and have the wavelength appropriate to that speed. We saw in Section 13.1 that the slowest waves on water have a wavelength of 17 mm and travel at 0.23 m/s. Waves of longer or shorter wavelength travel faster. For the shorter wavelengths, surface tension is the predominant flattening force, but for the longer waves it is gravity. The animals that we are concerned with in this section travel faster than 0.23 m/s, so they make waves. They are much more than 17 mm long, so these waves are gravity waves. Work is needed to give potential and kinetic energy to the water in the waves, so there is a force resisting the motion of the boat or animal due to the presence of the waves. It is called wave drag.

The importance of wave drag is illustrated by comparisons of the metabolic rates of animals swimming at the surface or deeply submerged. For example, Baudinette and Gill (1985) measured the oxygen consumption of little penguins (*Eudyptula minor*) swimming in a flume (Fig. 5.2B). They found that the metabolic rates of the penguins were 6.3 W/kg when they were resting, 8.4 W/kg when they were swimming under water at 0.72 m/s, and 12.4 W/kg when they were swimming at the surface at the same speed. Similar measurements on sea otters (*Enhydra*) showed that for them swimming at the surface required 70% more metabolic power than swimming submerged at the same speed (Williams 1999). Some wave drag acts even on animals that are completely submerged but close to the surface. To avoid it, it is necessary to swim deeper than about 2.5 body diameters (Hertel 1966).

A simple equation gives the speed of waves that have wavelengths long enough for surface tension to be unimportant. A wave of wavelength λ travels at speed v_{wave},

$$v_{wave} = \sqrt{\frac{g\lambda}{2\pi}} \tag{13.1}$$

where g is the gravitational acceleration (Denny 1993). If the wavelength of the waves matches the waterline length of the hull, l_{hull}, the bow and stern waves interfere constructively with each other and the waves become very large. The speed at which this happens is called the hull speed; it is $\sqrt{gl_{hull}/2\pi}$. For a duck with a waterline length of 0.33 m, this is 0.72 m/s, and for a duckling with a waterline length of 0.12 m it is 0.43 m/s. As a boat or animal approaches its hull speed, the wave drag becomes very large.

Prange and Schmidt-Nielsen (1970) measured the oxygen consumption of mallard ducks (*Anas platyrhynchos*) swimming in a flume. By increasing the speed of the flowing water, they were able to make the ducks swim at speeds up to 0.7 m/s, which is remarkably close to the calculated hull speed of 0.72 m/s. The metabolic rates of the ducks were 7 W/kg when they were resting on the surface of the water, and around 12 W/kg when they swam at speeds of 0.35–0.5 m/s. At higher speeds the metabolic rates rose steeply, reaching 22 W/kg at 0.7 m/s.

Speedboats travel faster than their hull speeds by hydroplaning. The boat rides up its bow wave, so that it slopes bow up and strikes the water at a positive angle of attack. This results in hydrodynamic lift. A boat that is traveling below its hull speed floats relatively low in the water and is supported entirely by buoyancy, but a hydroplaning boat rides higher in the water and is supported largely by lift. Drag on a hydroplaning boat is considerably lower than would be predicted by extrapolation from lower speeds, because less of its surface area is under water and because wave drag is largely avoided. Aigledinger and Fish (1995) found that ducklings with a hull speed of 0.43 m/s were capable of short bursts of swimming at 1.73 m/s by hydroplaning. When hydroplaning, they rose higher in the water and their bodies had an angle of attack of about 16°.

The siphonophore *Velella* floats on the surface of the sea and is blown along by the wind. The gas-filled float that gives it buoyancy has a triangular extension on its upper surface, which functions as a sail. This sail has a low aspect ratio (i.e. it is long and low) so its lift/drag ratio is low, at any angle of attack (Francis 1991). Consequently, it cannot sail at small angles into the wind, like yachts with high aspect ratio sails. Also unlike a yacht, it has no means of steering. It sails downwind at a small angle to the direction of the wind.

Though all the animals discussed in this chapter travel over the surface of water, their modes of propulsion and the phenomena that prevent them from sinking are very different. Fisher spiders are supported by surface tension, basilisk lizards by a combination of effects that depend on the foot being driven down into the water, and ducks and *Velella* by buoyancy. *Velella* sails like a yacht, and other surface swimmers are propelled by limb movements. Capillary waves were important in our discussion of spiders, and gravity waves in our discussion of ducks. I suspect that further research on surface swimmers might be profitable, looking for ways in which they may possibly minimize wave drag.

Chapter Fourteen

• •

Swimming with Oars and Hydrofoils

*T*HIS CHAPTER is about swimming by means of limbs or fins that remain more or less flat as they move through the water. It is not concerned with animals that swim by undulation either of a fin or of the whole body; they are dealt with in the next chapter.

A swimming animal can use its appendages to propel it in two different ways, making use either of drag or of lift. For example, ducks swim by spreading their webbed feet and moving them backward through the water. Drag on the backward-moving feet acts forward, providing the thrust that drives the bird forward. Penguins, however, swim by beating their wings up and down more or less as if they were flying, and are propelled by lift on the wings. Similarly, paddle steamers are driven by drag on the blades of the paddle, whereas more modern ships are driven by lift on the blades of a propeller.

14.1. FROUDE EFFICIENCY

Flying animals have to exert large vertical forces to support their weight, and generally much smaller horizontal forces to propel themselves. Most swimming animals, however, are close in density to the water they swim in, so any vertical forces they may have to exert are small. Their principal requirement is for thrust, not weight support. Whether the thrust is derived from lift or from drag, it is obtained by driving water backward. From this follows a principle that is important for all swimming animals. The thrust equals the rate at which momentum is added to the wake. Consider an animal swimming with velocity v. Let dm_{wake}/dt be the mass of water that it drives backward in unit time, accelerating this water from rest to velocity $-v_{\text{wake}}$. Then the thrust is $v_{\text{wake}}\,dm_{\text{wake}}/dt$. Notice that the same thrust can be obtained by accelerating a lot of water to a low velocity, or a little to a high velocity. The power that the animal has to exert to overcome the drag on its body at velocity v is the thrust multiplied by the velocity of the body, $vv_{\text{wake}}\,dm_{\text{wake}}/dt$; we can think of this as the useful power. If the animal were propelling itself by pushing on an immovable object, this is all the power it would need. However, while the body is being driven forward, water is driven backward. Additional power (induced power) is needed to give kinetic energy to this water. The rate at which kinetic energy is added to the water in the wake is $\frac{1}{2} v_{\text{wake}}^2 \, dm_{\text{wake}} / dt$. Thus the total of the power requirements we have considered so far is $(v + \frac{1}{2} v_{\text{wake}}) v_{\text{wake}} \, dm_{\text{wake}} / dt$. We can define an efficiency, known as the Froude efficiency, as the power expended against drag on the body divided by the sum of this power and the induced power:

$$\text{Froude efficiency} = \frac{v}{v + \frac{1}{2} v_{\text{wake}}} \tag{14.1}$$

This shows that for efficient propulsion, v_{wake} should be made as low as possible. It is better to accelerate a lot of water to a low velocity than a little to a high velocity.

The Froude efficiency is also called the theoretical efficiency, because there may be further power requirements of which it fails to take account. For example, in the case of an animal that propels itself by means of hydrofoils, work has to be done against profile drag on the hydrofoil. (The distinction between profile drag and induced drag was explained in Section 10.3.)

14.2. DRAG-POWERED SWIMMING

Water beetles and various aquatic bugs swim by means of oarlike legs. Some of them use two pairs of legs for swimming, and some just one pair (Nachtigall 1965). Figure 14.1 shows forces acting on a beetle that is rowing at speed v, moving the blades of its oarlike legs backward and forward at speed v_{oar} relative to its body. (It would, of course, be possible for the speeds of the oars in the power stroke and the recovery stroke to be different, but we will assume for the sake of simplicity that both strokes are made at the same speed.) In the power stroke (Fig. 14.1A) the oars move backward relative to the water at speed $v_{\text{oar}} - v$. Because they are moving backward, the drag on them acts forward. I have given these drag forces the symbol F_{thrust} because these are the forces that propel the beetle. In the recovery stroke (Fig. 14.1B) the oars move forward relative to the water at speed $v_{\text{oar}} + v$, so the drag on the oars now acts backward, resisting the beetle's motion. At this stage the oars are feathered so as to make this drag, F_{feather}, as small as possible. Throughout the cycle of movement, drag F_{hull} acts on the body. The net force on the body is $F_{\text{hull}} - 2F_{\text{thrust}}$ in the power stroke and $F_{\text{hull}} + 2F_{\text{feather}}$ in the recovery stroke. If the beetle is swimming at a steady speed, the forces acting over a complete cycle must balance, whence

$$F_{\text{hull}} = F_{\text{thrust}} - F_{\text{feather}} \tag{14.2}$$

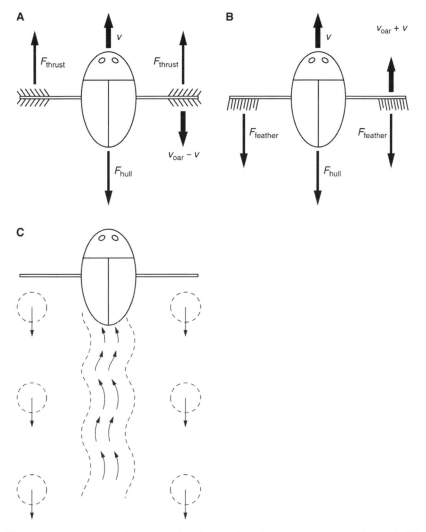

Fig. 14.1. Diagrams of a swimming beetle (A) during a power stroke and (B) during a recovery stroke. (C) Water movements in the beetle's wake.

The power required to move the oars relative to the body is $2v_{oar}F_{thrust}$ in the power stroke and $2v_{oar}F_{feather}$ in the recovery stroke, so the mean power output is $v_{oar}(F_{thrust} + F_{feather})$. The power required to move the body through the water is $vF_{hull} = v(F_{thrust} - F_{feather})$. Thus, the efficiency of rowing is

$$\text{Efficiency} = \left(\frac{v}{v_{oar}}\right)\left(\frac{F_{thrust} - F_{feather}}{F_{thrust} + F_{feather}}\right) \tag{14.3}$$

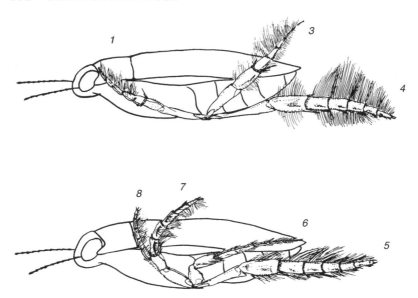

Fig. 14.2. Ventral and lateral views of the water beetle *Acilius* showing successive positions of the hind legs in swimming. The other legs are not shown. Redrawn from Nachtigall (1960).

The speed of the oars, v_{oar}, must be greater than v, because the oars must move backward relative to the water in the power stroke. Equation 14.3 tells us that for efficient swimming v_{oar} should be as little greater than v as possible, which implies that large oars are more efficient than small ones. This is the principle that we have already met in our discussion of Froude efficiency; it is more efficient to propel yourself by pushing large masses of water slowly than by pushing small masses fast. The equation also makes the obvious point that for efficient swimming the drag on the oars in the recovery stroke should be kept as low as possible.

The oars give backward momentum to the water near them, but the body drags water along with it, giving forward momentum to the water in its wake. The principle of conservation of momentum tells us that the beetle's movements cannot alter the total momentum of it and the water. If it is swimming at constant speed, the backward momentum given to the water by the oars equals the forward momentum in the wake of the body.

Nachtigall (1960) studied the water beetle *Acilius*, which is about 17 mm long and swims under water at up to about 0.5 m/s. It rows itself along with its second and third pairs of legs, which are fringed with long hairs (Fig. 14.2). You might suppose that a fringe of hairs would be ineffective as an oar blade, because water can pass through the gaps between the hairs. Cheer and Koehl (1987) have shown that very little water may pass between the hairs of fringed appendages moved through water if the hairs are closely spaced and if the Reynolds number is low enough. (In this context, the linear dimension used to calculate Reynolds number is hair diameter.) The Reynolds number for the hairs on *Acilius* swimming legs is higher than for the appendages of the smaller animals that Cheer and Koehl considered. Nevertheless, the fringes of hairs provide 54% as much thrust as a solid paddle of the same area would do (Nachtigall 1960).

In the power stroke (Fig. 14.2, positions 1–4) the hairs are fully spread in a vertical plane, making the effective area of the blade as large as possible. In the recovery stroke (positions 5–8) the hairs fold down so as to trail more or less horizontally behind the leg. In addition, the hair-bearing segments of the leg, which are oval in cross section, rotate about their long axes so as to move edge-on through the water in the recovery stroke, further reducing drag. These changes in hair and leg position are made automatically, due to the asymmetrical arrangement of hairs on the legs.

Nachtigall (1960) fixed *Acilius* legs in a water tunnel and measured the forces on them in the positions both of the power stroke and of the recovery stroke. He also analyzed film of *Acilius* swimming. His results show that the term v/v_{oar} in Equation 14.3 was about 0.7 and the term $(F_{thrust} - F_{feather})/(F_{thrust} + F_{feather})$ was about 0.8 (see also Alexander [1983]). Multiplied together, these give an efficiency of about 0.56. However, Nachtigall (1960) pointed out that there are further energy losses due to water being pushed sideways as the legs swing in their arcs. When account was taken of this, the overall efficiency with which work done by the legs was used to propel the body fell to about 0.45. I have not considered inertial work in this discussion because, in contrast to the situation for hovering flight, inertial work in rowing can generally be assumed to be small compared to hydrodynamic work. This implies that kinetic energy given to a limb and its added mass of water at the beginning of a stroke can generally be used to do work against drag at the end (see Section 3.6).

The energy cost of swimming depends not only on the efficiency, but also on the drag coefficient of the body. A poorly streamlined body will need a lot of power to propel it, even if the efficiency of rowing is high. Nachtigall (1960) measured drag on *Acilius* bodies in a water tunnel and found that the drag coefficient based on frontal area was 0.23. At the same Reynolds number, the drag coefficients of a well-streamlined body and of a sphere would be about 0.1 and 0.5, respectively, so *Acilius* is only moderately well streamlined.

Power strokes, in which the beetle accelerates, alternate with recovery strokes in which it decelerates. Consequently, v and F_{hull} fluctuate. To discover the consequences of this, we will calculate the work done against drag on a body that accelerates from an initial speed $v - \Delta v$ to a final speed $v + \Delta v$,

Posterior

Anterior

Locomotion direction

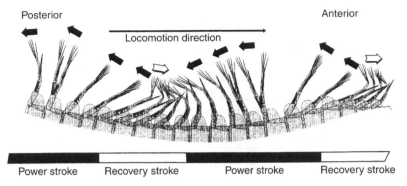

| Power stroke | Recovery stroke | Power stroke | Recovery stroke |

Fig. 14.3. Part of the trunk of the crustacean *Speleonectes*, showing the limb movements that power swimming. From Kohlhage and Yager (1994).

with constant acceleration a. Its speed at time t is $(v - \Delta v) + at$]. Drag is approximately proportional to the square of speed, and the power required to overcome drag is the drag multiplied by the speed, and so is proportional to speed cubed. Thus, the power required at time t is $k[(v - \Delta v) + at]^3$, where k is a constant. The work done against drag can be calculated by integrating the power with respect to time, from time 0 to time $2\Delta v/a$, when the final velocity is attained. This work is $(2kv^3\Delta v/a)[1 + (\Delta v/v)^2]$. The power is this work divided by the time $2\Delta v/a$ in which it is done:

$$\text{Power} = kv^3 \left[1 + \left(\frac{\Delta v}{v}\right)^2 \right] \tag{14.4}$$

Thus, the fluctuations of velocity increase the work that has to be done against drag by a factor $1 + (\Delta v/v)^2$. Note that the increase is relatively small, unless the amplitude of the velocity fluctuations is a large fraction of the mean velocity. In this analysis, I have ignored the work done giving kinetic energy to the body while it is accelerating, because this energy is used to do work against drag as the body decelerates.

The velocity fluctuations Δv could be reduced by making shorter oar strokes with higher frequency. This would tend to reduce the power required for swimming at the same mean speed. However, if the strokes were made too short, the inertial work done accelerating the oars would become too large, in comparison to the hydrodynamic work, to be recovered. Alternatively, $\Delta v/v$ could be reduced by having two sets of oars that worked out of phase with each other, so that each made its power stroke during the recovery stroke of the other. However, it is difficult to see how the second and third pairs of legs of a water beetle could work more than slightly out of phase without colliding; and if strokes of the left legs alternated with strokes of the right ones the beetle would yaw from side to side and drag would be increased.

Many small crustaceans swim by means of limbs fringed with bristles. Some of them have many pairs of swimming limbs, up to 32 in the case of *Speleonectes*. This animal is about 20 mm long and lives in caves in the Bahamas, where it swims with its ventral side uppermost. It usually beats its limbs in a metachronal rhythm, each pair of legs moving slightly after the next more posterior pair. Thus, waves of limb movement travel forward along the body (Fig. 14.3). This makes the animal move at about 7 mm/s. About half the limbs are making their power stroke at any instant, so it might be expected that the speed of swimming would be almost constant. However, the speed fluctuates through a range of a few millimeters per second; the peaks coincide with the power stroke of the longest limbs, in the middle of the trunk (Kohlhage and Yager 1994).

Some fish row with their pectoral fins and some others use their pectoral fins as hydrofoils, as we shall see in the next section. Blake (1979b, 1980a) analyzed film of an angelfish (*Pterophyllum*) rowing, calculating both the drag forces on the fins and the inertial forces needed to accelerate and decelerate the masses of water that moved with them. He found that in this case the inertial forces were greater than the drag, so that the fin muscles had to do work to accelerate the added mass of water and negative work to decelerate it. The conclusion from his calculations was that the overall efficiency, with which work done by the fins was used to propel the body, was only 0.16.

Frogs swim by a kicking action of the hind legs, similar to the human breaststroke. Thrust is provided principally by drag on the large hind feet, but additional thrust is obtained at the end of the stroke when the feet meet sole to sole, squeezing out a rearward jet of water (Gal and Blake 1988a, b).

Ducks row with their webbed feet, making power strokes with the left and right foot alternately. This causes little or no yawing, because the feet keep close to the median plane. Semiaquatic mammals such as mink (*Mustela*) and muskrats (*Ondatra*) also perform drag-based swimming by alternate movements of their hind legs (Fish 1996).

14.3. SWIMMING POWERED BY LIFT ON LIMBS OR PAIRED FINS

In the power stroke of drag-powered swimming, the limb or fin that serves as an oar moves backward through the water, with its surface more or less at right angles to its path (Fig. 14.4A). In contrast, in the power stroke of lift-powered swimming the limb or fin moves vertically or along a sloping path, with a relatively small angle of attack (Fig. 14.4B, C, D). Grebes (*Podiceps*) swim by means of their lobed feet. Johannson and Norberg

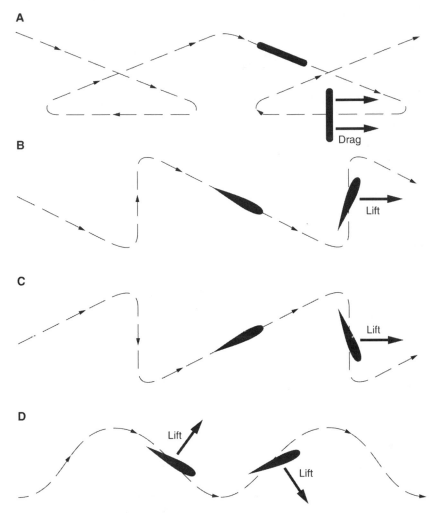

Fig. 14.4. Diagrams of (A) drag-powered swimming and (B–D) alternative techniques of lift-powered swimming. The animals are swimming toward the right of the page. Each diagram shows the path of the oar or hydrofoil relative to the water. The oar or hydrofoil is shown in section at two positions in each diagram, with the force that provides thrust indicated by arrows.

(2001) described how the feet move upward through the water in the power stroke, generating lift that serves to propel the bird, as shown in Fig. 14.4B. The bird wrasse (*Gomphosus*) swims by beating its pectoral fins more or less as shown in Fig. 14.4B (Walker and Westneat 1997). The body decelerates during the downstroke and accelerates during the upstroke, showing that at least most of the thrust is provided by the upstroke.

In the human crawl stroke, the hand moves downward during the first half of the power stroke, apparently providing lift to propel the swimmer as in Fig. 14.4C (D. I. Miller 1975). Sea lions (*Zalophus*) swim by beating their pectoral flippers, more or less as shown in Fig. 14.4C (English 1976). All of these examples could easily be mistaken for drag-powered swimming, because the foot, hand, or flipper moves posteriorly *relative to the body* in the power stroke, with its large surfaces more or less vertical. The observation that shows that the animal cannot be being propelled by drag is that the limb is not moving backward *relative to the water*.

Though *Gomphosus* obtains thrust principally from the upstroke, this is not the case for all fish that swim by using their pectoral fins as hydrofoils. Drucker and Lauder (1999) used particle image velocimetry (Section 5.4) to reveal the water movements driven by the pectoral fins of *Lepomis*. A 20-cm fish swimming slowly, at 10 cm/s, produced vortex rings only during the downstroke, showing that only the downstroke was generating thrust. At higher speeds both the downstroke and the upstroke produced vortex rings. The downstroke ring and the subsequent upstroke ring remained linked, like the vortex rings produced by flying moths (Fig. 12.1). The downstroke ring was angled slightly downward, providing weight support for the fish, which is a little denser than water. Another fish of very similar size and shape, *Embiotoca*, produces pairs of linked vortex rings even when swimming slowly (Drucker and Lauder 2000). Both species supplement the thrust from their pectoral fins by tail movements when they swim fast (Section 15.1). Drucker and Lauder (2001) used particle image velocimetry to investigate turning by *Lepomis*.

The hydrodynamic force on an oar or hydrofoil can be regarded as acting at a single point, the center of pressure. In the power stroke of rowing, the blade of the oar must be moved posteriorly, relative to the body, at a speed greater than the swimming speed. In the power stroke of lift-powered swimming by the techniques illustrated in Fig. 14.3B and C the hydrofoil moves vertically relative to the water, which implies that it must be moved posteriorly relative to the body at a speed equal to the swimming speed. These requirements may limit the speed of swimming (Alexander 1989b). Consider an animal that beats a limb backward and forward with frequency f, each forward or backward stroke taking time $1/2f$. If the center of pressure is at a distance r from the base of the limb, the distance it moves relative to the body in its backward stroke cannot exceed $2r$, so the speed of the backward movement (and so the speed of swimming) cannot exceed $4rf$. We will work out the consequence of this for penguins, if they swam as illustrated in Fig. 14.3B or C. A 4-kg Humboldt penguin (*Spheniscus*) had wings 0.16 m long, so the center of pressure was probably about 0.10 m lateral to the shoulder (Hui 1988). This bird beat its wings at 2.75 Hz when swimming at 1.7 m/s. It would probably have beaten them at higher frequencies when swimming faster, but seems unlikely to have been capable of frequencies above 4 Hz (Clark and Bemis 1979). At 4 Hz, $4rf$ would have been 1.6 m/s. If it swam as shown in Fig. 14.3B or C, it could not swim faster than this.

Penguins do not swim by these techniques, but more nearly as shown in Fig. 14.3D, and can swim much faster. Humboldt penguins have been filmed swimming at 4.5 m/s in a zoo aquarium (Hui 1987). Hui (1988) filmed Humboldt penguins swimming submerged in a long tank with a window in the side. Stripes painted on the wings enabled him to measure their angles of attack as well as to plot their paths through the water. The path of the distal part of the wing sloped at around $-50°$ and $+50°$ in the downstroke and upstroke, respectively. The angle of attack generally lay between 0 and $+10°$ in mid downstroke, and between -10 and $-25°$ in mid upstroke. Lift must act forward and up in the downstroke, forward and down in the upstroke. The upward and downward components cancel out over a cycle, so the mean force on the wings is forward thrust. Hui also made resin casts of penguin wings and measured the lift and drag on them in a wind tunnel at an appropriate Reynolds number. He obtained lift coefficients for the distal part of the wing ranging from a surprisingly low value of $+0.3$ at angles of attack around $+20°$, to -0.8 at $-20°$. Taking these measurements together with the angles measured from the films, he calculated that peak thrust was 4 N in the downstroke and 7 N in the upstroke for a 4-kg penguin swimming at 1 m/s. The mean calculated thrust force was approximately equal to the drag on the body, which Hui measured by towing a formaldehyde-fixed carcass (or a resin cast of the carcass) along a tank of water. These measurements gave drag coefficients based on wetted area of 0.011–0.015 for the cast, and slightly more for the preserved carcass. This is much less than for unstreamlined bodies such as spheres, but a well-streamlined body with the same ratio of thickness to length, at the same Reynolds number, would have had a drag coefficient based on wetted area of only about 0.003. Nachtigall and Bilo (1980) calculated drag coefficients for live penguins (*Eudyptula*) from measurements of their decelerations while they glided with their wings motionless relative to the body. They obtained a much lower wetted-area drag coefficient than Hui did, 0.0044. The measurements are not strictly comparable because Hui had removed the wings from his specimens, but this does not explain the discrepancy.

Hui (1988) considered the possibility that unsteady effects might be important in penguin swimming. The distance traveled in each cycle was only about 12 wing chords, so unsteady effects cannot be confidently discounted (Section 11.2). However, his attempt to apply an unsteady theory gave a much less good match of thrust to drag than the calculation (described above) that ignores unsteady effects.

We have already seen that for flying birds there is an optimum wing area that minimizes the gliding angle for any given speed (Equation 10.23). A similar argument could be formulated for birds that use their wings only for swimming. It would give a much smaller optimum area, mainly because density (which appears in Equation 10.23) is 800 times larger for water than for air at sea level. Appropriately, penguins have much smaller wings than flying birds of equal mass; for example, about 1/40 of the area of the wings of an eagle (Hui 1988). Guillemots (*Uria aalge*) use their wings both for flight and for swimming, so their wing area has to be a compromise. The wing loading of a 0.95-kg guillemot was 171 N/m^2, whereas a more typical bird of the same mass, a herring gull (*Larus argentatus*), had a wing loading of only 46 N/m^2 (Pennycuick 1997). Accordingly, guillemots fly fast, at about 19 m/s, whereas herring gulls fly at only about 10 m/s. Guillemots beat their wings at about 8.7 Hz when flying, but only 1.9–2.8 Hz when swimming (Lovvorn et al., 1999), so their muscles cannot be shortening at optimum speed in both modes of locomotion. J. M. V. Rayner tells me that flying herring gulls beat their wings at about 2.5 Hz.

Marine turtles swim by beating their fore flippers, apparently producing thrust as shown in Fig. 14.3D. At least some of them can swim quite fast. Davenport et al. (1984) recorded a young *Chelonia mydas*, with a carapace only 11 cm long, swimming at speeds up to 1.4 m/s.

There are some indications that swimming with limbs used as hydrofoils tends to be more economical of energy than swimming with limbs used as oars. Baudinette and Gill (1985) measured the oxygen consumption of penguins and a duck of closely similar mass swimming at the surface in a flume. Their resting metabolic rates were similar, but when they swam at speeds ranging from 0.3 to 0.7 m/s, the duck used oxygen about twice as fast as the penguins did at the same speed. Williams (1999) gives metabolic rates for a 20-kg sea otter and sea lions of similar mass swimming submerged. The comparison is less satisfactory than the one between penguins and ducks, because the mammals had different resting metabolic rates and swam at different speeds, but it did appear that the sea lions were the more economical swimmers.

One reason why lift-powered swimming may in some cases be more economical of energy than drag-powered swimming is that thrust may be obtained by pushing on larger volumes of water, giving higher Froude efficiencies (Section 14.1). A photograph of a duckling published by Aigledinger and Fish (1995) shows a vortex ring produced by a foot stroke, made visible by trapped bubbles of air. Its small diameter (similar to the width of the foot) confirms that the foot can push on only a small volume of water. The vortex rings produced by the fin movements of *Lepomis* and *Embiotoca* are also rather small, with diameters of 3–6 cm for 20-cm fish (Drucker and Lauder 2000). The rings produced by the left fin remain separate from those produced by the right one. The wings of penguins and guillemots, however, are much longer in proportion to the dimensions of the body than are the fins of the fish, and produce relatively larger vortex rings. The vortices are shown (again made visible by air bubbles) in a photograph of a pigeon guillemot (*Cepphus* [Rayner 1995]). This picture shows that each wing stroke produces a single vortex ring, with a diameter only a little less than the wing span.

Walker and Westneat (2000) wondered whether rowing might offer some advantage that would make it preferable to hydrofoil propulsion for some animals. They made computer simulations, using a model that resembled a fish propelled by a pair of pectoral fins. They compared two versions of the model, identical in body size (15 cm long) and fin dimensions (3 cm long). One beat its fins forward and back in a rowing motion, feathering them for the return stroke. The other beat its fins up and down with the same amplitude, adjusting their angle of attack so as to generate forward lift in both strokes. Walker and Westneat calculated the forces on the fins, taking account of unsteady effects and of the added mass of water that moved with them. In these calculations, they made use of the results of the experiments of Dickinson et al. (1999) on a model fly (Section 11.2). The model fly operated at lower Reynolds numbers than the simulated fish, but no comparable data for more appropriate Reynolds numbers were available.

In one set of simulations, Walker and Westneat (2000) set the model moving at a chosen speed, then adjusted the frequency of the fin beat to make the thrust generated by the fins match the drag on the body. They calculated the work done on the fins and hence the efficiency of propulsion. These simulations showed, as expected, that hydrofoil propulsion was more efficient than rowing at all speeds.

In another set of simulations, they made the fins beat with a fixed frequency of 10 Hz and calculated the thrust that they would generate with the body moving at various speeds. In these simulations, the thrust (averaged over a complete cycle) was generally greater or less than the drag, implying that the fish was accelerating or decelerating. At every speed except the highest (0.9 m/s), rowing gave more thrust than hydrofoil propulsion. These results indicate that if efficiency is the most important criterion, hydrofoil propulsion is preferable to rowing; but that if the ability to accelerate is more important, rowing is better. The high thrusts that oars can give will also be valuable for turning and braking. Hydrofoils are more efficient, but oars seem to be better for maneuverability.

14.4. SWIMMING WITH HYDROFOIL TAILS

Unlike penguins and sea turtles, which swim by beating paired hydrofoils, whales swim by beating a single hydrofoil on the tail. However, the principle of swimming is the same; Fig. 14.4D is as good a diagram of whale swimming as of penguin swimming. The swimming movements of whales have been described by (among others) Videler and Kamermans (1985) and Fish (1993, 1998). Tunas have vertical tail fins that they beat from side to side, unlike the horizontal flukes of whales that are beaten up and down, but their swimming action is otherwise similar (Fierstine and Walters 1968).

There is no sharp distinction between the swimming mechanism of whales and tunas and that of the more typical fishes described in Chapter 15 as swimming by undulation of the body. The significant difference is that whales and tunas have a narrow caudal peduncle between the main part of the body and the hydrofoil tail. To a first approximation, they can be thought of as rigid hulls propelled by a separate oscillating hydrofoil at the rear. When we consider more typical fishes we will have to think of body and tail as a single undulating unit.

The distance traveled by a whale in each cycle of tail movements is 12 or less times the chord length of the fluke, and as few as 4.5 chord lengths in the case of a killer whale (*Orcinus*) swimming slowly (Fish 1998). It must therefore be assumed that unsteady effects are important. For unsteady analyses of whale swimming, the fluke is generally assumed to move sinusoidally with amplitude h, frequency f, and forward speed v. The angle at which the plane of the fluke is tilted relative to the horizontal is also assumed to fluctuate sinusoidally. The fluke has plan area A and its chord, where it joins the caudal peduncle, is c. The analysis uses two dimensionless numbers, the reduced frequency and the feathering parameter. The reduced frequency is $2\pi fc/v$, that is, 2π divided by the number of chords traveled in each tail beat cycle. The feathering parameter is the ratio of the maximum angle of attack of the fluke to the maximum angle of the path of the fluke to the horizontal: it is zero if the fluke tilts so as to have zero angle of attack throughout the tail beat cycle, and one if it remains horizontal throughout the cycle. Fish (1998) calculated feathering parameters ranging from 0.4 to 0.7 for four species of toothed whales swimming at various speeds. For any combination of reduced frequency and feathering parameter, a thrust coefficient C_{thrust} can be calculated (Yates 1983):

$$\text{Thrust} = \frac{1}{2}\rho A v^2 C_{thrust}\left(\frac{h}{c}\right)^2 \qquad (14.5)$$

where ρ is the density of water. Fish (1993) calculated thrust in this way for several species of toothed whale, and went on to calculate drag coefficients. He obtained coefficients based on wetted area ranging from about 0.03 at a Reynolds number of 2×10^6 to about 0.003 at a Reynolds number of 3×10^7. At these Reynolds numbers, the drag coefficient of a thin, flat plate aligned with the direction of flow would be 0.004 and 0.002, respectively (assuming a turbulent boundary layer; see Alexander [1983]). Drag coefficients based on wetted area for well-streamlined bodies are only 20–30% higher than these, but drag on a swimming whale is higher, probably mainly because of the "recoil" movements of the body that inevitably occur when the animal beats its tail. Lang (1975) measured the deceleration of a dolphin (*Stenella*) gliding with its body straight between bursts of swimming. Its drag coefficient *based on (volume)$^{2/3}$* was 0.027, approximately the value expected for a streamlined body with a turbulent boundary layer.

Reduced frequency and feathering parameter can also be used to calculate the efficiency with which work done by the animal is used to overcome drag on the body. Fish (1998) calculated efficiencies between 0.75 and 0.9 for the toothed whales that he studied.

The swimming muscles of dolphins insert on the vertebrae of the tail through a very large number of long, slender tendons. Bennett et al. (1987) measured the elastic properties of the tendons and considered whether they might serve as energy-saving springs. The tail flukes of the dolphins that we studied were only around 1% of body mass, but the added mass of water calculated to move with them increased the total mass that had to be oscillated to 7% of body mass. Bennett et al. (1987) calculated that the inertial work was 0.3 of the hydrodynamic work for a dolphin (*Lagenorhynchus*) swimming at 5 m/s. Taking account of this, we estimated the effect of tendon elasticity on the work required of the muscles. We argued that the work required of the muscles is increased if negative work done at one stage of the tail beat cycle has to be balanced by additional positive work at another. We assumed that the metabolic energy cost of swimming would be least if the need to do negative work were eliminated, and concluded that the tendons were excessively compliant; the muscles would have to do more positive and negative work than if the tendons had been thicker. However, we failed to take account of unsteady effects in our estimate of hydrodynamic work. Blickhan and Cheng (1994) reanalyzed the data taking account of unsteady effects and concluded that the tendon compliance was close to the optimal value that would eliminate the need for muscles to do negative work.

Neither of these investigations took account of the relationship between the efficiency of a muscle and the rate at which it is shortening (Fig. 2.3). Alexander (1997b) did, in the theory illustrated by Fig. 3.5. This theory shows that to minimize metabolic energy costs, the tendons should be less compliant than would be required to eliminate negative work. When interpreted in the light of this theory, even Blickhan and Cheng's (1994) calculations indicate that the tendons are more compliant than would be optimal.

Tunas also have tendons running through the caudal peduncle, connecting the swimming muscles to the skeleton of the tail. Their possible role as energy-saving springs seems not to have been investigated. Knower et al. (1999) used a buckle transducer (Section 5.6) to record the forces transmitted by the tendons to the tails of yellowfin (*Thunnus*) and skipjack tuna (*Katsuwonus*) swimming at sustainable speeds in a large water tunnel. They also recorded electromyograms from the red (aerobic) muscle, which powers swimming at these speeds. In other experiments, the same group used sonomicrometry (Section 5.6) to record length changes of red muscle fibers in skipjack (Shadwick et al., 1999). Their results show that the period when the muscles of the left side of the body are shortening coincides precisely with the period when the tail is moving toward the left, and similarly for the right side. This suggests that the tendons are not stretching much, for the following reason. Peak inertial forces act when the tail is at the extremes of its motion to left and right. For this reason, force is first required in the muscles of the left side before the tail has finished moving to the right, and ceases to be required before the tail has finished moving to the left. The tendons must stretch in phase with the fluctuations of force, so if they were stretching much the length changes of the muscle fibers would be out of phase with the movements of the tail. Unfortunately, the published force records do not show the precise phase relationship between the forces and the tail movements: the force records were made to establish the relationship between electromyographic activity and force.

Dolphins and tunas are remarkably fast swimmers. The highest reliably recorded speed for a dolphin (or, indeed, for any swimming animal) seems to be 11 m/s, briefly attained by a trained *Stenella attenuata* in a lagoon in Hawaii. The dolphin, 1.9 m long, was chasing a lure towed by a variable-speed winch. Its sprinting speed in water is about the same as the peak speed attained on land by elite human athletes in the fastest part of a 100-m race. Dolphins are often seen swimming in the bow waves of ships that are traveling at speeds around 10 m/s, but this is not unaided swimming; an animal that positions itself in the front slope of the bow wave may be carried along passively at the speed of the ship (Hertel 1966).

Speeds up to 21 m/s have been claimed for *Thunnus* and *Acanthocybium*, measured by recording the speed at which hooked fish drew out a line (Walters and Fierstine 1964). As the line was drawn out, instruments detected magnetic markers spaced at regular intervals along it and recorded their passage as blips on a chart recorder. The published example of a record shows a great deal of electrical noise, and it seems possible that some of the blips that were counted as signals were merely noise. The highest speeds that have been recorded for tunas by other means are around 10 m/s (Magnuson 1978).

Tunas presumably owe their speed in part to the unusual arrangement of blood vessels that keeps their swimming muscles warmer than the water (see Section 15.2).

14.5. PORPOISING

Dolphins and penguins often leap repeatedly out of the water as they swim, in a behavior known as porpoising (Au and Weihs 1980; Hui 1987). This gives the impression of being a strenuous mode of locomotion, but Au and Weihs suggested that it may actually save energy if the animal is traveling fast. The essence of their argument is that drag increases with speed, so the work required to swim unit distance is greater at higher speeds. However, the work needed to jump unit distance is independent of speed; Equation 8.4 shows that the length of a jump is proportional to the square of speed, and so to the animal's initial kinetic energy. Therefore, there must be a critical speed, above which it becomes more economical to porpoise than to remain perpetually underwater.

Attempts to calculate the critical speed lead to much uncertainty (Au and Weihs 1980; Gordon 1980). The energy cost of swimming close under the surface is much higher than the cost of swimming deeper (Section 13.3), so conclusions are affected by the depth at which the animal is assumed to swim. It is also not easy to calculate the cost of a leap. The kinetic energy of the animal is the same when it reenters the water as it was at the start of the leap (ignoring the trivial effect of air resistance). However, the added mass of water that moves with the animal when it is swimming falls away when it leaps, so the kinetic energy of this added mass is lost. Energy is needed to accelerate a new added mass when the animal reenters the water. Further, a porpoising animal is traveling downward as it enters the water and upward as it leaves, so while it is in the water it needs an upward acceleration. To give itself this acceleration, it must drive water downward. This implies an energy cost. For example, if it uses lift on its flippers to give itself the upward acceleration it will have to do work against induced drag.

Au and Weihs (1980) estimated that the critical speed above which porpoising would save energy was 5 m/s for a 50-kg dolphin and 11 m/s (probably faster than the animal could swim) for a 5-tonne whale. These speeds would have been a little different if Au and Weihs had considered all the points made above. However, it does not seem very useful to revise the calculations, because of the uncertainties involved and because of the paucity of data about the speeds at which dolphins do or do not porpoise.

Hui (1987) observed that Humboldt penguins (*Spheniscus*) in the San Diego Zoo generally porpoise only at speeds of 3 m/s or more. He found that even when porpoising, they were out of the water for no more than 22% of the time, and argued that any resulting energy savings must be marginal. He suggested that the principal significance of porpoising might be that it enabled penguins to breathe without spending much time close below the surface, where drag is high. This suggestion is supported by the observations of Yoda et al. (1999), who attached data loggers to wild Adélie penguins (*Pygoscelis*) to record acceleration, depth, and, in some cases, speed. The penguins were caught at their nests so that the data loggers could be fitted, and again when the birds returned from foraging trips so that the loggers could be retrieved. The records showed zero depth and downward acceleration whenever the bird was out of the water. They showed that the birds porpoised only occasionally, and that even when porpoising they were out of the water only for a small fraction of the time.

To swim forward, animals must drive water backward. This chapter started by explaining Froude efficiency, which is higher for animals that swim by accelerating large masses of water to low velocities than for those that accelerate small masses to high velocities. Swimmers that use lift on hydrofoils to propel themselves generally have higher Froude efficiencies than others that depend on drag on oars. Topics that I would like to see investigated further include the possible role of tendons as energy-saving springs in whales and tunas, and the energetics of porpoising.

• •

Chapter Fifteen

Swimming by Undulation

FIGURE 15.1 shows twelve stills from a film of a dogfish swimming. It is passing waves of bending backward along its body; dots on frames 3 to 8 mark successive positions of the crest of one of the waves. This action drives the fish forward. Many fishes and some snakes and worms swim in this way, by undulating their bodies. In addition, many fish and cephalopods swim by passing waves of bending along fins. This chapter is about swimming by undulation, both of the body as a whole and of fins separately.

15.1. UNDULATING FISHES

In Section 14.4 we considered the tails of tunas and whales in isolation from the body. We thought of the body as a passive structure, pushed along by the beating of the hydrofoil tail. This seemed to be a reasonable simplification, because these animals bend the main bulk of their bodies only a little as they swim. Their tails are connected to the body by a slender caudal peduncle that is streamlined in cross section, so that the hydrodynamic forces that act on it as it moves from side to side must be small. If we had examined the swimming movements of tunas and whales more closely, however, we would have found waves on their bodies (see illustrations in Webb [1975]). There is no sharp distinction between the movements of these animals and those of fishes such as the dogfish.

There are substantial differences between the patterns of undulation that different fishes use for swimming (Fig. 15.2). In the eel the wavelength of the waves is about 0.6 of the length of the body, so at any instant the body forms 1.7 waves. In saithe the wavelength is about equal to the length of the body, and in scup it is 1.5 times the length of the body. The wavelike nature of the movements is less obvious in scup than in the others, because the body never forms a complete cycle of the waveform. In all three cases, the amplitude of the waves increases toward the tail. The stride length is the distance traveled in one cycle of undulation. For the waves to travel backward relative to the water, as required for propulsion, the stride length must be less than the wavelength. It seems usually to be at least half the wavelength.

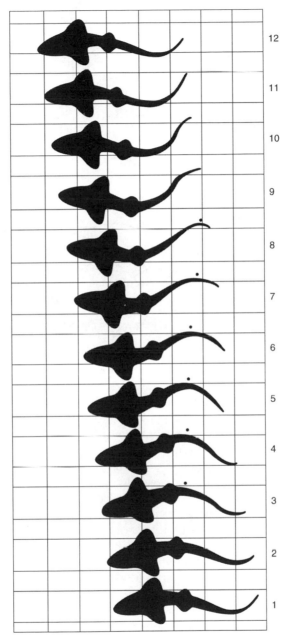

Fig. 15.1. Tracings at intervals of 0.1 s from a film of a dogfish (*Squalus*) swimming. The grid squares on the background have 3-inch (76-mm) sides. Sir James Gray, from whose work this picture is taken, was the most influential pioneer of research in animal locomotion. From Gray (1933).

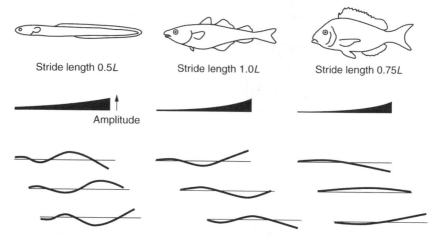

Fig. 15.2. Swimming movements of (left) an eel (*Anguilla*); (center) a saithe (*Pollachius*); and (right) a scup (*Stenotomus*). Below the outlines of the fish, diagrams show how the amplitude of the transverse movements increases from snout to tail. At the bottom of the figure, sets of three diagrams show how the center line of the body moves as the tail beats from one side (in the lowest of the three diagrams) to the other (in the highest). Throughout the figure, the fishes' heads face to the left. From Wardle et al. (1995).

Except at the lowest speeds, fish generally keep the amplitude of the tail beat more or less constant and increase speed by increasing the frequency of the beat. Stride length (the distance traveled in one tail beat cycle) equals speed/frequency. It changes little with increasing speed, and is about the same fraction of body length in different-sized members of the same species (Videler 1993). This implies that Strouhal numbers (Section 4.2) are more or less independent of speed and size; Strouhal number can be expressed as length divided by stride length. Triantafyllou et al. (1993) have argued that the Strouhal numbers that fish and cetaceans use are close to optimal for the energetics of swimming.

We have seen how the hydrofoil tails of tunas and whales must reverse their angles of attack at the end of each half cycle of movement, so that the lift always has a forward component (Fig. 14.4D). This implies that the direction of the circulation around the hydrofoil (Fig. 10.2D) must be reversed. Each half stroke must produce a vortex ring, and these rings are presumably linked to form a chain as shown in Fig. 15.3A. I write "presumably" because, so far as I know, no one has visualized the vortices in the wakes of tunas or whales.

The wakes of fishes that swim by more obvious undulation have been studied by various methods, most successfully by particle image velocimetry. Wakes like the one shown in Fig. 15.3A have been observed for mullet

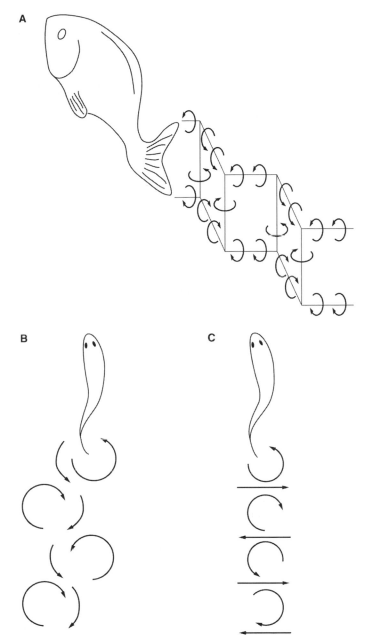

Fig. 15.3. (A) Linked vortex rings in the wake of a swimming fish. (B, C) Horizontal sections through swimming fishes and their wakes. The fish is accelerating in (B), and swimming at constant speed in (C).

(*Chelon* [Müller et al., 1997]) and *Danio* (Wolfgang et al., 1999). However, in these fish the undulations of the body as well as the side-to-side movements of the tail produce vortexlike movements in the water. To understand this effect, consider a point of inflection in the wave form, where a concave bend merges into a convex one. As the wave travels posteriorly, its amplitude increases while its wavelength remains unchanged. Therefore, the angle of the body to the direction of swimming at the point of inflection increases. This makes the water near the point of inflection rotate, forming an incipient vortex. The point of inflection moves posteriorly along the fish, carrying this body vortex with it, It reaches the rear edge of the tail fin as the tail reaches the end of a stroke and as the vortex ring formed by that stroke is being completed. The body vortex contributes to the vortex that the tail is then shedding.

That is what happens in *Chelon*, in *Danio*, and presumably also in other typically "fish-shaped" fishes. In eels (*Anguilla*), however, the point of inflection reaches the posterior end of the body not when the tail is at the end of a stroke, but halfway through a stroke. The body vortex is shed separately from the tail vortex, and the wake is more complicated (Videler et al., 1999).

The dorsal and anal fins of fishes such as *Danio* are just a little anterior to the caudal fin. The swimming movements of the body move them as well as the caudal fin from side to side; and they, like it, generate lift forces that help to propel the fish. If the tail were cut off immediately posterior to them, they would function like a caudal fin and propel the fish, leaving behind a wake of linked vortex rings. In the intact fish, the vorticity from these fins reaches the caudal fin before it has had time to roll up into vortex rings (Wolfgang et al., 1999). Lighthill (1970) has argued that if the gaps between the dorsal and anal fins and the caudal fin are small, the fins will function like a single continuous fin. However, if the gaps are large enough, the efficiency of swimming will be enhanced.

The law of conservation of momentum applies to the swimming of fish as well as of beetles. A water beetle leaves behind it forward-moving water associated with the drag on its body and backward-moving water associated with the thrust on its oars (Fig. 14.1C). Water movements associated with the thrust and drag on undulating fish, however, are not separate. In Fig. 15.3B, a jet of backward-moving water zigzags through the chain of linked vortex rings. The fish is giving backward momentum to the water, so by the law of conservation of momentum the fish itself must be gaining momentum; it must be accelerating. Fish often alternate bursts of swimming in which they accelerate with intervals of passive gliding in which they lose speed. They must produce wakes like Fig. 15.3B in the accelerating bursts. A fish swimming at constant velocity would presumably form a wake like Fig. 15.3C, in which water is driven alternately to the left

A.

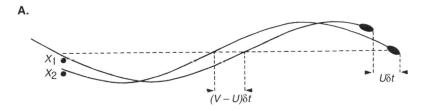

B.

Fig. 15.4. Diagrams of a fish swimming toward the right, which are explained in the text. From Alexander (1975b).

and to the right, but there is no backward jet. If the water is given equal momentum to either side, its net momentum gain is zero and the momentum of the fish will remain constant.

Lighthill (1960) devised a simple theory of swimming that calculates the Froude efficiency (see Section 14.1) by considering the sideways momentum left behind in the wake. Figure 15.4A represents a fish swimming to the right at two instants separated by a short time interval δt. The fish is swimming with velocity U, so it advances $U\delta t$ in the interval. Waves of bending are traveling backward along its body with velocity V, so in the interval δt these waves travel $(V - U)\delta t$ backward *relative to the undisturbed water*. A particle of water that is initially at position X_1 has moved to X_2 by the end of the interval. (Elsewhere in this book I give all velocities the symbol v with suitable subscripts, but here we will have four velocities to consider and it seems clearer to give each a different letter.)

Figure 15.4B shows the fish in the same two positions as in A. The tip of the tail, traveling forward with velocity U and transversely with velocity W, has moved $U\delta t$ forward and $W\delta t$ transversely in the interval of time δt. Meanwhile, the particle of water that was initially at X_1 has moved transversely a distance $w\,\delta t$. The similar triangles in Fig. 15.4B show us that

$$w = \frac{W(V - U)}{U + V - U} = \frac{W(V - U)}{V} \tag{15.1}$$

It is a fair approximation to think of the fish as leaving behind a cylinder of water moving transversely with velocity w. The diameter of this cylinder equals the height h of the tail fin. (We made a similar assumption about the air in the wake of flying animals, when we derived Equation 10.5.) The cross-sectional area of this cylinder is $\pi h^2/4$, it is growing with the velocity U of the fish, and the density of the water is ρ. Thus, the mass of water set moving in unit time is $\pi\rho Uh^2/4$. This water is being given velocity w, so in unit time the water is given momentum $\pi\rho Uwh^2/4$ and kinetic energy $\pi\rho Uw^2h^2/8$. By Newton's second law of motion, the transverse force on the tail equals the rate at which transverse momentum is being given to the water. The work done in unit time is this transverse force multiplied by the transverse velocity W of the tail; it is $\pi\rho UwWh^2/4$. Of this, $\pi\rho Uw^2h^2/8$ is used giving kinetic energy to the water in the wake, and the remainder, $\pi\rho Uw(W-\frac{1}{2}w)h^2/4$, serves to overcome the drag on the body. The Froude efficiency is the work done against drag divided by the total work:

$$\text{Froude efficiency} \;=\; \frac{\pi\rho Uw\,(W-\frac{1}{2}w)\,h^2/4}{\pi\rho UwWh^2/4} \;=\; \frac{W-\frac{1}{2}w}{W} \qquad (15.2)$$

From this and Equation 15.1

$$\text{Froude efficiency} \;=\; \frac{V+U}{2V} \qquad (15.3)$$

(This must be to some extent an overestimate, because we have ignored the tendency of the tail to wag the fish.)

No efficiency can be greater than 1, so the speed V of the waves along the body must be greater than the forward speed U of the fish. To make swimming as efficient as possible, V should be only a little greater than U. This can be achieved by making the height h of the tail fin large. This is another case where the principle we established in Section 14.1 applies; the required thrust can be obtained at lower energy cost by accelerating a large mass of water to a low velocity than a small mass to a high velocity.

The speed U is the stride length multiplied by the tail beat frequency, and the wave speed V is the wavelength multiplied by the frequency, so Equation 15.3 can be written

$$\text{Froude efficiency} \;=\; \frac{\text{Wavelength} + \text{Stride length}}{2 \times \text{Wavelength}} \qquad (15.4)$$

For the scup (Fig. 15.2) the stride length is half the wavelength and the Froude efficiency is 0.75. For the saithe the stride length is only a little less than the wavelength and the efficiency is even higher.

The theory we have used to calculate efficiency also provides us with an estimate of the drag on the body. We have seen that the work done against drag in unit time is $\pi\rho\,Uw\,(W-\frac{1}{2}w)\,h^2/4$. The drag is this work divided by the distance U traveled in unit time:

$$\text{Drag} = \frac{\pi\rho\,w\,(W-\frac{1}{2}w)\,h^2}{4} \tag{15.5}$$

$$= \frac{\pi\rho\,W^2\,(V^2-U^2)\,h^2}{8V^2}$$

(using Equation 15.1 to eliminate w).

Webb (1975) analyzed film of a 0.3-m rainbow trout (*Salmo gairdneri*) swimming in a water tunnel at 0.1–0.6 m/s. He measured U, V, W, and h, and used Equation 15.5 to calculate the drag. He also calculated the friction drag that would act on the fish if it were a rigid body moving at the same speed, using Equation 10.3. He found that throughout the range of speeds, the drag calculated from Lighthill's theory was about 4 times the friction drag. Pressure drag would act on a rigid body, as well as friction drag, but for a well-streamlined body such as a trout it is expected to add only about 20% to the friction drag. The drag on the swimming fish is thus around 3 times the drag on an equivalent rigid body. It has been suggested that the undulating movements of the body may prevent the boundary layer growing to the thickness that would be expected for a rigid body at the same Reynolds number, increasing friction drag (Lighthill 1971). The thinner the boundary layer for a given speed of movement through the water, the greater the friction drag; notice in Equation 3.7 that viscous force is inversely proportional to the thickness of a fluid layer. However, there is another reason for the drag calculated from Equation 15.5 being higher than for a rigid body. Side-to-side movements of the tail cause transverse "recoil" movements of the anterior end of the body; to some extent, the tail wags the fish. Work has to be done against the drag resisting the recoil movements, as well as against the drag resisting forward movement. Webb (1992) argued that this effect is more important than thinning of the boundary layer. However, Anderson et al. (2001) showed by particle image velocimetry that some thinning of the boundary layer occurs around swimming dogfish (*Mustelus*) and scup (*Stenotomus*).

Webb (1975) used Equation 10.3 (which assumes a turbulent boundary layer) to calculate the friction drag. He used it although the Reynolds numbers calculated from the length and speed of the fish were only $0.2–1.6 \times 10^5$, well within the range in which the boundary layer might be expected to remain laminar. He did this because there was a good deal of turbulence in the water tunnel, even in the absence of the fish. If he had used Equation 10.2 (for a laminar boundary layer) the drag calculated from the films would have been about 8 times the friction drag (Alexander 1977b).

Attempts to measure drag on fish carcasses have often given much higher drag coefficients than would be expected for a rigid streamlined body, apparently because drag is enhanced by the carcass flapping like a flag (Webb 1975). However, measurements from film of the deceleration of a cod (*Gadus*) gliding after a burst of swimming, with body and tail straight and fins folded, gave a drag coefficient based on wetted area of only 0.011 (Videler 1993). The fish was 0.3 m long, and it decelerated from 1.2 to 0.7 m/s, so its Reynolds numbers were around 3×10^5. The friction drag coefficient should have been about 0.004 for a laminar boundary layer or 0.007 for a turbulent one. This confirms (if any confirmation is needed) that the drag on a fish with its body held straight is similar to the drag on other streamlined bodies.

Webb (1971) performed an ingenious experiment to check that Lighthill's theory gives a good estimate of the power required for swimming. He measured both the mechanical power (calculated from Lighthill's theory) and the metabolic power (from oxygen consumption) of trout swimming in a water tunnel. From these data he calculated that the muscles were doing work with efficiencies around 0.1. He then attached small plates to the backs of the fish, increasing drag by amounts that could be calculated, and again measured oxygen consumption at a range of swimming speeds. As expected, the metabolic rate was higher, at any speed, than without the plates. From the difference he calculated that work was done against drag on the plates with efficiencies around 0.1. The efficiency with which work was done against the known drag on the plates agreed with the apparent efficiency of doing the work calculated from Lighthill's theory, tending to confirm that the theory gave reasonable estimates of power requirements.

Many other measurements have been made of the metabolic rates of swimming fishes (Videler 1993). Metabolic rate increases rapidly with increasing speed. As a general rule,

$$\text{Metabolic rate} \approx \text{Resting metabolic rate} + a(\text{Speed})^b \qquad (15.6)$$

where a and b are constants. The exponent b seems generally to be about 2.5 (Alexander 1974b). This is the exponent we would expect if the fish were a rigid body with a laminar boundary layer, whose motion was resisted only by friction drag. (This follows from Equation 10.2; remember that the Reynolds number, which appears in the equation, is proportional to the square root of speed, making the drag proportional to $(\text{Speed})^{1.5}$, and that the power requirement is the drag multiplied by the Speed.) We have, of course, just seen that the power requirement is much greater than for a rigid body, but it is greater by a fairly constant factor.

Fish have a maximum range speed, at which the energy cost of swimming unit distance is least. It can be found as in Fig. 7.11A, by drawing a tangent from the origin on a graph of metabolic rate against speed. Videler (1993) determined maximum range speeds from measurements of metabolic rates of teleost fishes of masses ranging from 5 mg (a larva) to 360 g. He found that maximum range speed (m/s) was about 0.47 (mass, kg)$^{0.17}$. At this speed, the cost of transport (J/kg m) was about 1.1(mass, kg)$^{-0.38}$. This is much less than the costs of transport for running or flying animals of the same mass (Figs. 7.12 and 12.3; note that costs of transport can be calculated from the metabolic powers in Fig. 12.3 by dividing by speed). However, the comparison of swimming with running and flight looks very different if only endotherms are considered (Williams 1999). Mammals and birds that swim by means of hydrofoils (penguins, seals, and whales) have costs of transport around ten times as high as fish of equal mass, approximately equal to the costs of running. (We saw in Section 14.3 that mammals and birds that swim by rowing have even higher costs of transport.)

This does not imply that penguins, seals, and whales are inefficient swimmers; they have higher costs of transport than fish because their maximum range speeds are higher, due to higher resting metabolic rates. Consider a fish whose swimming metabolic rate is given by Equation 15.6 with an exponent b of 2.5. Its cost of transport T at speed v is the metabolic rate divided by mv, where m is body mass:

$$T = \frac{(R/v) + av^{1.5}}{m} \qquad (15.7)$$

Here R is the resting metabolic rate and a is a constant. To find the maximum range speed at which T has its minimum value we will differentiate this equation.

$$\frac{dT}{dv} = \frac{-(R/v^2) + 1.5av^{0.5}}{m}$$

At the maximum range speed, $dT/dv = 0$, so

$$\text{Maximum range speed} = \left(\frac{2R}{3a}\right)^{0.4} \qquad (15.8)$$

By putting this speed into Equation 15.7 we get

$$\text{Cost of transport at maximum range speed} = \frac{2.0a^{0.4}R^{0.6}}{m} \qquad (15.9)$$

Birds and mammals generally have resting metabolic rates around 10 times as high as fish of equal mass (Alexander 1999). If they swim as efficiently as fishes, a will be the same as for fishes. Thus, Equations 15.8 and 15.9 tell us that we should expect the maximum range swimming speeds of mammals and birds to be $10^{0.4} = 2.5$ times as high as for fish of equal mass, and cost of transport at the maximum range speed to be $10^{0.6} = 4$ times as high as for fish. The maximum range speeds and costs of transport given for seals, whales, and penguins by Videler (1993) actually average about 2 and 10 times the predicted values for fish, respectively. This argument has not given us a full explanation of the high costs of transport of endotherms, but it shows that they are largely due to higher resting metabolic rate, rather than to less efficient swimming.

Fishes accelerating rapidly from rest (making "fast starts") start by bending the body into a C- or S-shape (reviewed by Domenici and Blake [1997]). In a C-start, the preparatory bend is followed by powerful extension of the body. The resulting direction of travel is very variable, but is commonly 120–180° to the direction in which the fish was originally facing. In S-starts, the movements following the preparatory S bend are broadly similar to the movements of steady swimming. The fish accelerates more or less in the direction in which it was initially facing. C-starts are commonly used as escape responses, and S-starts for predator strikes. Papers on fast starts commonly give peak instantaneous accelerations, but the final velocity and the time in which it is attained are probably generally more relevant to the life of the fish. Performances tabulated by Domenici and Blake (1997) include a 0.32-m trout (*Onchorhynchus*) reaching 2.8 m/s in 125 ms, and a 0.4-m pike (*Esox*) reaching 4.0 m/s in 108 ms. James and Johnston (1998) found that the speeds attained in fast starts by *Myoxocephalus* 5.5–32 cm long were roughly proportional to the square root of body length.

One might expect the speeds attained in fast starts to be limited by the work that muscle can perform in a single contraction. The pike mentioned above reached 4 m/s after one contraction of the muscle of each side of its body. At this speed, the kinetic energy per unit mass of its body was 8 J/kg. To this we must add the kinetic energy of the added mass of water that would move with it, giving a total of about 10 J/kg. Further kinetic energy would be given to the water in the wake. A clever experiment by McCutcheon (1977) showed that a small fish (*Brachidanio*) accelerating from a low speed by beating its tail once to left and right, pushed in each beat on a mass of water that was three or four times its body mass. If this water had been pushed directly backward, the kinetic energy given to the water would have been only a small fraction of the kinetic energy given to the fish. However, because the water was pushed at large angles to left and right, the kinetic energy that had to be given to it added about 80% to the work that had to be done to accelerate the fish. If this were true also for the pike fast start, the work required would be about 18 J/kg. The axial muscles of a pike make up about 55% of body mass, so this is 33 J/kg muscle. This may be compared to the 54 J/kg done by the leg muscles of a jumping bushbaby (Section 8.1). The comparison is probably unfair, because relatively little work may be done in the preparatory stroke of a fast start, leaving the muscle of one side of the body to do most of the work (James and Johnston 1998).

Ahlborn et al. (1997) used a physical model to support their contention that momentum given to the water in the preparatory stroke is used to contribute to the impulse accelerating the body in the return stroke. Wolfgang et al. (1999) studied the flow in the wake of a *Danio* making a 60° turn.

15.2. Muscle Activity in Undulating Fishes

Many records have been made of electromyographic activity in swimming fishes (see Altringham and Ellerby 1999). It has been found that waves of muscle activation travel backward along the body, like the waves of bending. However, in all cases the wave of activation travels faster than the wave of bending. In eels it travels only a little faster than the wave of bending, but in scup it travels so fast that activation is almost simultaneous all along the length of the fish. At the anterior end, activation of the muscle on one side of the body generally starts while that part of the body is completing a bend toward the other side. This allows time for the muscle to develop substantial force before it starts shortening and so bending the body toward its own side. Electromyographic activity continues through much of the time while the muscle is shortening, but ends while the muscle is still shortening, allowing time for the force to decline before the muscle starts being stretched again. Further posteriorly, because the wave of activation is traveling faster than the wave of bending, the electromyographic signal starts earlier in the period when the muscle is being stretched, and may end very soon after it has started to shorten. This suggests that posterior muscles may exert force largely while being stretched, and so may do as much negative as positive work (Hess and Videler 1984).

This possibility has been investigated for several species (Altringham et al., 1993; Wardle et al., 1995; Hammond et al., 1998). Electromyograms were recorded while the fish swam. At the same time, the length changes of muscles were determined, either by calculation from the curvature of the body, as seen in films, or by sonomicrometry. The fish were then killed and muscle fiber bundles were dissected out. Work loop experiments were performed, in which the muscles were subjected to length changes with frequencies and amplitudes imitating swimming, while being stimulated electrically for the part of the cycle in which electromyographic activity

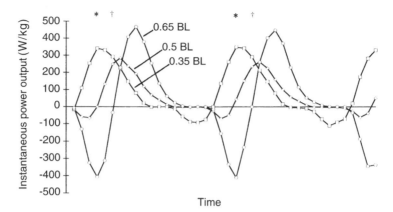

Fig. 15.5. Graphs of instantaneous power output against time for superficial white muscle 0.35, 0.5, and 0.65 body lengths (BL) from the snout of a swimming saithe (*Pollachius virens*). From Altringham et al. (1993).

had been observed. This procedure should produce work loops imitating the muscle's performance during swimming. These experiments have confirmed that posterior muscles do negative work when first activated (Fig. 15.5). As a result, these muscles may do little or no net work in steady swimming. Their principal function may be to transmit forces exerted by more anterior muscles to the tail.

This situation has been made more understandable by the theoretical work of Cheng et al. (1998). They undertook the very difficult task of calculating the forces all along the body of a swimming saithe. They started with detailed measurements of the body form and swimming movements of a 0.4-m saithe swimming at 1.2 m/s. They calculated not only the hydrodynamic forces, but also the inertial forces needed to accelerate and decelerate each part of the body in its side-to-side beating, and the forces needed to overcome the elasticity and viscosity of the body tissues. Taking account of all these forces, they calculated the bending moments that the muscles would have to exert, in all parts of the body and at all stages of a cycle of swimming movements. They found that to make the body move as it does, the wave of muscular bending moment would have to travel faster along the body than the wave of bending, and that the bending moments required in the posterior parts of the body were greatest while the muscles were still being stretched. It may seem wasteful to have the wave of muscle activation traveling so fast along the body that muscle near the posterior end does largely negative work, but if the wave did not travel so fast, the body would not move as it does. It would be interesting to extend the calculations to discover how different rates of propagation of the wave of activation would affect the pattern of movement, the speed of swimming, and the energy cost. We do not yet know whether the observed rate is optimal.

In most fishes, the bulk of the axial musculature is white but there is a superficial band of red muscle on each side of the body (Fig. 15.6A), and sometimes some pink muscle with intermediate properties. Bone (1966) showed that the white muscle is anaerobic and the red aerobic. Electromyographic records from fish swimming in water tunnels show that at low speeds, which can be sustained for long periods by aerobic metabolism, only the red muscle is active (see, for example, Rome et al., [1988]). At higher speeds that require anaerobic metabolism and cannot be sustained, the white muscle is brought into use. Burgetz et al. (1998) found that in *Onchorhynchus* white muscle is brought into use at speeds above 70% of the critical speed, defined as the maximum speed that could be maintained for 30 min.

The proportion of red muscle varies markedly between species. It is about 1% of the whole in cod (*Gadus*), 5% in saithe (*Pollachius*), 10–14% in herring (*Clupea*) and mackerel (*Scomber*), and even more in tunas (Videler 1993). In tunas and lamnid sharks it is not confined to the surface, but extends well in toward the vertebral column, and is kept up to 20 K warmer than the water (Carey 1982). Its elevated temperature increases the power output that can be obtained from it, as demonstrated by work-loop experiments with bundles of red muscle fibers from tuna (*Thunnus* [Altringham and Block 1997]). This should enable the fish to swim faster. In experiments in a water tunnel, Sepulveda and Dickson (2000) compared the swimming performance of juvenile tuna (*Euthynnus*), whose muscles were 1.0–2.3 K warmer than the water, and mackerel (*Scomber*), which were only 0.1–0.6 K warmer than the water. They found no significant difference in maximum sustainable swimming speed between fish of the same size. However, their tuna were only slightly warmer than the water, and there may have been some other difference between the species that gave the mackerel an advantage.

Fishes can swim at their peak sprinting speeds only for short times. Bainbridge (1960) measured the speeds of fish swimming in an annular tank and found, for example, that a 20-cm dace (*Leuciscus*) could be stimulated to swim at 2.2 m/s only for a short burst lasting less than one second, but would sustain 1.2 m/s for 5 s and 0.08 m/s for 20 s. He and Wardle (1988) found that 25-cm saithe could swim a short burst at 2.2 m/s, and could sustain 1.2 m/s for 2 min or 0.9 m/s for 200 min. The maximum speed that can be sustained aerobically is presumably a larger fraction of the speed attainable in a short burst for fish with a larger proportion of red muscle.

The red muscle has a lower maximum shortening speed (v_{max}, Equations 2.3) than white muscle from the same fish, by a factor of 2.8 in the case

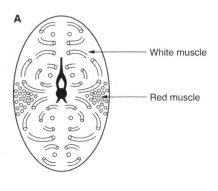

White muscle

Red muscle

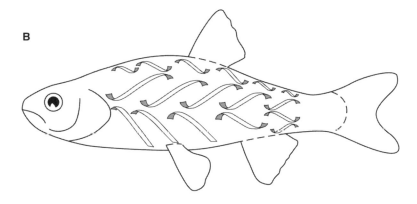

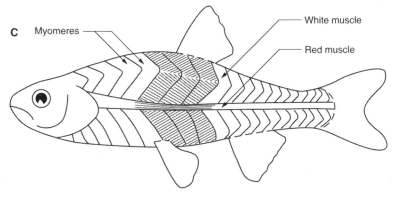

Fig. 15.6. Diagrams of the axial musculature of a typical teleost. (A) A thick transverse section, showing superficial red fibers running parallel to the long axis of the body and deeper white ones arranged helically. (B) A side view, showing a few of the helical trajectories formed by white fibres. (C) A few of the myomeres, as seen when the skin is removed.

of carp (*Cyprinus* [Rome et al., 1988]). In work loop experiments it gives maximum power at a lower cycle frequency, as shown, for example, by Fig. 2.5B. Altringham and Johnston (1990) found that the optimum frequency for power output from *Myoxocephalus* muscle was about 2 Hz for the red muscle and 6 Hz for the white. The red muscle was used in slow swimming with tail beat frequencies of 1–4 Hz, and the white muscle in faster swimming at 4–9 Hz. Thus, each muscle type was adapted to the range of frequencies in which it was required to operate. The maximum power output was about 5 W/kg for the red muscle, and 30 W/kg for the white. Suppose that a fish whose muscle is capable of these power outputs has 5% red muscle and 95% white. The power available for a short burst of swimming will be almost 120 times the power that can be sustained for prolonged swimming. If power requirements are proportional to (speed)$^{2.5}$, as suggested by the sentence following Equation 15.6, the maximum burst speed may be expected to be $120^{0.4} = 6.8$ times the aerobically sustainable speed. However, this may be an overestimate because no account has been taken of the power required to accelerate to high speed.

The red muscle fibers run parallel to the long axis of the body, but many of the white fibers make large angles with the long axis, up to a maximum of about 40° (van der Stelt 1968; Alexander 1969). For most of the length of the body of typical teleosts they are arranged in helices, as indicated in Fig. 15.6A, B, but a different pattern is found in the caudal peduncle and throughout the swimming muscle of selachians and some of the more primitive teleosts.

If all the white fibers ran parallel to the long axis of the body, fibers close to the median plane would shorten much less in a bend, than more lateral fibers, and so would do less work. Alexander (1969) presented mathematical models showing that the observed fiber arrangements might enable all the white fibers to shorten by equal fractions of their length. Rome and Sosnicki (1991) measured sarcomere lengths in bent carp (*Cyprinus*) and found as predicted that at several positions along the length of the body, sarcomere length was uniform throughout the white muscle of each side of the body. Katz et al. (1999) measured muscle length changes in the white muscle of swimming *Chanos*, by sonomicrometry. They checked by dissection after each experiment that the sonomicrometry crystals were aligned with the muscle fibers, so as to measure muscle strain accurately. They found that the strain in deep white muscle was about half the strain in superficial red muscle in the same cross section, but since they recorded from only one position in the white muscle, their results do not show whether strain was uniform throughout the white muscle as predicted by Alexander (1969). Wakeling and Johnston (1999) measured strain at several distances from the median plane in swimming carp, again by sonomicrometry. Their results seemed to show that muscle strains were proportional to distance from the median plane, as when a homogeneous beam is bent. It is not clear from their paper whether the pairs of sonomicrometry crystals were aligned with the muscle fibers or parallel to the long axis of the body; if the latter, their results are not inconsistent with the predictions of Alexander (1969).

If my theory is correct, strain in the helically arranged white muscle of typical teleosts is about one-quarter of red muscle strain for any particular radius of bending. If the muscle behaves like a homogeneous beam, white muscle strain ranges from almost equal to red muscle strain (in the most lateral white muscle) to very much less (close to the median plane). In either case, the average strain in the white muscle is much less than in the red. To take full advantage of the capacity of the muscle to do work, the fish should bend its body to smaller radii of curvature when swimming with the white muscles than when swimming with the red. Katz et al. (1999) measured red muscle strain in *Chanos*. They found that when it swam at speeds at which only red muscle was active, peak strains in red muscle about halfway along the body were about ±7.5%. When it swam faster, using the white muscle, accelerating bursts of a few tail beat cycles in which peak strains at the same location in the red muscle were ±13% alternated with decelerating intervals in which peak strain was ±6%. More posteriorly, strains were larger. Fast starts also are powered by white muscle. Teleosts bend to smaller radii of curvature (therefore, larger muscle strains) in fast starts than in steady swimming (Domenici and Blake 1997).

Thin sheets of collagen fibers (myosepta) divide the axial muscles of fishes into W-shaped myomeres (Fig. 15.6C; dissection shows that the shape is more complex than is apparent from this surface view). Successive myomeres are derived from different segments of the embryo and are separately innervated. Van Leeuwen (1999) pointed out that because the myomeres are not activated simultaneously, there may be differences between the forces exerted by the muscle in successive myomeres, making it necessary for forces to be transmitted from the muscle to the skin or axial skeleton. He presented a mathematical analysis that seems to show that the complex shape of the myosepta is optimally adapted to this force-transmitting role.

15.3. FINS, TAILS, AND GAITS

At low speeds, fish often swim by undulating a fin or fins, instead of the whole body. The underlying principle of this technique of locomotion is the same as for swimming by undulation of the whole body. A possible advantage of undulating the fins is that the increased drag that results from the undulations will act only on the fins, and not on the whole body.

A disadvantage is that all the power must come from the fin muscles, which are generally much smaller than the axial muscles.

In some cases median fins are used. Sea horses (*Hippocampus*) and the African electric fish *Gymnarchus* swim by undulating their dorsal fins (Blake 1980b). Both South American knife fishes (Gymnotidae) and African knife fishes (Notopteridae) swim by undulating their long anal fins (Blake 1983). In other cases, pectoral fins are more important, but it is not always easy to make a sharp distinction between the use of pectoral fins as oars or hydrofoils (Chapter 14) and swimming by undulation of the pectoral fins. The action of the greatly enlarged pectoral fins of rays is clearly undulation in the case of species with relatively low aspect ratio fins, such as the stingray *Taeniura* (Rosenberger and Westneat 1999), but is better regarded as the flapping of a hydrofoil in Myliobatidae (Daniel 1988; see also Rosenberger 2001). Many teleosts use undulations of both median and pectoral fins (Long et al., 1994).

Fish that propel themselves by means of several fins may be able to turn on the spot; in other words, they may be able to turn while keeping the center of gravity of the body stationary. Walker (2000) showed that the boxfish *Ostracion* can (almost) do this. He pointed out that this does not necessarily make the fish good at turning in confined spaces. *Ostracion* has 70% of the length of its body enclosed in a rigid carapace, so needs a space 0.7 body lengths wide to turn in. In contrast, Schrank et al. (1999) found that three species of fish with flexible bodies could turn around between vertical walls only 0.11–0.26 body lengths apart.

Webb (1994; and in Alexander 1989a) has shown how teleost fish change gaits as they increase speed. Typically, the following gaits are used, starting with the slowest:

1. Propulsion by median and paired fins. The muscles that operate these fins are predominantly red.

2. Burst and coast swimming by undulation of the body, powered by red muscle. In burst and coast swimming, groups of a few cycles of body undulation alternate with periods in which the body remains straight and glides passively forwards.

3. Swimming by continuous undulation of the body, still powered by red muscle.

4. Burst and coast swimming by undulation of the body, powered by white muscle.

5. Swimming by continuous undulation of the body, powered by white muscle.

Burst and coast swimming is used (gaits 2 and 4) in the ranges of speeds at which the muscles powering locomotion would be working well below their capacity for power output if the body were undulated continuously.

An advantage of burst and coast swimming is that, while coasting, the fish presumably avoids the increased drag associated with undulation. However, *Chanos* appear to forego this advantage. At speeds at which they might be expected to use gait 4 they alternate bursts of high-amplitude body undulation with intervals of lower amplitude undulation, instead of coasting with the body straight (Katz et al., 1999).

The bass (*Micropterus*) is an example of a fish that uses the full range of gaits, but many fish omit some of them (Webb 1994). For example, tunas and sharks do not use gait 1, and sea horses and rays use only gait 1. Hove et al. (2001) have described the sequence of gaits used by boxfish (*Ostracion*), which use both median fins and pectoral fins at low speeds.

Surfperch (*Embiotoca*) swim with their pectoral fins at low speeds and by undulating their bodies at higher speeds. Drucker and Jensen (1996) filmed surfperch of masses ranging from 5 to 500 g swimming in a water tunnel, and measured the speeds at which they made the gait transition. This speed rose from 0.2 m/s for 5-g fish to 0.4 m/s for 100-g fish, but did not rise further for larger fish. The frequency of the pectoral fin beat at the gait transition speed was proportional to (body mass)$^{-0.12}$. The exponent is close to the exponent of -0.14 observed for stride frequency at the trot–gallop transition in mammals (Heglund et al., 1974), but this may be coincidental.

Squids and cuttlefishes swim by fin undulation at low speeds and by jet propulsion (discussed in Chapter 16) at higher speeds.

15.4. Undulating Worms

The animals discussed so far in this chapter have been relatively large and fast. Consequently, the Reynolds numbers associated with their swimming were large enough for inertial forces to have been much more important than viscous forces for their propulsion. For example, even a fish larva 1 cm long swimming at 5 cm/s has a Reynolds number of 500.

Some animals that swim by undulation are much smaller and slower. For example, Gray and Lissman (1964) filmed a 0.8-mm nematode (*Turbatrix*) swimming at 0.7 mm/s with a Reynolds number of 0.6. Because swimming is driven by side-to-side movements of the slender body, it may be informative to calculate a Reynolds number using the diameter of the worm (0.03 mm) as the scale of length and the speed of the side-to-side movement of a point on the body (3.5 mm/s) as the velocity. That gives a Reynolds number of 0.1. With Reynolds numbers as low as these, inertial forces will be negligible and we need consider only viscous forces.

Gray and Hancock (1955) calculated the forces on a cylindrical, undulating organism swimming at low Reynolds numbers. They applied their

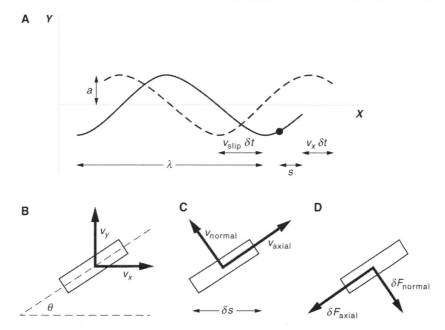

Fig. 15.7. Diagrams of a nematode worm swimming toward the right of the page. (A) Two successive positions of the worm, with a segment of the body marked by a dot. (B) The velocity of the marked segment resolved into forward and transverse components. (C) The velocity resolved differently into axial and normal components. (D) The axial and normal components of the hydrodynamic force on the segment. From Alexander (2001).

theory to spermatozoa, but it is equally applicable to small nematodes. It depends on the assumption (which greatly simplifies the mathematics) that the hydrodynamic force on any short segment of the body is equal to the force on an equal length of an infinite cylinder of the same diameter moving with the same velocity and orientation. Lighthill (1976) presented a more rigorous theory that avoids this assumption.

The worm shown in Fig. 15.7A is swimming toward the right. The segment of its body marked by a dot has components of velocity forward (i.e., toward the top of the diagram) and to the right (Fig. 15.7B). Alternatively, its velocity can be resolved into a component v_{axial} along the axis of the segment, and a component v_{normal} at right angles to it (Fig. 15.7C). Consequently, a hydrodynamic force acts on the segment, which has components δF_{axial} and δF_{normal} in the axial and normal directions (Fig. 15.7D). These forces could be resolved into a component in the direction of swimming and a component at right angles to it. At any instant, the transverse components to left and right will (more or less) cancel out. The thrust component on the segment may sometimes be positive and sometimes negative, but if the integral of the thrust components along the whole length of the body is positive, the worm will accelerate; and if the integral is negative, the worm will slow down.

At the low Reynolds number involved, hydrodynamic forces are proportional to velocity. The axial force per unit length on the segment will be $C_{axial}v_{axial}$, and the normal force per unit length will be $C_{normal}v_{normal}$, where C_{axial} and C_{normal} are constants. Let the waves have wavelength λ and amplitude a, and let them travel posteriorly along the body with velocity v_{wave}. Then Gray and Hancock's (1955) theory leads to

$$\frac{v}{v_{wave}} = \frac{1 - (C_{axial} / C_{normal})}{1 + (\lambda^2 / 2\pi^2 a^2)(C_{axial} / C_{normal})} \qquad (15.10)$$

This equation shows that backward-moving waves will propel the worm forward, only if C_{axial} is less than C_{normal}. This will be the case; C_{axial}/C_{normal} will be 0.5 or a little more, depending on the ratio of worm diameter to wavelength (Lighthill 1976). If the amplitude is one-fifth of the wavelength and the diameter is $1/25$ of the wavelength (these are typical values), Equation 15.10 gives $v/v_{wave} = 0.22$, which lies within the range of 0.2–0.3 observed by Gray and Lissmann (1964).

Alexander (2001) used Gray and Hancock's (1955) theory to estimate the mechanical power required for swimming. Making reasonable assumptions about the efficiency of nematode muscle, I estimated that the metabolic power required for swimming was likely to be between 1 and 10% of the animal's metabolic rate. In contrast, the metabolic rate of a fish swimming at its maximum range speed is typically three times its resting metabolic rate (Alexander 1998b). Locomotion is probably far less important in the energy budgets of these small worms than in those of larger swimmers.

Equation 15.10 shows that if C_{axial} is greater than C_{normal}, the animal will be propelled in the same direction as the waves. This point was made initially by Taylor (1952), a distinguished mathematician who did not at the time realize its application to real animals. The parapodia of polychaete worms have the effect of making C_{axial} greater than C_{normal}, and these worms swim forward by means of forward-moving waves. However, the parapodia are not simply passive appendages. They beat forward and back as the animal undulates, enhancing their propulsive effect (Clark and Tritton 1970).

This chapter has shown how most fish and some other animals swim by undulating either the whole body or the fins alone. It has shown that the drag on an undulating body seems to be greater than for a rigid body of the same size and shape traveling at the same speed. It has shown how waves of muscle activation travel along the body of a swimming fish and how the red and white muscles are used, and that fish use different gaits at different speeds. Finally, we have seen that the hydrodynamics of swimming of small worms that swim by undulation is quite different from the hydrodynamics of fish that swim by similar movements, because the Reynolds numbers are much lower.

The invention of particle image velocimetry (Section 5.4) has given a great boost to research on swimming by undulation, making it very much easier than before to study the water movements around a swimming animal. Much has been achieved with it, but much remains to be done. There is scope also for much more research on gaits. Are gait changes as speed increases driven simply by the need to recruit more muscle, or do fish (like horses, Fig. 7.11) save energy by changing gaits? Another topic that I would like to see explored further is the significance of the helical arrangement of the white swimming muscles of fishes. My theory (Alexander 1969) has been challenged, and may be wrong.

Chapter Sixteen

....

Swimming by Jet Propulsion

*T*HIS CHAPTER is about animals that propel themselves through water by squirting a jet of water out of a contracting cavity. They include squids and other cephalopods that drive water out of the mantle cavity by contraction of its muscular wall (Fig. 16.1A); a few bivalve molluscs, such as *Pecten*, that squirt jets of water out of their mantle cavities by adducting the valves of the shell (Fig. 16.1B); and medusae, which contract to expel water from the space enclosed by their bell. Other examples of jet-propelled swimmers include salps, which draw water in at the anterior end of the body and expel it at the rear (Madin 1990); and dragonfly larvae, which squirt water from the rectum (Mill and Pickard 1975). Jet propulsion by squids, scallops, and medusae has been studied more thoroughly than the jet propulsion of other animals. Accordingly, this chapter concentrates on them.

16.1. EFFICIENCY OF JET PROPULSION

Figure 16.2 shows two designs for jet-propelled animals. Figure 16.2A represents an animal such, as a jellyfish, that draws water in from the rear, then ejects it toward the rear. Figure 16.2B shows an animal, such as a salp, that takes water in from in front and ejects it to the rear. In each case the animal is swimming at velocity v, and the water has velocity v_{intake} (forward) as it enters the animal and v_{jet} (backward) as it is ejected. These velocities are defined *relative to the undisturbed water*. In case A, $v_{intake} > v$ (water entering from the rear must be traveling faster than the animal). In case B, however, $v_{intake} < v$ (the entering water is being pushed forward by the approaching animal, but it is traveling less fast than the animal and so is overtaken by it).

In previous chapters, we used Equation 14.1 to calculate Froude efficiencies for rowing, hydrofoil propulsion, and swimming by undulation. In swimming by those methods, the water is accelerated just once, toward the rear. In jet-propelled swimming, however, water is accelerated first forward and then backward, and we need a different equation, as Anderson and De Mont (2000) have pointed out. The analysis that follows is not the same as theirs.

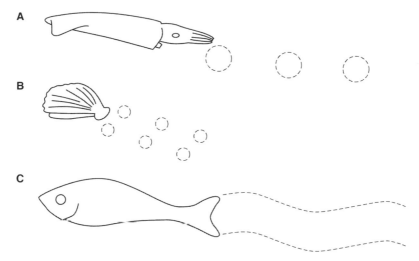

Fig. 16.1. Diagrams showing the masses of water pushed on by swimming animals: (A) a squid and (B) a scallop swimming by jet propulsion; and (C) a teleost fish swimming by undulation. All the animals are swimming toward the left of the diagram, by driving water toward the right.

Let a mass m_{jet} of water pass through the animal in unit time. This water starts at rest, is accelerated to velocity v_{intake} (forward), decelerated, and finally accelerated to v_{jet} (backward). To calculate the net rate of change of momentum we need to consider only the initial and final velocities; the rate of change of momentum is $m_{\text{jet}}v_{\text{jet}}$. The force propelling the animal equals the rate of change of momentum, so the rate at which work is being done against drag (the useful power) is $m_{\text{jet}}vv_{\text{jet}}$. Kinetic energy is given to the entering water at a rate $\frac{1}{2}\,m_{\text{jet}}v_{\text{intake}}^2$, and though this kinetic energy is subsequently taken from the water there is no apparent mechanism for it to be recovered and reused. Kinetic energy is finally lost in the wake at a rate $\frac{1}{2}\,m_{\text{jet}}v_{\text{jet}}^2$. The total power requirement is $m_{\text{jet}}vv_{\text{jet}} + \frac{1}{2}\,m_{\text{jet}}(v_{\text{intake}}^2 + v_{\text{jet}}^2)$:

$$\text{Efficiency} = \frac{\text{Useful power}}{\text{Total power}} = \frac{2vv_{\text{jet}}}{2vv_{\text{jet}} + v_{\text{intake}}^2 + v_{\text{jet}}^2} \qquad (16.1)$$

If water is taken in at the rear as in Fig. 16.2A, $v_{\text{intake}} > v$ and the efficiency has an upper limit at which $v_{\text{intake}} = v$:

$$\text{Maximum efficiency (rear intake)} = \frac{2vv_{\text{jet}}}{2vv_{\text{jet}} + v^2 + v_{\text{jet}}^2} \qquad (16.2)$$

$$= \frac{2vv_{\text{jet}}}{(v + v_{\text{jet}})^2}$$

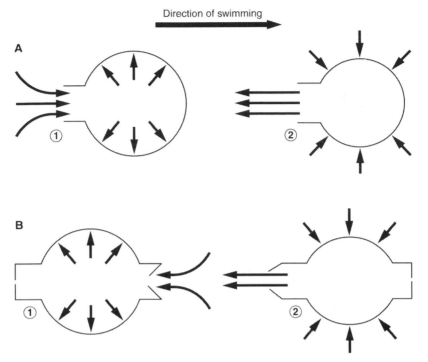

Fig. 16.2. Diagrams of jet propulsion by animals that take water in (A) from the rear and (B) from in front. In each case (1) shows water being taken in and (2) shows it being ejected.

However, if water is taken in at the front as in Fig. 16.2B, $v_{intake} < v$, and a higher maximum efficiency is possible. It is obtained when $v_{intake} = 0$:

$$\text{Maximum efficiency (front intake)} = \frac{2vv_{jet}}{2vv_{jet} + v_{jet}^2} \qquad (16.3)$$

$$= \frac{v}{v + \frac{1}{2}v_{jet}}$$

This equation is identical with Equation 14.1, apart from the difference in the subscript. We have not so far considered the possibility that v_{intake} may be negative. In that case, the efficiency is as given in Equation 16.3, because the intake velocity and the jet velocity are in the same direction, and kinetic energy given to the incoming water is retained in the jet.

There is a hidden assumption in our discussion so far. We have assumed that the swimming velocity v is constant. However, when an animal is swimming by jet propulsion its velocity fluctuates. The animal accelerates while it is expelling a jet of water. Unless v_{intake} is strongly negative, it decelerates while water is being taken in to refill the cavity from which the water is expelled. The drag on the animal fluctuates not only because the velocity fluctuates, but also because the animal is bigger when the cavity is full than when it is empty. Further, the effective mass of the animal fluctuates; an animal with its cavity full has greater mass than when its cavity is empty. Equations that took account of all these fluctuations would be rather complicated. However, if the frequency of the jet-propulsion cycle is sufficiently high, these fluctuations are small enough to be ignored, and Equations 16.1 to 16.3 apply.

Now make the very different assumption that the intervals of time between jets are so long that the animal glides (almost) to a halt between one jet and the next. Assume also that the thrust, during the brief jet pulses, is very much greater than the drag. During each pulse, a mass m_{jet} of water that was initially at rest is ejected at velocity v_{jet}, and the body of mass m is accelerated from velocity 0 to a maximum velocity v_{max}. By the law of conservation of momentum,

$$mv_{max} = m_{jet}v_{jet} \qquad (16.4)$$

Because the animal is almost stationary when it takes in water, even water drawn in from the rear can be drawn in at low velocity, so little energy need be wasted imparting forward velocity to the water. The work required for one cycle of swimming is the sum of the kinetic energies given to the body and to the jet, $\frac{1}{2}(mv_{max}^2 + m_{jet}v_{jet}^2)$. Of this, $\frac{1}{2}mv_{max}^2$ can be thought of as useful work. It provides the work that is done against drag as the animal decelerates in the interval between jets. Thus,

$$\text{Efficiency} = \frac{mv_{max}^2}{mv_{max}^2 + m_{jet}v_{jet}^2}$$

From this and Equation 16.4,

$$\text{Efficiency} = \frac{v_{max}}{v_{max} + v_{jet}} \qquad (16.5)$$

Equations 16.3 and 16.5 show that, both for steady front-intake jetting and for intermittent jetting, it is more efficient to swim by means of a low-velocity jet than with a high-velocity jet. To attain the same swimming speed, the mass of water ejected in unit time must, of course, be greater if the jet velocity is low than if it is high. However, the efficiency of rear-intake jetting, given by Equation 16.2, is greatest when $v_{jet} = v$.

Daniel (1983) has made a more realistic (and correspondingly more complicated) analysis of animal swimming by jet propulsion. His mathematical model represents a hydrozoan medusa of diameter 10 mm, which alternately expels water at a constant rate for 0.1 s, and draws water in at a constant rate for 0.2 s. Figure 16.3 shows how it moves in the first six

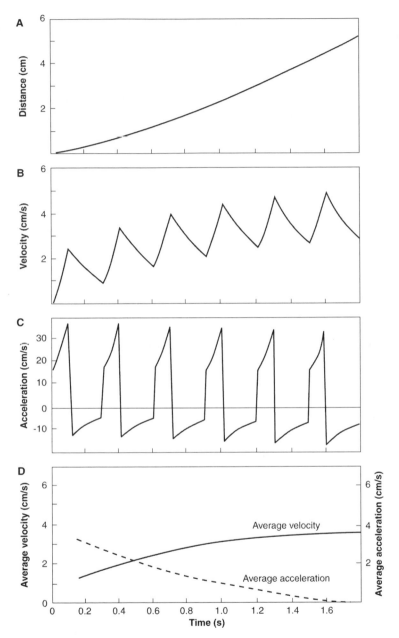

Fig. 16.3. Graphs showing the movements predicted by a mathematical model of a hydrozoan medusa, in the first few cycles of swimming, starting from rest. (A) Distance traveled; (B) instantaneous velocity; (C) instantaneous acceleration; and (D) velocity and acceleration averaged over a complete swimming cycle. From Daniel (1983).

cycles of jetting, starting from rest. It accelerates in each contraction phase, and decelerates in each refilling phase. The acceleration is greater at the end of each contraction phase than at the beginning, although the rate of ejection of water is constant, because the effective mass of the animal falls as water is ejected. The mean velocity (averaged over a cycle) increases during the first few cycles starting from rest, but then settles down at an almost constant value. The pattern of movement predicted by the model agrees well with observations of living hydrozoan medusae (see also Daniel 1985).

Each time the medusa's muscles contract, kinetic energy is given to the animal and the added mass of water that moves with it. In the intervals between contractions, some of this kinetic energy is lost, doing work against drag as the animal decelerates. In the first few cycles of jetting, the kinetic energy gained in the acceleration exceeds that lost in the deceleration. There is a net increase in kinetic energy in each cycle, and a large proportion of the work that the muscles have to do is required to supply it. If swimming continues until a steady state is reached, there is no net gain in the kinetic energy of the body and added mass. Work is then required only to overcome drag, to give kinetic energy to the water in the jet, and to deform the bell of the medusa. For the present, we will ignore the work required to deform the bell: it is discussed in the next section.

Hydrozoan medusae generally alternate bouts of a few cycles of jet propulsion with intervals of rest. For example, Daniel (1985) observed that *Gonionemus* typically made 5 to 10 bell contractions in a 1- to 3- bout, followed by a rest of 10–90 s. It is only in the later stages of a bout that the animal approaches a steady state. Daniel's (1983) model predicted that for these cycles the Froude efficiency was only 0.09; in other words, the work done giving kinetic energy to the jet was ten times the work done against drag. Daniel (1985) measured the oxygen consumption of swimming medusae and calculated that the metabolic cost of transport was an order of magnitude higher than predicted for fish of the same mass. The animals were tethered and may not have been using oxygen at the same rate as if they had been swimming freely, but it seems clear that medusan locomotion is costly.

Thus, the swimming performance of medusae is unimpressive. We will now ask whether the same is true for other animals that swim by jet propulsion. Squids swim by alternately expanding and contracting the mantle cavity. They draw water into the cavity through a wide slit at its anterior end and expel it through a tube called the siphon. Squids appear well streamlined and are reasonably fast. For example, *Illex* with masses of 0.4–0.5 kg can attain 2.8 m/s in a short burst of swimming, and can sustain a speed of 0.76 m/s aerobically (O'Dor and Webber 1986). In its resting position, the siphon faces anteriorly, so a jet from it propels the animal backward, with the posterior end leading. In that case, although water is taken in at the anterior end, this is the rear end relative to the direction of travel. Figure 16.2A can represent a squid swimming backward. However, the siphon can be bent to face posteriorly and propel the animal forward, in which case the animal functions as in Fig. 16.2B.

How efficient is squid swimming? Webber and O'Dor (1986) inserted a cannula connected to a pressure transducer into the mantle cavity of *Illex*, so that they could record the pressure there as the squid swam in a water tunnel. The faster it swam, the larger the pressures that they recorded. For example, when a 0.3-kg specimen was swimming steadily at 0.58 m/s, the peak pressure in the mantle cavity in each jetting cycle was 5300 Pa above the pressure in the water outside the animal. Thus, the pressure of the water fell by 5300 Pa as it was ejected from the cavity. By applying Bernoulli's equation (Equation 10.6), we can estimate that it must have been ejected at a velocity of 3.2 m/s relative to the animal. The velocity v_{jet} of the jet relative to the undisturbed water would have been $3.2 - 0.58 \approx 2.6$ m/s. Anderson and De Mont (2000) have pointed out that calculations like this may be inaccurate, because there may be differences of pressure within the mantle cavity. However, the errors are probably small enough not to concern us here. By putting the swimming speed of 0.58 m/s and jet velocity of 2.6 m/s in Equation 16.2, we can calculate that the efficiency cannot have been greater than 0.29. This estimate is a little pessimistic because, although most of the water was presumably ejected while the pressure was near its peak, some must have been ejected while it was lower. If we use the mean pressure during ejection, 2500 Pa, instead of the peak pressure, we get an efficiency of 0.38. Even this is far lower than the efficiency of 0.75 that we calculated in Section 15.1 for a typical teleost swimming by undulation.

Webber and O'Dor (1986) also measured the oxygen consumption of squid swimming in their water tunnel. They compared an *Illex* that had a mass of 0.4 kg with its mantle cavity empty to a 0.5-kg salmon (*Onchorhynchus*). This comparison seemed fair because the mass of water in the mantle cavity fluctuated during the jetting cycle about a mean value of about 0.1 kg. At 0.76 m/s, the maximum speed that could be sustained aerobically, the metabolic rate of the squid was 1.6 W more than when the animal was stationary. The salmon could swim aerobically at 1.35 m/s, using 1.2 W more than when stationary. Thus, the squid swam more slowly than the fish, but nevertheless used more power. This must have been largely due to its lower Froude efficiency.

The efficiency is lower because the mass of water that the squid ejects in unit time from its mantle cavity is much less than the mass that the fish pushes on with its tail. At 0.76 m/s, the squid jetted with a frequency of 0.89 Hz, expelling about 0.2 kg water each time. Thus, it was ejecting puffs of water at a mean rate of 0.18 kg per second. In contrast, the fish pushed on a continuous cylinder of water of diameter equal to the height of its caudal fin (Fig. 16.1C; see also Section 15.1). The fish was 0.37 m long, and the height of its caudal fin must have been about 0.08 m. A cylinder of water of this diameter, 0.76 m long, has a mass of almost 4 kg. This is the mass of water accelerated each second. Thus, the squid is accelerating much smaller masses of water than the fish, and has to accelerate them to much higher velocities to obtain the same thrust.

Squid such as *Illex* swim continuously in aquaria and undertake long migrations in the wild (Webber and O'Dor 1986). In contrast, scallops spend most of their time resting on the bottom and swim only a few meters at a time (Dadswell and Weihs 1990). They escape from predators such as starfish by bursts of swimming in which they climb at a steep angle. They swim by repeatedly opening their shells and clapping them shut. As the shell opens, water is drawn in from all sides. When it closes, a flap of tissue around the edge of the shell directs the outflowing water through openings on either side of the hinge (Fig. 16.1B). Cheng et al. (1996) used a starfish to stimulate *Placopecten* to swim. These scallops had shells about 65 mm long, and swam at around 0.25 m/s, climbing at angles around 25°. As they climbed, they opened and closed their valves through angular ranges of around 10°, at frequencies of about 3.6 Hz. Dadwell and Weihs (1990) described level swimming at speeds up to 0.79 m/s. In level swimming, the scallop's tendency to sink is apparently counteracted by the shell functioning as a hydrofoil.

Marsh *et al.* (1992) used miniature pressure transducers to record pressures inside the mantle cavities of slightly smaller (50 mm) scallops of other species, again using predators to stimulate swimming. They recorded peak pressures of about 3000 Pa as the valves closed in each swimming cycle. Bernoulli's equation (Equation 10.6) implies that water ejected when the pressure was at this peak would be accelerated to 2.5 m/s, many times faster than the scallops can have been swimming (the swimming speed was not reported). Hence, the Froude efficiency must have been low. Cheng and De Mont (1996) estimated Froude efficiencies of around 0.3 for *Placopecten*.

As well as the pressure transducers, Marsh et al. (1992) attached two sonomicrometry crystals to the shells of their scallops, one to each valve. This enabled them to record the changes of distance between the crystals (see Section 5.6) as the scallops swam. Knowing how far the crystals were from the hinge, and the dimensions of the valves, it was easy to calculate from these records the changes of volume of water in the mantle cavity. Thus, pressures and volume changes were measured simultaneously. When a pump moves a volume V of fluid against a pressure difference Δp, it does work $V\Delta p$. Marsh et al. (1992) calculated the work done on the water by the adductor muscle (Fig. 16.4A). They found that this was about 20 J/kg muscle in the first cycle of a burst of swimming, falling to about

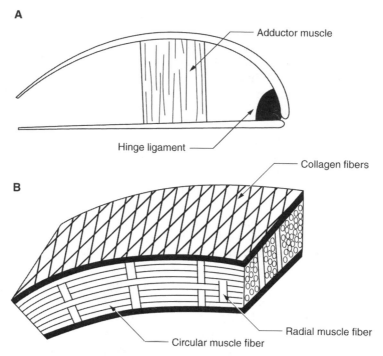

Fig. 16.4. (A) A diagrammatic section through a scallop, showing the adductor muscle and the hinge. (B) A diagram of the mantle of a squid, showing the arrangement of collagen fibers and muscle fibers.

10 J/kg after about 20 cycles. The muscle has to do work on the elastic hinge that connects the valves, as we will see in the next section, but this is small compared to the work done on the water.

Marsh and Olson (1994) performed work loop experiments (see Section 2.3) on bundles of fibers isolated from *Argyropecten* adductor muscle. In these experiments they imitated the pattern of length change seen in swimming scallops, and stimulated the fibers electrically at the appropriate stage of the cycle. The work that the muscle did in these experiments (17 J/kg) was approximately equal to the work per unit mass of muscle recorded in the experiments on intact animals.

16.2. ELASTIC MECHANISMS IN JET PROPULSION

Scallops have a large muscle to close the shell, but no muscle to open it. Instead, the shell is opened by elastic recoil of a block of abductin (a rubberlike protein). The two valves of the shell are held together at the hinge by a strip of flexible but relatively inextensible protein, the outer hinge ligament. The abductin forms the inner part of the hinge ligament (Fig. 16.4A). It is compressed as the shell closes, and recoils to open the shell. Thus, some of the work done by the adductor muscle as it closes the shell is stored as elastic strain energy in the abductin and later released by elastic recoil to do the work of opening. Alexander (1966) showed by a simple experiment that 91% or more of the work done on the abductin as the shell closes is returned in the elastic recoil as it opens.

This work can be estimated from the records of Marsh et al. (1992), which show the pressure in the mantle cavity falling to 300 Pa below ambient while the shell is opening. We have already seen that the pressure rises to 3000 Pa above ambient while it is closing. The volume of water that enters the shell during opening equals the volume that leaves during closing, so these pressures indicate that the work done on the water during opening is about one-tenth of the work of closing. This work is low because the area of the gape through which the water enters is larger than that of the openings through which the jet leaves; and because the forward movement of the scallop, with the opening gape in front, will help to fill the mantle cavity.

If the hinge ligament were stiffer, it would store more strain energy when the shell was closed. More work would have to be done on it to close the shell, but correspondingly more energy would be available to open it, and it would open faster, enabling the scallop to make more swimming cycles in unit time. This might enable it to swim faster. Alternatively, if it swam at the same speed as before, its efficiency would be improved, because it would expel more water in its jets, per unit time. These arguments may seem attractive, and suggest that it would be possible to predict an optimum stiffness for the hinge by balancing the benefits of faster opening against the increased work of opening. However, a simple mathematical model on those lines would predict that the shell should start opening as soon as closing is complete. It does not do that. Instead, the swimming cycle consists of three approximately equal parts; one-third of the period is spent opening, one-third closing, and one-third gliding forward with the shell closed (Cheng et al., 1996).

We have not yet considered the inertia of the valves or of the water immediately outside them. The valves gain and lose angular velocity twice in each cycle of opening and closing. This implies fluctuations of their kinetic energy, and of the kinetic energy of the added mass of water that moves with them. Are these large enough to be important for the work of swimming? Cheng et al. (1996) showed that they are not; the peak kinetic energies given to the valves and to the added mass of water are very small compared to the kinetic energy given to the jets of water that propel the animal. They made a proper hydrodynamic analysis of swimming, but a simple argument will confirm that the inertial work of swimming must be small. The scallops they studied opened and closed their shells through a range of about 0.2 radians, opening and closing each taking about 0.1 s. Thus, the angular velocities of opening and closing were about 2 rad/s. Each valve approached and moved away from the median plane at about 1 rad/s. The lengths of the valves (measured at right angles to the hinge) were about 60 mm, so the peak velocities of opening and closing, at the edge of the shell, were about 0.06 m/s. This is only one-fortieth of the velocity of 2.5 m/s, which we estimated for the jet. The added mass of water is much greater than the mass of the valves themselves, and can be estimated as the mass of a sphere of water of 60 mm diameter, or about 100 g. A pair of circular valves of 60 mm diameter, opening through 0.2 rad, would draw in almost 20 g water, so the added mass is about five times the jet mass. If five times the mass is given one-fortieth of the velocity, the kinetic energy given to the added mass (twice in each cycle) is $5/40^2 = 0.003$ of the kinetic energy given (once in each cycle) to the jet. Critical readers will find many gross approximations in that calculation. However, the result is not too different from the conclusion of the far more sophisticated calculations of Cheng et al. (1996) that the inertial work of opening and closing the shell is about 1% of the work required to power the jet.

Energy savings by elastic mechanisms have been a recurring theme in this book, starting with the discussion of basic principles in Section 3.6. It appears that in the case of the scallop the potential for savings of this kind is negligible, because the inertial work estimated above is so small. The hinge ligament makes it unnecessary for scallops to have shell-opening muscles, but it does not save them useful amounts of energy.

Just as scallops depend on the elasticity of the hinge ligament to open the shell, medusae depend on the elastic properties of their mesoglea to enlarge the bell. De Mont and Gosline (1988) measured the elastic properties both of the intact bell of the medusa *Polyorchis* and of isolated blocks of mesoglea, and calculated the strain energy stored during swimming movements. They also performed experiments on living *Polyorchis* that were tethered by gluing the apex of the bell to a support. They used a pressure transducer to record pressure changes within the bell of tethered animals as they made swimming movements. At the same time, they made video records from which they were able to calculate the changing volume of the bell. They plotted the pressures and volumes as a work loop, from which they were able to calculate the work done pumping water to produce the jet. This was presumably not exactly the same as if the medusae had been swimming freely, but is probably a reasonable approximation. It would have been extremely difficult to make the measurements on free-swimming medusae.

One of the tethered medusae habitually made single contractions, after which its bell was left vibrating passively. The frequency of these free vibrations matched the frequency of contractions of the bell of freely swimming medusae of the same size, showing that the swimming frequency is the animal's resonant frequency and that the bell can be refilled by elastic recoil of the mesoglea. In this, the medusa resembles scallops, in which the shell is refilled by elastic recoil of the hinge ligament.

The work loops showed that *Polyorchis* of 30 mm diameter did about 50 μJ of work in each swimming cycle, driving the jet. The measurements of elastic properties showed that the strain energy stored in each cycle was about 30 μJ. As the medusae swam at their resonant frequency, the inertial work required for the swimming movements must have matched the elastic work, 30 μJ. We saw that in scallops the inertial work is very small compared to the hydrodynamic work of driving the jet. In contrast, in medusae the inertial work is similar in magnitude to the hydrodynamic work. The explanation for this difference is that scallops expel water through small openings at high velocity, whereas medusae expel water through a wide opening at low velocity. Both in scallops and in medusae, energy is saved by swimming at the resonant frequency of the system; but whereas the savings are trivial in scallops, they are substantial in medusae. In scallops, the elasticity of the hinge ligament is important only as a mechanism for opening the shell, but in medusae the mesoglea also has an important energy-saving role.

Though important, the mesoglea is rather inefficient in this role. The mechanical tests on isolated blocks of mesoglea showed that only 60% of the work done deforming it was returned in its elastic recoil. Observations of the vibrations of a tethered medusa as they died away after a single contraction gave the same result (De Mont and Gosline 1988). In contrast, we have already seen that the hinge ligament of scallops gives 91% energy return.

An elastic mechanism also has a role in squid swimming. The mantle consists of a thick layer of muscle sandwiched between two sheets of collagen fibers (Fig. 16.4B). Most of the muscle fibers run circumferentially, so when they contract, the diameter of the mantle is reduced and water is driven out of the mantle cavity. The volume of the muscle remains constant, so the circumferential contraction must be accompanied by lengthening of the mantle, or thickening of its wall, or both. The collagen fibers in the inner and outer sheets run at small angles to the animal's long axis and more or less prevent lengthening, so the mantle must thicken as it contracts. In addition to the inner and outer sheets, there are some collagen fibers running through the thickness of the muscle layer that are stretched as the mantle contracts and thickens. Refilling of the mantle cavity is powered largely by elastic recoil of these fibers (Gosline and Shadwick 1983; MacGillivray et al., 1999). However, their elastic recoil does not do all the work of refilling. In addition to the circular muscle fibers, there are some radial muscle fibers (Fig. 16.3B). When they contract, they thin the mantle and stretch the circular fibers. Gosline et al. (1983) showed that the radial muscle fibers are active especially in the later part of the refilling phase of the jetting cycle.

The (negative) pressure in the mantle cavity during refilling is small compared to the (positive) pressure in jetting (Webber and O'Dor 1986). This shows that the work of refilling is small compared to the work of jetting. Also, the rates of change of radius of the mantle are small, compared to the velocity of the jet (calculated from data in Gosline et al., 1983). By an argument like the one for scallops (above), this shows that the inertial work required to drive the radial movements of the mantle wall, and of the water immediately around it, is relatively small.

The principal message of this chapter is that jet propulsion is a relatively inefficient swimming technique. This is because the volume of the animal limits the volume of water that can be ejected to provide thrust, in each cycle of movement. One of the major gaps in our knowledge that I would like to see filled concerns gait change and maneuverability in squids.

Chapter Seventeen

....

Buoyancy

FRESHWATER has a density of 1000 kg/m^3, and seawater about 1026 kg/m^3. Animals' bodies consist largely of materials that are denser than either. For example, the muscles of fishes (both selachians and teleosts) have densities between about 1040 and 1080 kg/m^3, and teleost guts have densities around 1040 kg/m^3 (Alexander 1993b). The soft parts of *Nautilus*, removed from the shell, have a density of about 1060 kg/m^3 (Denton and Gilpin-Brown 1973). Skeletal materials are generally denser, for example, 1060–1180 kg/m^3 for selachian cartilage, 1300–2000 kg/m^3 for teleost bone, and 2700 kg/m^3 for mollusc shell (Alexander 1993b; Wainwright et al., 1976). Consequently, aquatic animals that lack adaptations that would give them buoyancy are denser than the water they live in. For example, the densities of fish that lack swimbladders or other buoyancy adaptations are generally between 1050 and 1090 kg/m^3 and the densities of typical squids about 1070 kg/m^3 (Jones and Marshall 1953; Denton and Gilpin-Brown 1973). These animals will sink if they stop swimming. This chapter is about the means by which aquatic animals avoid sinking, either by means of low-density organs that reduce their density to that of the water, or by generating upward hydrodynamic forces.

17.1 BUOYANCY ORGANS

Many different materials serve as buoyancy aids in animals. Gases give buoyancy to many siphonophores, which have gas filled floats; to a few cephalopods (*Nautilus*, *Sepia* and *Spirula*) that have gas-filled chambers in their shells; and to most teleost fishes, which have gas-filled swimbladders. The densities of gases are negligible in comparison with the densities of water and of animals' bodies, but the material that encloses the gas may have a high density. For example, the walls of a *Nautilus* shell have a density of 2700 kg/m^3, and when the chambers are completely full of gas the density of the complete shell is about 910 kg/m^3, only a little less than the density of water (Denton and Gilpin-Brown 1973). The gas-filled shell of *Sepia* is much more lightly built, and has a density of only about 500 kg/m^3 in the most buoyant individuals. The densities of *Nautilus*, *Sepia*, and most teleosts that have well-developed swimbladders are very close to the density of the water they live in (Denton and Gilpin-Brown 1973; Alexander 1993b).

Some other animals gain buoyancy from low-density lipids. The livers of a few sharks contain very large quantities of the hydrocarbon squalene (Corner et al., 1969) or of wax esters (Van Vleet et al., 1984), and the densities of these sharks are very close to that of seawater. Wax esters predominate in the adipose tissues of the coelacanth *Latimeria* and permeate its remarkably oily muscles (Nevenzel et al., 1966). Many lantern fishes (Myctophidae) have gas-filled swimbladders but some do not. Some that have no gas accumulate wax esters around the vestigial swimbladder and in their muscles, and have densities of 1025–1037 kg/m^3, close to or only a little more than the density of seawater (Capen 1967). Triglycerides have densities around 930 kg/m^3, but squalene and wax esters have lower densities, around 860 kg/m^3.

In some other animals, the body fluids have remarkably low densities, due to peculiarities in their ionic composition. Many squids that swim at substantial depths contain large quantities of coelomic fluid, which has the same osmotic concentration as seawater or the blood, but has the sodium ions largely replaced by ammonium (Denton and Gilpin-Brown 1973). The densities of these fluids are about 1010 kg/m^3, and the intact animals have almost exactly the density of seawater. Scyphozoan jellyfish, some siphonophores, and ctenophores have mesogloea that is slightly less dense than seawater, due to exclusion of sulfate ions (Bidigare and Biggs 1980).

Some deep-water teleosts have no swimbladder and contain only modest quantities of lipids, but have poorly ossified bone and watery tissues, and have less than half as much protein in their bodies as coastal teleosts of equal mass (Denton and Marshall 1958). Two species had densities of 1032 and 1039 kg/m^3, substantially less than most teleosts without swimbladders, but still more than seawater.

We will calculate the quantities of low-density materials that are needed to match the densities of typical fish to the water they live in. Consider a fish that, without a buoyancy organ, would have volume $V_{without}$ and density $\rho_{without}$. We will give it a buoyancy organ of density ρ_{buoy}. What must the volume V_{buoy} of this organ be, to match the density of the animal to that of the water (ρ_{water})? The total mass of the animal is $V_{without}\rho_{without} + V_{buoy}\rho_{buoy}$, so for it to have the same density as the water

$$\frac{V_{without}\,\rho_{without} + V_{buoy}\,\rho_{buoy}}{V_{without} + V_{buoy}} = \rho_{water}$$

By rearranging this equation we find that the volume of the buoyancy organ, as a fraction of the total volume of the animal, is

Table 17.1.

The volumes and masses of buoyancy organs of various densities required to match the densities of animals to seawater (1026 kg/m^3) or freshwater (1000 kg/m^3)

	Density of buoyancy organ (kg/m^3)	In seawater		In freshwater	
		Volume	Mass	Volume	Mass
Swimbladder	Negligible	0.05	0	0.07	0
Sepia shell	600	0.09	0.04	0.13	0.07
Squalene or wax esters	860	0.23	0.19	0.35	0.30
Nautilus shell	910	0.30	0.26	0.45	0.41
Ammoniacal fluids	1010	0.75	0.74	Impossible	Impossible

Note. The volumes and masses of buoyancy organs are expressed as fractions of total body volume or mass. It is assumed that, without the buoyancy organ, the density of the animal would be 1075 kg/m^3. Two entries are "impossible" because a fluid that is denser than freshwater cannot give buoyancy in freshwater.

$$\frac{V_{\text{buoy}}}{V_{\text{without}} + V_{\text{buoy}}} = \frac{\rho_{\text{without}} - \rho_{\text{water}}}{\rho_{\text{without}} - \rho_{\text{buoy}}} \tag{17.1}$$

It is often more convenient to measure the mass of a buoyancy organ, than its volume. Let the mass of the body excluding the buoyancy organ be m_{without}, and let the mass of the buoyancy organ be m_{buoy}. Then $V_{\text{buoy}} = m_{\text{buoy}} / \rho_{\text{buoy}}$, and if the animal has the same density as the water $(V_{\text{without}} + V_{\text{buoy}}) = (m_{\text{without}} + m_{\text{buoy}})/(\rho_{\text{water}})$. By substituting these equations in Equation 17.1 we find

$$\frac{m_{\text{buoy}}}{m_{\text{without}} + m_{\text{buoy}}} = \frac{(\rho_{\text{buoy}}/\rho_{\text{water}}) (\rho_{\text{without}} - \rho_{\text{water}})}{\rho_{\text{without}} - \rho_{\text{buoy}}} \tag{17.2}$$

Table 17.1 has been calculated from Equations 17.1 and 17.2. It shows how large buoyancy organs need to be to match the density of a typical animal to freshwater or seawater. As it predicts, teleosts with densities close to seawater or freshwater generally have swimbladders occupying about 5 or 7%, respectively, of their volume (Jones and Marshall 1953; Alexander 1959b). Other buoyancy organs are much denser, so have to be much larger.

17.2. Swimming by Dense Animals

A fish that is denser than water may prevent itself from sinking simply by swimming with its body tilted, so that the thrust generated by its swimming movements has a vertical as well as a horizontal component. The horizontal component must equal the drag F_{drag}, so the vertical component must be $F_{\text{drag}} \tan \alpha$. This vertical component of the thrust may be augmented by hydrodynamic lift (F_{lift}), due to the body functioning as a hydrofoil. The ratio of these two effects is

$$\frac{\text{Lift}}{\text{Vertical component of thrust}} = \frac{F_{lift}}{F_{drag} \tan \alpha} \qquad (17.3)$$

The ratio F_{lift}/F_{drag} will depend on the shape of the fish. Flatfish (Pleuronectiformes) are shaped like low-aspect ratio aerofoils, so their bodies will be more effective as hydrofoils than those of most other fish. Weihs (1973) estimated that lift/drag ratios up to 5 should be possible for them. This implies that when the body is tilted at small angles, the lift should be much larger than the vertical component of thrust.

Most other fish have narrow bodies, much less well shaped for generating lift. Rather than swimming with the whole body inclined at an angle of attack, these fish can be expected to do better by swimming with the body horizontal and using fins as hydrofoils to counteract their tendency to sink.

Sharks obtain some of the lift that they need from the tail, which has the asymmetrical form that is described as heterocercal (Fig. 17.1). It drives water downward as well as backward as it beats from side to side, and so provides an upward component of force as well as forward thrust. Alexander (1965) demonstrated the upward force in experiments with severed tails. By releasing streams of dye into a water tunnel in which a leopard shark (*Triakis*) was swimming, Ferry and Lauder (1996) showed that the tail drives water downward as well as backward.

For a swimming shark to be in equilibrium, not only must the upward forces balance the downward ones, but the moments of the forces about a transverse axis must balance. Until recently, it was believed that the balance of forces was as shown in Fig. 17.1A. The hydrodynamic force on the tail exerts an anticlockwise moment about the center of mass. The weight of the fish acts (by definition) at the center of mass, and so exerts no moment about it. However, the Archimedes upthrust acts at the center of buoyancy, which is slightly anterior to the center of mass because the posterior parts of the body are denser than the anterior parts. Consequently, the upthrust exerts a clockwise moment about the center of mass. However, because the centre of buoyancy is so close to the center of mass, this moment is far too small to balance the anticlockwise moment from the tail. It was believed that the upward force needed to balance the moments was provided by lift on the pectoral fins. These fins cannot be folded, but project on either side of the body like the wings of aircraft. Alexander (1965) calculated that for the moments on a swimming dogfish (*Scyliorhinus*) to be balanced, the pectoral fins must supply about 70% of the required lift, and the tail 30%.

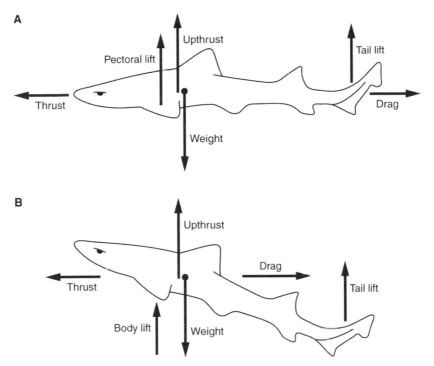

Fig. 17.1. Diagrams showing the forces acting on a shark swimming at constant depth (A) if lift is obtained from the pectoral fins and (B) if lift is provided by the tilted body. In each case, the heterocercal tail also contributes to the lift.

That interpretation had to be changed when Wilga and Lauder (2000) used particle image velocimetry to examine the flow behind the pectoral fins of a leopard shark swimming in a water tunnel. They found no vortices behind the fins except when the shark was initiating a climb or descent. There were no vortices behind the fins in level swimming, so they were providing no lift. Indeed, the fins were held at an angle of attack at which no lift could be expected. However, the fish swam with their bodies tilted nose-up. The angle of tilt was about 11° when the shark swam slowly, and less at higher speeds. It appeared that lift was acting on the body, as shown in Fig. 17.1B. Presumably, the center of lift on the body was anterior to the center of mass, as required to balance the moments. It would be expected to be well forward, because the centers of lift of aerofoils are much nearer their leading than their trailing edges, and because the posterior parts of the body are much narrower than the head. It seems likely that other species of shark also rely on body lift, but this has not yet been demonstrated. The finding that *Triakis* swims so as to get lift from its body rather than its fins is unexpected. The body is much longer than it is wide, so its aspect ratio is very low. The lower the aspect ratio of a hydrofoil, the greater the induced drag (Equation 10.13).

Many tunas lack swimbladders, and so are denser than the water in which they swim. For example, *Euthynnus* has a density of about 1086 kg/m^3 (Magnuson 1970). They swim with their pectoral fins extended and have been believed, like sharks, to depend mainly on lift on these fins to counteract their weight in water (Magnuson 1970). This point needs reexamination in the light of the experiments on *Triakis*. For the present, we will suppose that the fins are the main source of lift. If the fish swam too slowly, they would not be able to provide enough lift. For example, a 1-kg (0.38-m) *Euthynnus* needs about 0.6 N lift from its pectoral fins (Magnuson 1970). The plan area of these fins (measured as indicated for bird wing areas, in Fig. 10.1A) is about 0.003 m^2. We have no measurements to tell us the maximum lift coefficient that these fins can have without stalling, but it seems likely to be about 1.0. Equation 10.4 tells us that with this lift coefficient, the pectoral fins would give the required lift at a swimming speed of 0.6 m/s. This suggests that a 0.38-m *Euthynnus* should be capable of swimming at speeds down to 0.6 m/s (1.6 body lengths per second). *Euthynnus* observed in an aquarium by Magnuson (1973) never swam slower than 2.0 body lengths per second.

Rather than swim a straight, horizontal path, it may be more economical of energy for a fish that is denser than water to alternate powered swimming on an upward slope and a gliding descent (Weihs 1973). This can save energy because swimming by beating the tail requires more power than would be needed to propel a rigid body at the same speed (Section 15.1). While a fish is gliding, the additional power is not required. Weihs (1973) estimated that *Euthynnus* should be able to glide at a minimum angle of about 11°. This would enable it to save 20% of the energy that would be needed for level swimming, by climbing and gliding. However, this estimate has not been checked by experiment.

As we have seen, fish that are denser than water and depend on their pectoral fins for lift cannot swim below a critical speed if they use these fins as fixed hydrofoils. However, they may be able to hover if they beat the fins backward and forward, like the wings of a hovering hummingbird. If the fish is substantially denser than the water, the power required is high (Section 17.3), probably more than the pectoral fin muscles of most teleosts can provide. Jones (1952) and Bishai (1961) altered the densities of various perciform fishes by reducing the ambient pressure, making the swimbladder expand. Initially, the fish had densities very close to that of the water. When the pressure was reduced, they compensated for their reduced density by beating their pectoral fins. This enabled them to remain stationary in mid water, in freshwater of density 1000 kg/m^3, until their densities fell to about 975 kg/m^3. If the pressure and so their density were reduced further, the fish had to swim downward by beating their tails to avoid being carried to the surface. The converse experiment, of testing the ability of fish to hover when their density was increased, does not seem to have been performed, but it seems likely that most teleosts would be able to compensate only for small increases of density by beating their pectoral fins. In this respect, the mandarin fish (*Synchropus*) seems to be exceptional (Blake 1979a). This small coral reef fish has an unusually high density of 1150 kg/m³, but spends much of its time hovering close above the bottom by beating its pectoral fins.

17.3. ENERGETICS OF BUOYANCY

A fish that is denser than water must swim or hover actively, using metabolic energy, to prevent itself from sinking. One with a buoyancy organ may be able to float almost motionless in mid water, but expends metabolic energy growing and maintaining the buoyancy organ. Also, because the buoyancy organ makes its body bulkier than it would otherwise be, it needs more energy to swim at any given speed. Alexander (1990) estimated the energy costs of different buoyancy strategies and discussed their relative merits for animals with different ways of life.

Consider first an animal of mass m whose body has a density ρ that is greater than the density (ρ_{water}) of the water. The weight of the animal is mg, where g is the gravitational acceleration. Its volume is m/ρ, so the upthrust that acts on it, by Archimedes' principle, is $mg\rho_{water}/\rho$. The lift F_{lift} needed to prevent sinking is the difference between the weight and the upthrust:

$$F_{lift} = \frac{mg\,(\rho - \rho_{water})}{\rho} \qquad (17.4)$$

The animal may hover, keeping itself stationary in the water by beating its fins. If it beats a pair of fins of length r through an angle ϕ, the power required can be estimated (using Equation 11.8) as

$$\text{Power} = \left(\frac{F_{lift}^{3}}{2\rho_{water}\,r^{2}\phi}\right)^{0.5} \qquad (17.5)$$

$$= \left[\frac{m^{3}g^{3}\,(\rho - \rho_{water})^{3}}{2\rho^{3}\rho_{water}r^{2}\phi}\right]^{0.5}$$

(This estimate is of induced power. The total power required, including also profile power, would be higher.)

To get a rough idea of the magnitude of this power, consider a 1-kg fish of density 1075 kg/m³, in water of density 1026 kg/m³. Teleosts of this

$V^{0.33}v$, and drag coefficients for streamlined bodies with laminar boundary layers are about proportional to $1/(\text{Reynolds number})^{0.5}$, or to $V^{-0.17}v^{-0.5}$. Thus, we expect the power required for swimming to be

$$P = kV^{0.5}v^{2.5} \tag{17.8}$$

where k is a constant. We started this argument with the assumption that the power was the same as would be needed to propel a rigid body, but we saw in Section 15.1 that the power used by a swimming fish is several times higher. Equation 17.8 will still hold, provided the power is the same multiple of the rigid body power for different fish and different speeds.

Now consider an animal that has a buoyancy organ of volume V_{buoy} in addition to its basic volume V. Because it is bigger, it needs additional power P_{buoy} to swim:

$$
\begin{aligned}
P + P_{buoy} &= kv^{2.5}(V + V_{buoy})^{0.5} \tag{17.9}\\
&= kv^{2.5}(V^{0.5} + 0.5\,V_{buoy}V^{-0.5} + \ldots)
\end{aligned}
$$

(using the binomial theorem, which is explained in mathematical textbooks). If V_{buoy} is not too large, we can ignore later terms in the series and subtract Equation 17.8 from 17.9 to obtain

$$P_{buoy} \approx 0.5\,kv^{2.5}V_{buoy}V^{-0.5} \tag{17.10}$$

$$\frac{P_{buoy}}{P} \approx \frac{0.5\,V_{buoy}}{V} \tag{17.11}$$

Thus, the increased size of the animal, due to the presence of the buoyancy organ, adds about 2.5% to the power needed for swimming by a fish with a swimbladder that adds 5% to its volume, and 11.5% for a fish with squalene or wax esters that add 23% to its volume (Table 17.1).

Equation 17.6 showed that for animals that are denser than the water, the power required to generate lift falls as the swimming speed increases. Equation 17.10 shows that for animals with buoyancy organs, the additional power required for swimming increases as speed increases. This tells us that for animals that swim slowly, it is more economical to have a buoyancy organ; and for those that swim fast, it is more economical to depend on hydrodynamic lift. The speed at which the balance of advantage changes from a buoyancy organ to lift should be higher for animals with swimbladders (which add only a little to the volume of the body) than for animals with denser buoyancy organs (which have to be larger).

However, we have not yet considered the cost of growing and maintaining the organ. A fish with a swimbladder must expend energy secreting gas into the swimbladder, both to enlarge the swimbladder as it grows and to replace gases lost by diffusion. Alexander (1972) attempted to calculate the energy cost of replacing diffusion losses for fish living at different depths in the sea. I was unable to reach a reliable quantitative conclusion, because the efficiency of gas secretion had not been measured. Even now there seem to be no measurements of the energy cost of secreting swimbladder gases, though the mechanism of gas secretion is well understood (Pelster and Scheid 1992). However, it seems clear that the cost of replacing diffusion losses from the swimbladder will be significant only for fish that live at substantial depths where the hydrostatic pressure is very high. Even for them it may not be very large. The walls of swimbladders are generally made remarkably impermeable to diffusing gases by deposits of ribbonlike guanine crystals, especially in deep sea fishes (Denton et al., 1970; Lapennas and Schmidt-Nielsen 1977).

Animals, such as ammoniacal squids, that make themselves buoyant by modifying the composition of their body fluids must use energy to secrete ions in and out of this fluid to maintain its composition. Cephalopods, such as *Nautilus* and *Sepia*, that depend on gas-filled shells for buoyancy must expend energy maintaining the osmotic pressure differences that prevent the cavities from filling with liquid (Denton and Gilpin-Brown 1973). I see no easy way of calculating the energy costs in these cases. However, we can make a very simple calculation of the energy cost of accumulating squalene or wax esters.

Consider an animal of mass m containing a mass m_{buoy} of low-density lipid. If it has a relative growth rate G (i.e., if its mass increases at a rate Gm), it must accumulate lipid at a rate Gm_{buoy} to keep its density constant. If the enthalpy of combustion of the lipid is H, this ties up energy that would otherwise be available for metabolism at a rate $Gm_{buoy}H$. For a typical 1-kg fish, G would be about 10^{-8} s^{-1} (Calder 1984). With this growth rate, the fish would add 37% to its body mass in a year. Also m_{buoy} would be about 0.19 kg (Table 17.1). The enthalpies of combustion of squalene and wax esters must each be about 40 MJ/kg (see data for similar compounds in Weast [1987]). Hence, we can estimate that the energy cost of accumulating the lipid would be 0.08 W. This is only half of the energy cost that we estimated for hovering by a 1-kg fish without a buoyancy organ.

17.4. BUOYANCY AND LIFESTYLE

We have seen that some swimming animals have densities very close to that of the water they live in, and so can remain almost motionless in mid water, with little tendency to rise or sink. Others are considerably denser than the water, and sink quite rapidly if they stop swimming. Some of the buoyant animals have gas-filled floats, while others depend for their buoyancy on low-density lipids or on body fluids of unusual composition. We have assessed as far as we could the energy costs associated with different buoyancy strategies. This section is about the relative merits of the different strategies for animals with different ways of life.

If an animal is denser than the water it lives in, frictional forces will help to hold it in place when it rests on the bottom. For this reason, there may be a positive advantage in being denser than water for animals that spend a lot of time resting on the bottom. These include selachians, such as *Scyliorhinus* and *Raia*, which have densities of about 1075 kg/m³ (Jones and Marshall 1953); teleosts, such as the flatfishes (Pleuronectiformes), which have lost the swimbladder and have similar densities; and octopus. Indeed, it might be an advantage to some of these fishes to be even denser. Webb (1989) has shown that the lift that acts on *Raia* and *Pleuronectes* when they rest on the bottom in even quite a slow current may be enough to make them lose their frictional grip.

Some animals that swim perpetually are also denser than water. They include pelagic sharks; tunnies, such as *Euthynnus* and *Katsuwonus*, which have lost their swimbladders and have densities of 1080 to 1100 kg/m³ (Magnuson 1973); and squids, such as *Loligo* (about 1070 kg/m³ [Denton and Gilpin-Brown 1973]). These animals depend on hydrodynamic lift to prevent sinking. Calculations in Section 17.3 showed that the energy cost of swimming fast might be less for animals that rely on lift than for those that have buoyancy organs. These dense animals may swim fast enough for buoyancy organs to be disadvantageous.

Our theoretical discussion showed that animals that swim slowly or hover in mid water can save energy by evolving a buoyancy organ. A large proportion of teleosts have well-developed swimbladders that give them densities very close to that of water (Jones and Marshall 1953; Alexander 1959b), and are capable of resting almost motionless in mid water. Whereas most sharks are denser, the basking shark *Cetorhinus* has enough squalene in its body to match its density almost exactly to seawater (Bone and Roberts 1969). It swims slowly, filter feeding. The cephalopods that have gas-filled shells that match their densities closely to the water (*Nautilus*, *Sepia*, and *Spirula*) are more sluggish in their behavior than squids such as *Loligo*. These animals may spend most of their time swimming slowly enough for the balance of advantage to favor buoyancy organs.

The less dense the buoyancy organ, the smaller it need be, and the less the extra energy cost of swimming given by Equation 17.10. This suggests that swimbladders or similar gas-filled floats should be the preferred buoyancy organs. However, evolution is constrained by ancestry, and we should not expect every animal to have the type of buoyancy organ that seems in theory to be the most advantageous. Also, gas-filled floats present problems for animals that make large changes of depth.

Pressure under water increases by one atmosphere for every ten meters of depth. The gases in the swimbladders of most fish are at the same pressure as the surrounding water, and are compressed or expand as the pressure changes, according to Boyle's law (Alexander 1959c). Thus, a swimbladder that contains enough gas to give the fish neutral buoyancy when it is close to the surface is reduced to half the required volume at 10 m depth (where the pressure is 2 atm), and one-tenth of the required volume at 90 m (10 atm). Whenever the fish descends from the surface, it becomes denser than the water and must swim to prevent itself from sinking further; unless, of course, it can secrete gas fast enough to keep the volume of the swimbladder constant. Similarly, a swimbladder that contains enough gas to give a fish neutral buoyancy at 90 m will expand so greatly if the fish ascends too far that the fish may be carried helpless to the surface and the swimbladder may burst.

The range of depths at which the density of a fish with a swimbladder is close enough to the density of the water for it to hover by fin movements alone may be very restricted. Jones (1952) found that perch (*Perca*) could not hover at pressures more than 16% below the pressure to which the swimbladder was adapted. Fishes of the order Cypriniformes (carps, characins, catfish, etc.) have swimbladders with less extensible walls than other teleosts, inflated with gas at pressures up to 0.14 atm above ambient. These swimbladders change volume less for small changes of ambient pressure than do the swimbladders of other fish, possibly enabling these fish to hover over a greater range of depths (Alexander 1959a, 1961).

Teleosts can keep their density constant if they change depth sufficiently slowly, by secreting gas into the swimbladder or allowing it to diffuse out into the blood. Jones and Scholes (1985) measured the rate at which cod (*Gadus morhua*) can compensate for depth changes by keeping them in a pressure tank with a viewing window. They were able to vary the pressure in the tank from 1 to 7.5 atm, simulating depths from 0 to 65 m. After each change of pressure, they monitored the rate of compensation by occasional brief tests in which they adjusted the pressure to find the one at which the fish just floated. They found that the cod adjusted to simulated increases of depth at a rate of only 1 m per hour at 12°C, or less at lower temperatures. The rate of compensation for simulated decreases of depth depended on the depth but not on the temperature; it was 1 m/h at a simulated depth of 5 m, and 20 m/h at 65 m.

Within the simulated range of depths, the rate of compensation for increased depth was constant. This is what we should expect. Water and fish tissues are effectively incompressible, so the swimbladder volume required for neutral buoyancy is the same at all depths. However, the mass of gas required to fill that volume increases in proportion to the pressure, which increases linearly with depth. If the mass of gas that can be secreted in unit time is independent of depth, the rate of compensation for increased depth should also be independent of depth. However, like other processes driven by metabolism, secretion can be expected to proceed more slowly at lower temperatures. Compensation for decreased depth is by passive diffusion from the swimbladder to the blood, driven by a difference in partial pressure that increases with depth. Accordingly, it is faster at greater depths.

Some other fish can secrete gas faster than cod. For example, *Pomatomus* can secrete fast enough to compensate for depth increases at a rate of 2.5 m/h (Wittenberg et al., 1964). However, it seems clear that any fish that is required to keep its swimbladder volume constant would be restricted to slow rates of change of depth.

Some fish make very rapid daily changes of depth. Notable among them are the lantern fishes (Myctophidae), which are extremely plentiful in the oceans. They spend the nights near the surface but the days much deeper. Barham (1966), in a submersible vehicle, observed lantern fishes commuting between the top 50 m of the sea and depths of around 300 m, at rates of the order of 100 m/h. Fish changing depth as fast as that cannot keep their swimbladder volumes constant. One option for them would be to keep enough gas in the swimbladder to match their density to the water at the nighttime depth, and to tolerate compression of the swimbladder to a small fraction of its nighttime depth by day. The converse strategy of keeping enough gas in the swimbladder for neutral buoyancy at the daytime depth would not be an option, because it would expand to unmanageable volumes as they ascended in the evening.

In these circumstances, swimbladders may lose their advantage over other types of buoyancy organs. Compare a fish with a swimbladder with one that relies on wax esters for buoyancy. At night, when they are near the surface, both fish have the same density as the water, but the swimbladder gives the same buoyancy for less volume. By day, when they are much deeper in the water, the fish with the swimbladder has to generate hydrodynamic lift to prevent itself from sinking further, but the one with wax esters still has almost exactly the same density as before. The shortcomings of the swimbladder at depth may outweigh its advantage by day. Lantern fishes that have lost their swimbladders and rely instead on wax esters for buoyancy (Capen 1967) may benefit from the change.

The shells of *Nautilus*, *Sepia*, and *Spirula* are sufficiently rigid to keep the volumes of gas within them constant as the animal changes depth. Consequently, large pressure differences develop across the wall of the shell as the animal descends, and if it goes too deep the shell implodes. *Spirula* has the strongest shell, capable of withstanding 170 atm, or a depth of about 1700 m (Denton 1974).

This chapter started with the observation that animals without buoyancy organs are denser than water. This may be an advantage if they spend most of their time on the bottom. Buoyancy organs that match the density of the animal to the water reduce the energy cost of slow swimming. They increase the cost of fast swimming because, by increasing the volume of the animal, they increase drag. Because gases have negligible density, a gas-filled float can be smaller than any other, but unless it has a rigid wall, its volume is changed by the changes of pressure when the animal swims to different depths. Our discussion of the costs and benefits of buoyancy organs has been incomplete, because we do not yet know the metabolic energy costs of maintaining buoyancy or changing depth by secretion of gas into a swimbladder or by withdrawing fluid from a cephalopod shell.

Chapter Eighteen

....

Aids to Human Locomotion

*U*NLIKE ANIMALS, we humans make a great deal of use of manufactured aids to locomotion. We wear shoes. Scuba divers carry
gas cylinders and wear fins on their feet. We ride bicycles and row
boats. And we make a great deal of use of vehicles with engines, including
cars, ships, and aircraft. It seems inappropriate to discuss engine-powered
vehicles in this book, but it seems interesting to ask how devices that do
not incorporate engines enable us to make more effective use of our own
muscles. Why, for example, is it faster and less tiring to cycle than to run?
This chapter attempts to answer questions like that.

18.1. SHOES

Shoes protect our feet and make walking more comfortable on hard,
rough surfaces. As well as protecting us from immediate injury from sharp
objects, running shoes with compliant soles may protect us from the cu-
mulative injuries that may result in osteoarthritis, by cushioning the im-
pacts of the foot with the ground. The foot is still moving when it hits
the ground, typically at speeds around 0.7 m/s. Both the ground and the
foot itself deform a little, but the foot is brought to rest in a very short
distance. The brief peak in force plate records such as Fig. 7.8D, immediately following impact with the ground, represents the force that decelerates the foot and lower leg (Ker et al., 1989). Most human runners hit the
ground first with the heel, and the fatty pad under the heel helps to cush-
ion the impact (Aerts et al. 1995). The force peak at impact may nevertheless be large, especially on artificial surfaces, such as concrete, that are less
compliant than most natural ground. Compliant heels on running shoes
supplement the natural cushioning. Dickinson et al. (1985) found that a
man running barefoot across a force plate produced much larger impact
peaks than when he wore trainers.

Bennett and Ker (1990) and Aerts et al. (1995) used a dynamic testing
machine to squeeze heel pads taken from amputated feet. They found that
briefly applied loads of 1500 N, simulating the impact peaks that occur in
running, deformed the pad by about 4 mm. Alexander and Bennett (1989)
made similar tests on the heels of running shoes and found that the same

force caused 7–15 mm deformation. Thus, the compliance of the shoe heels is a very substantial supplement to the compliance of the natural heel pad. The heels of some brands of shoe are twice as compliant as the heels of others, suggesting that some shoes should give much more effective cushioning than others. However, Nigg et al. (1987) made force plate records of athletes running at the same speed in shoes of different compliances, and found no significant difference in impact forces. The athletes must have adapted their running style when they changed shoes, but it is not clear how they did this. There was no apparent difference in the velocity of the foot or in the angle of the knee at impact. However, there seemed to be a tendency for runners in less compliant shoes to hit the ground first with the lateral edge of the heel. This may have increased the effective deceleration distance, due to the foot rotating about its long axis as it settled on the ground.

We saw in Section 7.4 that the elastic properties of the arch of the foot, as well as of the Achilles tendon, save energy in human running. Kinetic energy lost by the body in the first half of the foot's period of contact with the ground is stored as elastic strain energy and restored by elastic recoil in the second half. The strain energy stored in the heel of a running shoe cannot help here, because the heel leaves the ground, losing its strain energy, while the body is still decelerating. However, when the load has been transferred to the ball of the foot, strain energy is stored in the sole of the shoe. This energy is returned by elastic recoil just before the foot leaves the ground. This suggests that the compliance of the sole may have a significant energy-saving function. Alexander and Bennett (1989) applied forces of 2000 N to the soles of running shoes, simulating the peak force that would act on the sole in running, and found that they deformed by 9–12 mm. This should enable the sole of the shoe to store as much strain energy as the arch of the foot, which flattens by about 10 mm.

There is evidence that extra compliance, additional to the natural compliance of the tendons and foot, could be beneficial. McMahon and Greene (1978) designed a sprung indoor running track that was depressed and recoiled by about 9 mm during each footfall of an adult runner. Athletes regularly ran 3% faster in races on it than on conventional tracks. In principle, it should be possible to obtain the same advantage on a rigid track from shoes with compliant soles. Manufacturers have been reluctant to make soles more compliant than those currently used, because of the danger of the foot rocking to one side, resulting in a sprained ankle.

It is, of course, important that running shoes not be too heavy, because a shoe adds to the mass that has to be accelerated and decelerated as the foot swings forward and back (Section 7.5). However, a sole that incorporates gas bubbles can be highly compliant and reasonably light.

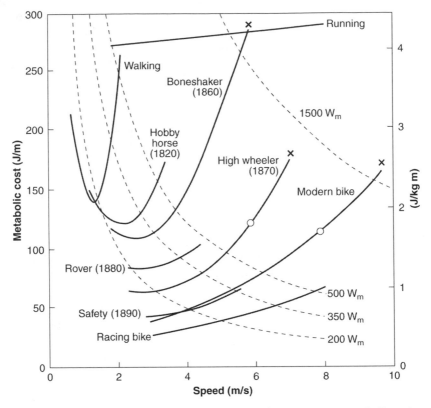

Fig. 18.1. Graphs of metabolic energy cost (joules per meter traveled) against speed, for adult humans walking, running, and riding various bicycles. The graphs for the Boneshaker, High Wheeler, and Safety bicycle have been extrapolated beyond the speeds used in the experiments. The crosses and circles mark record speeds achieved in one-hour and 24-hour trials, respectively, on various dates between 1870 and 1894. The broken lines are contours of equal metabolic power. From Minetti et al. (2001).

18.2 BICYCLES

Figure 18.1 shows energy costs of human locomotion, calculated from measurements of oxygen consumption. Running uses about 280 J for every meter traveled, at all speeds from a slow jog to the highest speed that can be sustained aerobically (about 6 m/s). The cost of cycling increases with increasing speed. It is only about 70 J/m on an ordinary bicycle or 40 J/m on a racing one, at 6 m/s. Cycling is very much more economical than running.

To understand why this should be, we need to remind ourselves of how energy is used in running. A runner loses and regains external kinetic energy and gravitational potential energy in every step. Some of the lost energy is stored as elastic strain energy and returned in an elastic recoil, but much of it has to be replaced by muscular work, at the cost of metabolic energy (Sections 7.3 and 7.4). In cycling at constant speed on level ground, however, the external kinetic energy and gravitational potential energy (both of the rider and of the bicycle) are constant. The internal kinetic energy of a runner fluctuates as the legs swing forward and back, demanding work from muscles (Section 7.5). Fluctuations of internal kinetic energy occur in cycling as the legs turn the pedals, but these fluctuations are much smaller than in running because the feet move more slowly. Metabolic energy is also needed in running to develop the large muscle forces that are needed to support the weight of the body (Section 7.6). In cycling, the legs do not have to support the weight of the body, and the forces they exert on the pedals are much smaller than the forces that the feet of runners exert on the ground. (The forces involved in cycling have been measured by means of bicycles with instrumented pedals [Davis and Hull 1981].)

Cyclists have to do some work to overcome friction in the bearings of the bicycle, but if the machine is well lubricated, that work is small enough to be ignored. Nearly all the work that cyclists have to do at constant speed on level ground is needed to overcome rolling resistance and air resistance (Pugh 1974). Rolling resistance arises because the tire and the ground are both distorted by the load at the point of contact. As the wheels revolve, different parts of the tires and of the ground are distorted and recoil. Much of the work done on them is returned in the recoil, but some is lost as heat, which is why tires get warm as you travel. In addition, energy is dissipated by vibrations of the bicycle. The energy lost in these ways has to be replaced by muscular work. Because every part of a tire is loaded and unloaded once in each revolution of the wheel, the work required per meter traveled is more or less independent of speed. Work is also required to overcome air resistance, but because aerodynamic drag increases roughly in proportion to the square of speed, the work required per unit distance is proportional to speed squared. In Pugh's (1974) experiments, the amounts of work needed to overcome rolling resistance and air resistance were roughly equal at 6 m/s, but at 12 m/s air resistance was four times as important as rolling resistance.

Figure 18.1 shows measurements of the metabolic cost of riding old bicycles as well as modern ones (Minetti et al., 2001). The old bicycles were genuine antiques, with the exception of the Hobby Horse, which was a faithful modern copy. The graph shows progressive reduction of the energy cost of cycling from the Hobby Horse of the 1820s to a modern racing bicycle. The Hobby Horse had no pedals. The rider sat on a saddle low enough for his feet to reach the ground, propelling himself by means of his feet. In comparison with running, fluctuations of external kinetic and gravitational potential energy were more or less eliminated, and the legs were relieved of the need to support the weight of the body, but the leg movements involved substantial fluctuations of internal kinetic energy. On this machine it is possible to travel at a moderate running speed, for little more than half the energy cost of running.

The Boneshaker of the 1860s has pedals attached directly to the front wheel. The leg movements of pedaling require smaller fluctuations of internal kinetic energy than those of riding the Hobby Horse, and the energy cost of riding the Boneshaker is slightly lower. The Boneshaker has iron rims on its wheels (hence its name). The front wheel has a diameter of 0.89 m (which is larger than modern bicycle wheels). Because the pedals are fixed to it, the bicycle advances only one front wheel circumference (2.8 m) for each revolution of the pedals. Consequently, the rider has to pedal at high frequency to travel fast.

This problem was alleviated in the High Wheeler of the 1870s (also known as the Penny Farthing) by making the front wheel much larger. The example used for the experiments has a front wheel diameter of 1.27 m. The pedals are still attached directly to the wheel, but the machine travels further for each revolution, so the pedaling frequency is lower at any given speed. Figure 2.3 shows that any muscle has an optimum rate of shortening, at which it does work most efficiently. The effect of different rates of pedaling on efficiency has been investigated by measuring the rates of oxygen consumption of people on bicycle ergometers. At fairly high power outputs, the efficiency is near-maximal at pedaling rates of 0.6 to 1.8 revolutions per second, and lower both at higher and at lower rates (Vandewalle et al., 1987). At a speed of 6 m/s, the pedaling rate is 2.1/s on the Boneshaker and 1.5/s on the High Wheeler. Cyclists on modern machines generally prefer rates of about 1.3/s, and select their gears accordingly.

Following the High Wheeler, the Rover bicycle of the 1880s improved safety rather than economy of energy. It has wheels of diameter only 0.75 m, making it less painful to fall off than a High Wheeler. The pedaling frequency at any given speed is nevertheless slightly lower than for the High Wheeler, because a chain drive (as in modern bicycles) makes the wheels rotate at a higher frequency than the pedals. The energy cost of cycling on it was a little higher than for the High Wheeler, due to higher rolling resistance. Both it and the High Wheeler had solid rubber tires. The Rover vibrated more severely than the High Wheeler because its spokes were shorter than in the large load-bearing wheel of the High Wheeler, giving it stiffer suspension. The Safety bicycle of the 1890s had pneumatic tires, making the suspension less stiff and greatly reducing the rolling resistance. Figure 18.1 shows that it is more economical of energy than the High Wheeler, and that a modern racing bicycle is still more economical.

We have seen that at high speeds, air resistance accounts for the great majority of the work of cycling. It can be reduced a little by using tubes with streamlined sections instead of round ones to build the frame of the bicycle. Capelli et al. (1993) found that riders on a bicycle with a stream-lined frame used 4.5% less oxygen, at a speed of 11 m/s, than on one with a traditional round-sectioned frame.

The air resistance can be reduced much more by enclosing the bicycle and rider in a streamlined shell, but this is forbidden by the rules of bicycle racing. Many enclosed, streamlined bicycles have been built (Abbott and Wilson 1995). One model (the Cheetah) has done a flying 200 m at 29 m/s. This equals the speed of a real cheetah (*Acinonyx jubatus*) over the same distance (Sharp 1997), and is very much faster than the world record for a flying 200 m on a machine that conformed to the rules (20 m/s).

Bicycles are fast and economical of energy on man-made roads. Mountain bikes, with thick tires and a wide range of gear ratios, are effective on softer or rougher natural surfaces. However, walking and running are practicable on much natural terrain that is unsuitable for wheeled vehicles (LaBarbera 1983).

18.3. SCUBA

Scuba (self-contained underwater breathing apparatus) enables people to swim underwater. The vital part of the equipment is the gas supply, but scuba divers also wear fins on their feet. When swimming at the surface, we depend mainly on our arms for power. We keep our arms submerged for the power stroke but (in strokes such as the crawl) lift them above the water for the recovery stroke. Divers cannot lift their arms out of the water, so their arms are less effective than in surface swimming. Accordingly, scuba divers depend on their legs for propulsion. Naked feet are too small to work well as paddles, so fins are used to enlarge them. Fins increase the mass of water that the feet push on, and so improve Froude efficiency (Section 14.1).

Pendergast et al. (1996) measured the oxygen consumption of scuba divers swimming 1.25 m below the surface in an annular pool. At the speeds that they investigated, the energy cost of scuba swimming was a

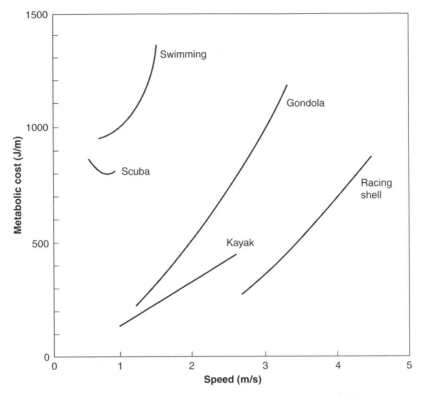

Fig. 18.2. Graphs of metabolic energy cost (joules per meter traveled) against speed for adult men swimming at the surface, scuba diving, sculling a gondola, paddling a kayak, and rowing a racing shell. Data are from Capelli et al. (1990) and Pendergast et al. (1989, 1996).

little less than for surface swimming, although the divers carried gas cylinders and the surface swimmers with whom I am comparing them did not (Fig. 18.2). The principal reason for this is that the divers avoided wave drag by swimming well below the surface (Section 13.3).

18.4. BOATS

Human-powered boats are generally faster than swimming, and less metabolic energy is needed to propel a small boat than to swim at the same speed. For example, Capelli et al. (1990) measured the oxygen consumption of men sculling a Venetian gondola and found that in the range of speeds at which swimming is possible, gondoliers used only about one-quarter as much energy as swimmers (Fig. 18.2).

Gondolas are heavy, for example 450 kg in the case of the one used for the experiment. Consequently, by Archimedes' principle, a large volume must be below the water surface and friction drag must act on a large area. Less drag will act on lighter boats of similar shape at the same speed. Pendergast et al. (1989) measured the oxygen consumption of elite men paddling a solo kayak with a mass of only 12.5 kg in a canal. At low speeds they used energy at about the same rate as gondoliers, but at higher speeds the kayak was much more economical than the gondola (Fig. 18.2).

Both gondolas and kayaks are propelled by lift on a hydrofoil. A gondolier moves his oar from side to side, adjusting the angle of attack of the blade appropriately, so as to generate thrust by the same principle as the side-to-side movements of a tuna's tail (Section 14.4). The blades of a kayak paddle are driven downward through the water in the power stroke, propelling the boat by the principle illustrated in Fig. 14.4C. In contrast, rowing boats are propelled by drag on the oars. The principle is the same as for water beetles (Section 14.2) but, unlike the beetle, a human rower lifts the oars clear of the water as he or she brings them forward in preparation for the next stroke.

Rowing boats designed for practical purposes such as fishing or crossing rivers are strongly built and correspondingly heavy. They are broad in the beam, making them stable and relatively safe. The seats are fixed, and rowing has to be powered by the muscles of the arms and trunk without assistance from the rower's legs. Racing shells are very different. They are extremely light, typically about 14 kg for single sculls and 93 kg for eights (Abbott and Wilson 1995). They are very long and narrow; a single sculls shell is typically 8 m long and only 0.3 m wide. This design keeps wave drag low. Sliding seats enable the leg muscles to be used, and indeed to provide most of the power. Athletes on ergometers that simulate sliding-seat rowing can produce almost as much power as on a bicycle ergometer (Abbott and Wilson 1995).

The world records for 2000-m races represent speeds of 5.0 m/s for single sculls and 6.2 m/s for eights. A single scull 8 m long has a hull speed (see Section 13.3) of 3.5 m/s, and an eight 18 m long has a hull speed of 5.3 m/s. It is only because wave drag is so low that they can exceed their hull speeds.

Energy costs of rowing as well as of kayaking are shown in Fig. 18.2. Unfortunately, the data are not strictly comparable because the rowing shell carried two oarsmen and a cox (each oarsman expended the energy shown), whereas the kayak carried only one man. Record speeds for kayaks are just a little slower than for racing shells, for example, 4.6 m/s for a solo kayak over 1000 m. This suggests that the difference in energy costs is small at high speeds.

McMahon (1971) explained why eights can go faster than boats with fewer crew. He assumed, realistically, that racing shells for different numbers of oarsmen are geometrically similar to each other, and that their weights are proportional to the number of oarsmen. Then, by Archimedes' principle, the submerged volume must be proportional to the number n of oarsmen, and the submerged surface area must be proportional to $n^{2/3}$. Most of the resistance to the boat's motion is friction drag, which at the high Reynolds numbers involved is very nearly proportional to the area multiplied by the square of the speed v (Equation 10.3). Thus, the drag is proportional to $v^2 n^{2/3}$, and the power required to $v^3 n^{2/3}$. The power available is proportional to n, so for the maximum speed

$$v^3 n^{2/3} \propto n \qquad\qquad (18.1)$$
$$v \propto n^{1/9}$$

This tells us that eights should travel $8^{1/9} = 1.26$ times as fast as single sculls. The ratio of the record speeds given above is very close to this, 1.24.

The fastest human-powered boats have submerged hydrofoils that lift the hull clear of the water at speed (Abbott and Wilson 1995). They are propelled by pedal-driven propellers, either a small submerged propeller or a large one operating in the air. Speeds up to 5.7 m/s have been recorded in 2000-m trials, and 9.5 m/s for a flying 100 m, for one-man craft.

Much higher speeds over water are attained without an engine by yachts and sailboards, but these depend on strong winds.

18.5. AIRCRAFT WITHOUT ENGINES

Sailplanes and hang gliders enable humans to soar, slope soaring and soaring in thermals in the same ways as birds (Section 10.6). These aircraft vary in performance, but some examples are compared, in Fig. 18.3, with two of the soaring birds that were represented in Fig. 10.7. The sailplanes glide faster than the animals, as their larger size should lead us to expect. Minimum sink speed and maximum range speed are both proportional to the square root of wing loading (Equations 10.20 and 10.21), and geometrically similar aircraft of equal density have wing loadings proportional to $(\text{mass})^{1/3}$. Therefore, optimum gliding speeds are expected to be proportional to $(\text{mass})^{1/6}$. It must, of course, be admitted that sailplanes are not geometrically similar to animals and do not have the same density, but they do nevertheless have higher wing loadings.

The sailplanes also differ from the animals in having lower minimum gliding angles. (The minimum gliding angles can be found in Fig. 18.3

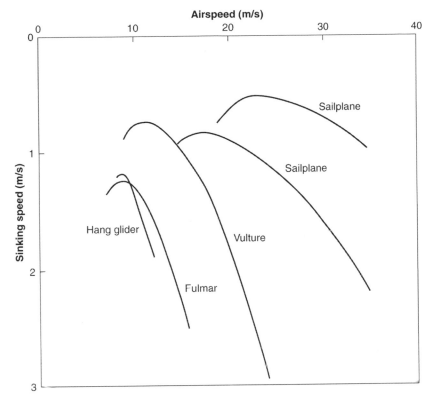

Fig. 18.3. Graphs of sinking speed against airspeed for two sailplanes, a hang-glider, a vulture (*Gyps africanus*), and a fulmar petrel (*Fulmarus glacialis*). Modified from Brower (1983).

by drawing tangents from the origin, as explained in Section 10.4.) Drag coefficients generally fall as Reynolds number increases (Equations 10.2 and 10.3, where the drag coefficients are the terms in parentheses). Hence, Equation 10.22 suggests that larger aircraft should have slightly smaller minimum gliding angles. However, the difference in minimum gliding angle between the better of the two sailplanes in Fig. 18.3 (1.4°) and the vulture (4°) is considerably larger than this argument can explain.

A pilot in a sailplane cannot take off unaided, but needs to be towed by a powered aircraft, a land vehicle, or a winch. Hang gliders, however, need no such assistance to take off from the top of a sufficiently steep hill. To make this possible and to make landing safe, hang gliders must be capable of gliding very slowly, which implies that their wing loading must be low. Figure 18.3 includes one example of a hang glider. Its minimum sink speed is only 9 m/s, about the same as for a fulmar petrel. Above the minimum sink speed, the hang glider performs less well than the bird.

One reason for this is that optimal wing loading increases with increasing speed (Equation 10.23). The bird can adjust its wing area to suit its speed by extending its wings fully at low speeds and partially folding them at high speeds, but the wing area of a hang glider is fixed.

As well as gliding and soaring, powered flight is possible without an engine. Many pedal-powered aircraft have been built (Abbott and Wilson 1995). In 1979, Bryan Allen achieved the remarkable feat of flying the *Gossamer Albatross* across the English Channel against a headwind. In 1988, *Daedalus* was flown 119 km from Crete to Santorini. These aircraft were necessarily very light and correspondingly fragile, with masses, excluding the pilot, of only 27 and 32 kg, respectively. With huge wings, their wing loadings were remarkably low, making slow flight possible. The *Gossamer Albatross* cruised at 5 m/s, and *Daedalus* at 7 m/s.

We will estimate the power needed to fly *Daedalus*. Its mass, with the pilot on board, was 104 kg, so the lift needed to keep it airborne was 1020 N. Its wingspan was 34 m. Assume that the induced drag factor was 1.0 and that the density of the air was 1.2 kg/m^3. Then, at an airspeed of 7 m/s, the induced drag calculated from Equation 10.9 is 10 N. If this is its maximum range speed (it would be unstable at lower speeds; Section 10.5), the profile drag equals the induced drag and the total drag is 20 N. The power requirement is the drag multiplied by the speed, 140 W. This is probably an overestimate, because the aircraft flew low enough to benefit from ground effect (Section 17.3). However, allowance has to be made for the Froude efficiency of the propeller being less than 100%. Abbott and Wilson (1995) give an estimate of 200 W for the power requirement of *Daedalus*, which is presumably based on informed calculation. Athletes in training can sustain 300 W on a bicycle ergometer for several hours. If you compare this with the metabolic power requirements for cycling shown in Fig. 18.1, please remember that 300 W mechanical power output requires about 1200 W metabolic power consumption.

It seems extremely unlikely that human-powered aircraft will ever be robust enough to be useful as a means of transport. To be robust they would have to be much heavier, but human power is adequate only for flying very light craft.

This chapter has shown how manufactured equipment can aid human locomotion, even if it has no motor. Running shoes protect our feet and enhance their elastic compliance. Bicycles enable us to travel much faster and more economically over reasonably smooth surfaces than we could do by running. Similarly, human-powered boats are faster and more economical than swimming. Human-powered aircraft are possible, but not practical.

Chapter Nineteen

....

Epilogue

S O FAR, I have discussed running, swimming, and flight separately. In this short concluding chapter I consider them together, and attempt some generalizations about locomotion. Dickinson et al. (2000) is a longer review with similar aims.

19.1. METABOLIC COST OF TRANSPORT

The metabolic cost of transport is the metabolic energy cost of moving unit mass of animal unit distance. The gross cost is (Metabolic rate of moving animal)/(Body mass × Distance traveled). The net cost is (Metabolic rate of moving animal – Metabolic rate of stationary animal) / (Mass × Distance). Both are generally smaller for larger animals.

For running, the net cost of transport (in J/kg m) is approximately $10.7(\text{body mass, kg})^{-0.32}$ (Fig. 7.12). This relationship was originally determined for mammals and running birds, but the graph shows that it is approximately true for reptiles, amphibians, and arthropods. The points for individual species are scattered on either side of the line, and in some cases species of equal mass have costs of transport differing by factors as large as three. It nevertheless seems to me that the relationship holds remarkably well for very diverse runners of an extremely wide range of sizes.

In the case of swimming, Videler (1993) found for fishes that the gross cost of transport (J/kg m) at the maximum range speed is approximately $1.1(\text{mass, kg})^{-0.38}$. (The maximum range speed is the speed at which the gross cost of transport is least.)

The metabolic rates (W/kg) of birds and bats flying at or near their maximum range speeds are approximately $57(\text{body mass, kg})^{-0.17}$ (Rayner 1995). Figure 11.3 shows that this expression also works reasonably well for flying bumblebees. Alexander (1998b) showed that maximum range speeds (m/s) for birds are expected to be about $16(\text{body mass, kg})^{0.14}$. By dividing the expression for metabolic rate per unit mass by the one for speed, we can estimate that the gross cost of transport for flying animals (J/kg m) is about $3.6(\text{body mass, kg})^{-0.31}$.

Notice that the exponents in the three expressions for cost of transport are very similar: −0.32 for running, −0.38 for swimming, and −0.31 for flight. However, the factors (10.7, 1.1 and 3.6, respectively) are very different. This seems to point to a general conclusion that costs of transport are high for running, moderate for flight, and low for swimming (Tucker 1970). However, our conclusion should not be as simple as that.

First, we should note that the expression for flight refers to flapping flight. Many of the larger birds travel by soaring, at much lower energy cost (Section 10.6). Secondly, the expression for swimming applies only to fish. Videler's (1993) data show that penguins and marine mammals, swimming submerged at their maximum range speeds, have costs of transport about ten times as high as for fish of equal mass. I explained in Section 15.1 that this results from higher maximum range speeds, a consequence of higher resting metabolic rates. Thirdly, animals that swim at the surface have costs of transport several times higher than similar animals that swim well below the surface (Section 13.3).

The comparison of metabolic and mechanical costs of transport in Fig. 7.12 shows that in running, the muscles of small animals work more efficiently than those of large ones. Figure 12.3C shows the same thing for swimming. This is why costs of transport are lower for large animals in both cases. The question, why the muscles of smaller animals should be less efficient, has not been satisfactorily answered.

19.2. Speeds

Figure 7.1A shows (unsurprisingly) that large mammals can generally sprint faster than small ones. A line drawn by eye through the cloud of points shows that mammal sprinting speeds (m/s) tend to be approximately $6.0(\text{body mass, kg})^{0.23}$. A 0.8-g cockroach had a maximum sprinting speed of 1.5 m/s (Full and Tu 1991), approximately as might be predicted by extrapolation of the relationship for mammals. Figure 7.1B shows that maximum aerobic speeds, like sprinting speeds, tend to be higher for larger mammals. However, the data points are so scattered that it does not seem useful to fit a line to them. Mammals of different sizes change from trotting to galloping at speeds (m/s) of about $1.5(\text{body mass, kg})^{0.24}$ (Heglund et al., 1974). These data suggest that mammalian running speeds are approximately proportional to $(\text{mass})^{0.24}$, but if we were to plot the speeds at which mammals do most of their traveling against body mass we would probably find a lower exponent; large mammals usually travel at a walk, but small mammals generally use running gaits.

Videler (1993) found that the maximum range speeds (m/s) of fish tend to be approximately $0.47(\text{body mass, kg})^{0.17}$. The maximum range speeds that he found for swimming penguins, seals, and whales average 1.8 times the predicted speed for a fish of the same mass. These data suggest that swimming animals should generally travel at speeds proportional to (mass)$^{0.17}$, but we do not know whether they do so. Maximum sprinting speeds of fishes (m/s) are generally about [0.4 + 7.4(body length, m)] (Videler 1993).

Alexander (1998b) argued that the maximum range speeds of birds (m/s) could be expected to be around 16(body mass, kg)$^{0.14}$. However, the speeds at which birds generally travel show no clear relationship with body mass; small birds may travel at speeds close to their maximum range speeds, but large ones fly at speeds well below their (higher) maximum range speeds (Pennycuick 1997). When traveling distances of a few hundred meters or more, most birds fly at speeds between 9 and 19 m/s. Insects fly more slowly. Only a few large insects seem to be capable of flying faster than 5 m/s, and it is doubtful whether any can exceed 10 m/s in still air (Dudley 2000).

These data about speeds cannot be summarized as easily as the data for costs of transport. However, it is clear that flying is generally much faster than running or swimming. The maximum aerobic running speeds of most animals fall well below the range of speeds (9–19 m/s) at which birds generally fly, and only the most remarkable endurance athletes, such as dogs and pronghorn antelope (*Antilocapra*), can sustain speeds in this range. Bees fly much faster than cockroaches can run. Dolphins and tunas seem to be the only swimming animals capable of exceeding 10 m/s, even in a short burst (Section 14.4).

We saw in the previous section that costs of transport for flight are moderate. However, moderate cost of transport with high speed implies a high power requirement. We saw in the previous section that the metabolic rate (W/kg) required for flight is approximately 57(body mass, kg)$^{-0.17}$. Taylor et al. (1981) found that the metabolic rates of mammals running at their maximum aerobic speeds were about 38(body mass, kg)$^{-0.19}$. Most mammals other than bats are incapable of the aerobic metabolic rates that enable birds of similar size to fly.

Alexander (1998b) argued that the advantages of long migrations for animals depend on their speeds and costs of transport. The potential benefit of migration may be a better supply of food energy in winter, but if too much time and/or energy are used on the journey, there may be no net benefit. My very rough calculations seemed to show that only flying animals, seals, and whales could be expected to benefit from very long migrations (round trips of 10,000 kg). All the animals known to make very long migrations belong to these groups.

19.3 GAITS

Gait change is a feature of the locomotion of many different groups of animals. Whether they run, swim, or fly, these animals use distinctly different patterns of movement at different speeds of locomotion. Quadrupedal mammals walk at low speeds, trot at moderate speeds, and gallop to travel fast (Section 7.2). Birds and people walk at low speeds and run or hop to go faster. In flight, many birds and bats use the vortex ring gait at low speeds and the continuous vortex gait at higher speeds (Section 12.2). Some birds use undulating flight at low speeds and bounding flight at high speeds. Fisher spiders row over the surface of water at low speeds and "gallop" when they move fast (Section 13.1). Many teleosts propel themselves at low speeds by fin movements and at high speeds by undulation of the body, and perform burst and coast swimming at some intermediate speeds (Section 15.3). Squids and cuttlefishes swim by fin undulation at low speeds and by jet propulsion when they go fast.

We saw in Section 1.10 that a gait change can be expected if a small change in speed results in an abrupt change in the optimum pattern of movement. It seems possible to explain the gait changes of mammals in this way. Measurements of the oxygen consumption of horses show that each gait (walk, trot, and gallop) requires less energy than either of the others in the range of speeds in which it is used (Fig. 7.11A). Similar measurements on humans have led to a similar conclusion, and a mathematical model seems to explain why this should be (Fig. 7.13). Similarly for flying birds, aerodynamic calculations seem to explain why the balance of advantage, in terms of energy cost, may shift abruptly from one gait to the other at a critical speed (Fig. 12.4). This needs confirmation by measurements of oxygen consumption. In the case of fish, swimming by fin movements alone may be more economical of energy than undulation at low speeds, because of the effects of boundary layer thinning and recoil movements (Section 15.1). However, the small fin muscles cannot supply enough power for fast swimming, so undulation has to be used at high speeds.

An argument based on the concept of dynamic similarity led to the prediction that legged animals of different sizes should make corresponding gait changes at equal Froude numbers (Section 4.2). This implies that they should change gaits at speeds proportional to the square roots of their leg lengths. The prediction is reasonably successful. We might reasonably hope that some similar rule would predict the speeds at which birds change from the vortex ring gait to the continuous vortex gait, but no such rule has yet been established. We also have no established means of predicting the speeds at which different fish change gaits. Our understanding of gait change in flying and swimming animals is still very imperfect.

19.4. Elastic Mechanisms

We have discussed the possibility of energy being saved by elastic mechanisms in the contexts of running (Section 7.4), flight (Sections 11.3 and 12.2), and swimming (Sections 14.4 and 16.2). The possibility of such savings arises in any mode of locomotion involving large fluctuations of kinetic energy, as explained in a general theory of oscillatory movements in Section 3.6. The role of elastic mechanisms in running is firmly established and seems well understood.

Many attempts have been made to discover whether energy is saved by elastic mechanisms in flight by calculating muscle efficiencies, assuming perfect or no elastic storage. The results have been inconclusive, but Dickinson and Lighton's (1995) ingenious experiment indicates modest savings in *Drosophila*. The elastic properties of the cuticle of the thorax, and especially of resilin structures, may be important. It seems likely that the muscles themselves are the important springs in the advanced insects that have fibrillar flight muscles. The possible role of the elastic compliance of the tendons of the flight muscles in bats and small birds does not seem to have been investigated. Thus, our understanding of elastic mechanisms in flight is very incomplete.

Our discussion in Section 16.2 concluded that any savings made by elastic mechanisms were trivial in the jet-propelled swimming of scallops and squids. However, the elastic compliance of the mesoglea may give useful savings in medusae. It seems likely that the long tendons of the tail muscles of dolphins may serve as energy-saving springs (Section 14.4). Tunas have similar tendons, but their possible role does not seem to have been investigated. As for flight, there is scope for more research on elastic mechanisms in swimming.

19.5. Priorities for Further Research

In this short concluding chapter, I have made some comparisons between running, swimming, and flight, and highlighted some of the issues that have arisen repeatedly in this book. In doing this, I have pointed to several topics that I regard as priorities for further research. We do not understand why the apparent efficiencies of running and flight are lower for small animals than for large ones. We would like to know more about why flying birds and bats and swimming fishes change gaits at particular speeds. We know regrettably little about the possible roles of elastic mechanisms in flight and swimming.

I would like to make two more points that apply to locomotion on land, in the air, and in water. First, research has concentrated on locomotion at constant velocity. We are relatively ignorant of the mechanics and energy costs of acceleration, deceleration, turning, and the maintenance of stability. Secondly, the great majority of investigations of locomotion have used captive animals in laboratories. Excellent use has been made of the telemetric techniques described in Section 5.7, especially in recent research on free-ranging birds and marine mammals. However, there are still rather few species for which we have good knowledge of their movements, speeds, and energy use for locomotion, during normal activity in their natural environments.

I have made many other suggestions for future research at the ends of earlier chapters, but the most informative research of the next few years may take quite different directions. We thought for many years that we understood snail crawling, until Denny destroyed the old theory and presented a new one (Section 9.3). Similarly, many of the current explanations of animal movement that are given in this book may soon be shown to be wrong. Unfortunately, we do not know which they will be.

References

Abbott, A. V. and Wilson, D. G. (editors) (1995) *Human-Powered Vehicles.* Human Kinetics, Champaign, IL.

Adams, N. J., Brown, C. R. and Nagy, K. A. (1986) Energy expenditure of free-ranging wandering albatrosses *Diomedea exulans. Physiological Zoology* **59**, 583–591.

Aerts, P. (1998) Vertical jumping in *Galago senegalensis*: the quest for an obligate mechanical power amplifier. *Philosophical Transactions of the Royal Society B* **353**:1607–1620.

Aerts, P., Ker, R. F., De Clercq, D., Ilsley, D. W. and Alexander, R. McN. (1995) The mechanical properties of the human heel pad: A paradox resolved. *Journal of Biomechanics* **28**, 1299–1308.

Ahlborn, B., Chapman, S., Stafford, R., Blake, R. W. and Harper, D. G. (1997) Experimental simulation of the thrust phases of fast-start swimming of fish. *Journal of Experimental Biology* **200**, 2301–2312.

Aigledinger, T. L. and Fish, F. E. (1995) Hydroplaning by ducklings: Overcoming limitations to swimming at the water surface. *Journal of Experimental Biology* **198**, 1567–1574.

Alexander, R. McN. (1959a) The physical properties of the swimbladder in intact Cypriniformes. *Journal of Experimental Biology* **36**, 315–332.

Alexander, R. McN. (1959b) The densities of Cyprinidae. *Journal of Experimental Biology* **36**, 333–340.

Alexander, R. McN. (1959c) The physical properties of the swimbladders of fish other than Cypriniformes. *Journal of Experimental Biology* **36**, 347–355.

Alexander, R. McN. (1961) The physical properties of the swimbladder of some South American Cypriniformes. *Journal of Experimental Biology* **38**, 403–410.

Alexander, R. McN. (1965) The lift produced by the heterocercal tails of Selachii. *Journal of Experimental Biology* **43**, 131–138.

Alexander, R. McN. (1966) Rubber-like properties of the inner hinge ligament of Pectinidae. *Journal of Experimental Biology* **44**, 119–130.

Alexander, R. McN. (1967) *Functional Design in Fishes.* Hutchinson, London.

Alexander, R. McN. (1969) The orientation of muscle fibres in the myomeres of fish. *Journal of the Marine Biological Association* **49**, 263–290.

Alexander, R. McN. (1971) *Size and Shape.* Arnold, London.

Alexander, R. McN. (1972) The energetics of vertical migration by fishes. *Symposia of the Society for Experimental Biology* **26**, 273–294.

Alexander, R. McN. (1974a) The mechanics of jumping by a dog (*Canis familiaris*). *Journal of Zoology* **173**, 549–573.

Alexander, R. McN. (1974b) *Functional Design in Fishes*, ed. 3. Hutchinson, London.

Alexander, R. McN. (1975a) *Biomechanics.* Chapman & Hall, London.

Alexander, R. McN. (1975b) *The Chordates.* Cambridge University Press, Cambridge, UK.

Alexander, R. McN. (1976) Mechanics of bipedal locomotion. In P. S. Davies (editor), *Perspectives in Experimental Biology* **1**, 493–504. Pergamon, Oxford, UK.

Alexander, R. McN. (1977a) Mechanics and scaling of terrestrial locomotion. In T. S. Pedley (editor) *Scale Effects in Animal Locomotion.* Academic Press, London.

Alexander, R. McN. (1977b) Swimming. In R. McN. Alexander and G. Goldspink (editor) *Mechanics and Energetics of Animal Locomotion*, Chapman & Hall, London, pp. 222–248.

Alexander, R. McN. (1981) The gaits of tetrapods: adaptations for stability and economy. *Symposia of the Zoological Society of London* **48**, 269–287.

Alexander, R. McN. (1982) *Locomotion of Animals.* Blackie, Glasgow.

Alexander, R. McN. (1983) *Animal Mechanics*, ed. 2. Blackwell, Oxford, UK.

Alexander, R. McN. (1984) Walking and running. *American Scientist* **72**, 348–354.

Alexander, R. McN. (1987) The spring in your step. New Scientist **114** (1558), 42–44.

Alexander, R. McN. (1988) *Elastic Mechanisms in Animal Movement.* Cambridge University Press, Cambridge, UK.

Alexander, R. McN. (1989a) Optimization and gaits in the locomotion of vertebrates. *Physiological Reviews* **69**, 1199–1227.

Alexander, R. McN. (1989b) *Dynamics of Dinosaurs and Other Extinct Giants.* Columbia University Press, New York.

Alexander, R. McN. (1989c) Mechanics of fossil vertebrates. *Journal of the Geological Society, London* **146**, 41–52.

Alexander, R. McN. (1990) Size, speed and buoyancy adaptations in aquatic animals. *American Zoologist* **30**, 189–196.

Alexander, R. McN. (1991a) Energy-saving mechanisms in walking and running. *Journal of Experimental Biology* **160**, 55–69.

Alexander, R. McN. (1991b) Elastic mechanisms in primate locomotion. *Zoological Morphology and Anthropology* **78**, 315–320.

Alexander, R. McN. (1992a) A model of bipedal locomotion on compliant legs. *Philosophical Transactions of the Royal Society B* **338**, 189–198.

Alexander, R. McN. (1992b) The work that muscles can do. *Nature* **357**, 360–361.

Alexander, R. McN. (1992c) *Exploring Biomechanics: Animals in Motion.* Freeman, New York.

Alexander, R. McN. (1993a) Legs and locomotion of Carnivora. *Symposia of the Zoological Society of London* **65**, 1–13.

Alexander, R. McN. (1993b) Buoyancy. In D. H. Evans (editor), *The Physiology of Fishes*, pp. 75–96. CRC Press, Boca Raton, FL.

Alexander, R. McN. (1995a) Simple models of human movement. *Applied Mechanics Reviews* **48**, 461–470.

Alexander, R. McN. (1995b) Leg design and jumping technique for humans, other vertebrates and insects. *Philosophical Transactions of the Royal Society B* **347**, 235–248.

Alexander, R. McN. (1995c) Hydraulic mechanisms in locomotion. In G. Lanzavecchia, R. Valvassori and M. D. Candia Carnevali (editors), *Body Cavities: Function and Phylogeny*, pp. 187–198. Mucchi, Modena.

Alexander, R. McN. (1996) *Optima for Animals*, ed. 2. Princeton University Press, Princeton, NJ.

Alexander, R. McN. (1997a) A minimum energy cost hypothesis for human arm movements. *Biological Cybernetics* **76**, 97–105.

Alexander, R. McN. (1997b) Optimum muscle design for oscillatory movement. *Journal of Theoretical Biology* **184**, 253–259.

Alexander, R. McN. (1997c) The U, J and L of bird flight. *Nature* **390**, 13.

Alexander, R. McN. (1998a) All-time giants: the largest animals and their problems. *Palaeontology* **41**, 1231–1245.

Alexander, R. McN. (1998b) When is migration worthwhile for animals that walk, swim or fly? *Journal of Avian Biology* **29**, 387–394.

Alexander, R. McN. (1999) *Energy for Animal Life*. Oxford University Press, Oxford, UK.

Alexander, R. McN. (2000a) Walking and running strategies for humans and other mammals. In P. Domenici and R. W. Blake (editors), *Biomechanics in Animal Behaviour*. Pp. 49–57. BIOS, Oxford, UK.

Alexander, R. McN. (2000b) Hovering and jumping: Contrasting problems in scaling. In J. H. Brown and G. B. West (editors), *Scaling in Biology*, pp. 37–50. Oxford University Press, Oxford, UK.

Alexander, R. McN. (2001) Locomotion. In D. L. Lee (editor), *The Biology of Nematodes*, pp.343–350. Taylor & Francis, London.

Alexander, R. McN. (In press) Stability and manoeuvrability of terrestrial vertebrates. *American Zoologist*.

Alexander, R. McN. and Bennett, M. B. (1989) How elastic is a running shoe? *New Scientist* 15 July 1989, 45–46.

Alexander, R. McN. and Jayes, A. S. (1978) Vertical movements in walking and running. *Journal of Zoology* **185**, 27–40.

Alexander, R. McN. and Jayes, A. S. (1980) Fourier analysis of forces exerted in walking and running. *Journal of Biomechanics* **13**, 383–390.

Alexander, R. McN. and Jayes, A. S. (1983) A dynamic similarity hypothesis for the gaits of quadrupedal mammals. *Journal of Zoology* **201**, 135–152.

Alexander, R. McN. and Maloiy, G. M. O. (1984) Stride lengths and stride frequencies of primates. *Journal of Zoology* **202**, 577–582.

Alexander, R. McN. and Pond, C. M. (1992) Locomotion and bone strength of the white rhinoceros (*Ceratotherium simum*). *Journal of Zoology* **227**, 63–69.

Alexander, R. McN. and Vernon, A. (1975a) Mechanics of hopping by kangaroos (Macropodidae). *Journal of Zoology* **177**, 265–303.

Alexander, R. McN. and Vernon, A. (1975b) The dimensions of knee and ankle muscles and the forces they exert. *Journal of Human Movement Studies* **1**, 115–123.

Alexander, R. McN., Langman, V. A. and Jayes, A. S. (1977) Fast locomotion of some African ungulates. *Journal of Zoology* **183**, 291–300.

Alexander, R. McN., Jayes, A. S., Maloiy, G. M. O. and Wathuta, E. M. (1979a) Allometry of the limb bones of mammals from shrews (*Sorex*) to elephant (*Loxodonta*). *Journal of Zoology* **190**, 155–192.

Alexander, R. McN., Maloiy, G. M. O., Hunter, B., Jayes, A. S. and Nturibi, J. (1979b) Mechanical stresses in fast locomotion of buffalo (*Syncerus caffer*) and elephant (*Loxodonta africana*). *Journal of Zoology* **189**, 135–144.

Alexander, R. McN., Maloiy, G. M. O., Njau, R. and Jayes, A. S. (1979c) Mechanics of running of the ostrich (*Struthio camelus*). *Journal of Zoology* **187**, 169–178.

Alexander, R. McN., Jayes, A. S., Maloiy, G. M. O. and Wathuta, E. M. (1981) Allometry of the leg muscles of mammals. *Journal of Zoology* **194**, 539–552.

Alexander, R. McN., Maloiy, G. M. O., Ker, R. F., Jayes, A. S. and Warui, C. N. (1982) The role of tendon elasticity in the locomotion of the camel. *Journal of Zoology* **198**, 293–313.

Alexander, R. McN., Dimery, N. J. and Ker, R. F. (1985) Elastic structures in the back and their role in galloping in some mammals. *Journal of Zoology A* **207**, 467–482.

Altringham, J. D. and Block, B. A. (1997) Why do tuna maintain elevated slow muscle temperatures? Power output of muscle isolated from endothermic and ectothermic fish. *Journal of Experimental Biology* **200**, 2617–2627.

Altringham, J. D. and Ellerby, D. J. (1999) Fish swimming: Patterns in muscle function. *Journal of Experimental Biology* **202**, 3397–3403.

Altringham, J. D. and Johnston, I. A. (1990) Modelling muscle power output in a swimming fish. *Journal of Experimental Biology* **148**, 395–402.

Altringham, J. D., Wardle, C. S. and Smith, C. I. (1993) Myotomal muscle function at different locations in the body of a swimming fish. *Journal of Experimental Biology* **182**, 191–206.

Anderson, B. D., Schultz, J. W. and Jayne, B. C. (1995) Axial kinematics and muscle activity during terrestrial locomotion of the centipede *Scolopendra heros*. *Journal of Experimental Biology* **198**, 1185–1195.

Anderson, E. J. and De Mont, M. E. (2000) The mechanics of locomotion in the squid *Loligo pealei*: Locomotory function and unsteady hydrodynamics of the jet and intramantle pressure. *Journal of Experimental Biology* **203**, 2851–2863.

Anderson, E. J., McGillis, W. R. and Grosenbaugh, M. A. (2001) The boundary layer of swimming fish. *Journal of Experimental Biology* **204**, 81–102.

Arnold, E. N. (1995) Identifying the effects of history on adaptation: Origins of different sand-diving techniques in lizards. *Journal of Zoology* **235**, 351–388.

Askew, G. N. and Marsh, R. L. (1998) Optimal shortening velocity (V/V_{max}) of skeletal muscle during cyclical contractions: Length–force effects and velocity-dependent activation and deactivation. *Journal of Experimental Biology* **201**, 1527–1540.

Åstrand, P.-O. and Rodahl, K. (1986) *Textbook of Work Physiology: Physiological Bases of Exercise*. McGraw–Hill, New York.

Au, D. and Weihs, D. (1980) At high speeds dolphins save energy by leaping. *Nature* **284**, 548–550.

Autumn, K., Liang, Y. A., Hsieh, S. T., Zesch, W., Chan, W. P., Kenny, T. W., Fearing, R. and Full, R. J. (2000) Adhesive force of a single gecko foot-hair. *Nature* **405**, 681–685.

Bainbridge, R. (1960) Speed and stamina in three fish. *Journal of Experimental Biology* **37**, 129–153.

Ballreich, R. and Kuhlow, A. (1986) *Biomechanik der Leichtathletik.* Enke, Stuttgart.

Barclay, C. J. (1994) Efficiency of fast- and slow-twitch muscles of the mouse performing cyclic contractions. *Journal of Experimental Biology* **193**, 65–78.

Barham, E. G. (1966) Deep scattering layer migration and composition: observations from a diving saucer. *Science* **151**, 1399–1403.

Basmajian, J. V. and de Luca, C. J. (1985) *Muscles Alive,* ed. 5. Williams & Wilkins, Baltimore.

Bateson, P. and Bradshaw, E. L. (1997) Physiological effects of hunting red deer (*Cervus elaphus*). *Proceedings of the Royal Society B* **264**, 1707–1714.

Baudinette, R. V. and Gill, P. (1985) The energetics of "paddling" and "flying" in water: Locomotion in penguins and ducks. *Journal of Comparative Physiology B* **155**, 373–380.

Baudinette, R. V. and Schmidt-Nielsen, K. (1974) Energy cost of gliding flight in Herring gulls. *Nature* **248**, 83–84.

Bekker, M. G. (1956) *Theory of Land Locomotion.* University of Michigan Press, Ann Arbor.

Bennet-Clark, H. C. (1975) The energetics of the jump of the locust *Schistocerca gregaria. Journal of Experimental Biology* **63**, 53–83.

Bennet-Clark, H. C. (1977) Scale effects in jumping animals. In T. J. Pedley (editor), *Scale Effects in Animal Locomotion,* pp. 185–201. Academic Press, London.

Bennet-Clark, H. C. and Alder, G. M. (1979) The effect of air resistance on the jumping performance of insects. *Journal of Experimental Biology* **82**, 105–121.

Bennet-Clark, H. C. and Lucey, E. C. A. (1967) The jump of the flea: A study of the energetics and a model of the mechanism. *Journal of Experimental Biology* **47**, 59–76.

Bennett, M. B. (1987) Fast locomotion of some kangaroos. *Journal of Zoology* **212**, 457–464.

Bennett, M. B. (1989) A possible energy-saving role for the major fascia of the thigh in running quadrupedal mammals. *Journal of Zoology* **219**, 221–230.

Bennett, M. B. and Ker, R. F. (1990) The mechanical properties of the human subcalcaneal fat pad in compression. *Journal of Anatomy* **171**, 131–138.

Bennett, M. B., Ker, R. F. and Alexander, R. McN. (1987) Elastic properties of structures in the tails of cetaceans (*Phocaena* and *Lagenorhynchus*) and their effect on the energy cost of swimming. *Journal of Zoology* **211**, 177–192.

Bennett, M.B., Ker, R. F., Dimery, N. J. and Alexander, R. McN. (1986) Mechanical properties of various mammalian tendons. *Journal of Zoology A* **209**, 537–548.

Bennett, N. C. (1991) Behaviour and social organisation of the Damaraland mole rat *Cryptomys damarensis. Journal of Zoology* **220**, 225–248.

Berger, M. and Hart, J. S. (1972) Die Atmung beim Kolibri *Amazilia fimbriata* während des Schirrfluges bei verscheidenen Umgebungstemperaturen. *Journal of Comparative Physiology* **81**, 363–380.

Berrigan, D. and Lighton, J. R. B. (1993) Bioenergetic and kinematic consequences of limblessness in larval Diptera. *Journal of Experimental Biology* **179**, 245–259.

Berrigan, D. and Pepin, D. J. (1995) How maggots move: allometry and kinematics of crawling in larval Diptera. *Journal of Insect Physiology* **41**, 329–337.

Bertram, J. E. A., Ruina, A., Cannon, C. E., Chang, Y. C. and Coleman, M. J. (1999) A point-mass model of gibbon locomotion. *Journal of Experimental Biology* **202**, 2609–2617.

Bevan, R. M., Butler, P. J., Woakes, A. J. and Prince, P. A. (1995) The energy expenditure of free-ranging black-browed albatrosses. *Philosophical Transactions of the Royal Society B* **350**, 119–131.

Bidigare, R. R. and Biggs, D. C. (1980) The role of sulfate exclusion in buoyancy maintenance by siphonophores and other gelatinous zooplankton. *Comparative Biochemistry and Physiology A* **66**, 467–471.

Biebach, H. (1998) Phenotypic organ flexibility in garden warblers *Sylvia borin* during long-distance migration. *Journal of Avian Biology* **29**, 529–535.

Biewener, A. A. (1989) Scaling body support in mammals: Limb posture and muscle mechanics. *Science* **245**, 45–48.

Biewener, A. A. (1990) Biomechanics of mammalian terrestrial locomotion. *Science* **250**, 1097–1103.

Biewener, A. A. (1992) In vivo measurement of bone strain and tendon force. In A. A. Biewener (editor) *Biomechanics: Structures and Systems*, pp.123–147. IRL Press, Oxford, UK.

Biewener, A. A. (1998) Muscle-tendon stresses and elastic energy storage during locomotion in the horse. *Comparative Biochemistry and Physiology B* **120**, 73–87.

Biewener, A. A. and Blickhan, R. (1988) Kangaroo rat locomotion: design for elastic energy storage or acceleration? *Journal of Experimental Biology* **140**, 243–255.

Biewener, A. A. and Full, R. J. (1992) Force platform and kinematic analysis. In A. A. Biewener (editor) *Biomechanics: Structures and Systems*, pp.45–73. IRL Press, Oxford, UK.

Biewener, A. A., Alexander, R. McN. and Heglund, N. C. (1981) Elastic energy storage in the hopping of kangaroo rats (*Dipodomys spectabilis*). *Journal of Zoology* **195**, 369–383.

Biewener, A. A., Corning, W. R. and Tobalske, B. W. (1998a) In vivo pectoralis muscle force–length behavior during level flight in pigeons (*Columba livia*) *Journal of Experimental Biology* **201**, 3293–3307.

Biewener, A. A., Konieczynski, D. D. and Baudinette, R. V. (1998b) In vivo muscle force–length behavior during steady-speed hopping in tammar wallabies. *Journal of Experimental Biology* **201**, 1681–1694.

Birch, H. L., Wilson, A. M. and Goodship, A. E. (1997) The effect of exercise-induced localised hypothermia on tendon cell survival. *Journal of Experimental Biology* **200**, 1703–1708.

Birch, J. M. and Dickinson, M. H. (2001) Spanwise flow and the attachment of the leading-edge vortex on insect wings. *Nature* **412**, 722–725.

Bishai, H. M. (1961) The effect of pressure on the distribution of some Nile fish. *Journal of Experimental Zoology* **147**, 113–124.

Blake, R. W. (1979a) The energetics of hovering in the mandarin fish (*Synchropus picturatus*). *Journal of Experimental Biology* **82**, 25–33.

Blake, R. W. (1979b) The mechanics of labriform locomotion, I: Labriform loco-motion in the angelfish (*Pterophyllum eimekei*): an analysis of the power stroke. *Journal of Experimental Biology* **82**, 255–271.

Blake, R. W. (1980a) The mechanics of labriform locomotion, II: An analysis of the recovery stroke and the overall fin-beat cycle propulsive efficiency in the angelfish. *Journal of Experimental Biology* **85**, 337–342.

Blake, R. W. (1980b) Undulatory median fin propulsion of two teleosts with dif-ferent modes of life. *Canadian Journal of Zoology* **58**, 2116–2119.

Blake, R. W. (1983) Swimming in the electric-eels and knifefishes. *Canadian Jour-nal of Zoology* **61**, 1432–1441.

Blickhan, R. (1989) The spring–mass model for running and hopping. *Journal of Biomechanics* **22**, 1217–1227.

Blickhan, R. and Cheng, J.-Y. (1994) Energy storage by elastic mechanisms in the tail of large swimmers—a re-evaluation. *Journal of Theoretical Biology* **168**, 315–321.

Blickhan, R. and Full, R. J. (1987) Locomotion energetics of the ghost crab, II: Mechanics of the centre of mass during walking and running. *Journal of Experi-mental Biology* **130**, 155–174.

Bobbert, M. F., Gerritsen, K. G. M., Litjens, M. C. A. and van Soest, A. J. (1996) Why is countermovement jump height greater than squat jump height? *Medicine and Science in Sports and Exercise* **28**, 1402–1412.

Bone, Q. (1966) On the function of the two types of myotomal muscle fibre in elasmobranch fish. *Journal of the Marine Biological Association* **46**, 321–349.

Bone, Q., Pulsford, A. and Chubb, A. D. (1981) Squid mantle muscle. *Journal of the Marine Biological Association* **61**, 327–342.

Bone, Q. and Roberts, B. L. (1969) The density of elasmobranchs. *Journal of the Marine Biological Association of the UK* **49**, 913–937.

Bonine, K. E. and Garland, T. (1999) Sprint performance of phrynosomatid liz-ards, measured on a high-speed treadmill, correlates with hindlimb length. *Journal of Zoology* **248**, 255–265.

Brackenbury, J. (1991) Kinematics of take-off and climbing flight in butterflies. *Journal of Zoology* **224**, 251–270.

Brackenbury, J. (1997) Caterpillar kinematics. *Nature* **390**, 563.

Brackenbury, J. and Wang, R. (1995) Ballistics and visual targeting in flea-beetles (Alticinae). *Journal of Experimental Biology* **198**, 1931–1942.

Bramble, D. M. and Jenkins, F. A. (1993) Mammalian locomotor–respiratory inte-gration: Implications for diaphragmatic and pulmonary design. *Science* **262**, 235–240.

Brower, J. C. (1983) The aerodynamics of *Pteranodon* and *Nyctosaurus*, two large pterosaurs from the Upper Cretaceous of Kansas. *Journal of Vertebrate Paleontol-ogy* **3**, 84–124.

Brown, J. H. and West, G. B. (editors) (2000) *Scaling in Biology*. Oxford University Press, New York.

Bryant, D. M. and Westerterp, K. R. (1980) The energy budget of the house mar-tin (*Delichon urbica*). *Ardea* **68**, 91–102.

Bryant, J. D., Bennett, M. B., Brust, J. and Alexander, R. McN. (1987) Forces exerted on the ground by galloping dogs (*Canis familiaris*). *Journal of Zoology* **213**, 193–203.

Burgetz, I. J., Rojas-Vargas, A., Hinch, S. G. and Randall, D. J. (1998) Initial recruitment of anaerobic metabolism during sub-maximal swimming in rainbow trout (*Onchorhynchus mykiss*). *Journal of Experimental Biology* **201**, 2711–2721.

Burkholder, T. J. and Lieber, R. L. (2001) Sarcomere length operating range of vertebrate muscles during movement. *Journal of Experimental Biology* **204**, 1529–1536.

Calder, W. A. (1984) *Size, Function and Life History.* Harvard University Press, Cambridge, MA.

Cannon, C. H. and Leighton, M. (1994) A comparative locomotor ecology of gibbons and macaques—selection of canopy elements for crossing gaps. *American Journal of Physical Anthropology* **93**, 505–524.

Capelli, C., Donatelli, C., Moia, C., Valier, C., Rosa, G. and di Prampero, P. E. (1990) Energy cost and efficiency of sculling a Venetian gondola. *European Journal of Applied Physiology* **60**, 175–178.

Capelli, C., Rosa, G., Butti, F., Ferretti, G., Veicsteinas, A. and di Prampero, P. E. (1993) Energy cost and efficiency of riding aerodynamic bicycles. *European Journal of Applied Physiology* **67**, 144–149.

Capen, R. L. (1967) Swimbladder morphology of some mesopelagic fishes in relation to sound scattering. *US Navy Electronics Laboratory, San Diego, California, Report* **1447**, 1–25.

Carey, F. G. (1982) Warm fish. In C. R. Taylor, K. Johansen and L. Bolis (editors), *A Companion to Animal Physiology*, pp. 216–232. Cambridge University Press, Cambridge, UK.

Cartmill, M. (1985) Climbing. In M. Hildebrand, D. M. Bramble, K. F. Liem, and D. B. Wake (editors), *Functional Vertebrate Morphology*, pp. 58–79. Belknap Press, Cambridge, MA.

Casey, T. M. (1976) Flight energetics of sphinx moths: power input during hovering flight. *Journal of Experimental Biology* **64**, 529–543.

Casey, T. M. (1991) Energetics of caterpillar locomotion: biomechanical constraints of a hydraulic skeleton. *Science* **252**, 112–114.

Casey, T. M., May, M. L. and Morgan, K. R. (1985) Flight energetics of euglossine bees in relation to morphology and to wing stroke frequency. *Journal of Experimental Biology* **116**, 271–289.

Cavagna, G. A., Saibene, F. P. and Margaria, R. (1964) Mechanical work in running. *Journal of Applied Physiology* **19**, 249–256.

Cavagna, G. A., Heglund, N. C. and Taylor, R. C. (1977) Mechanical work in terrestrial locomotion: Two basic mechanisms for minimizing energy expenditure. *American Journal of Physiology* **233**, R243–R261.

Cavagna, G. A., Willems, P. A. and Heglund, N. C. (1998) Walking on Mars. *Nature* **393**, 636.

Cavanagh, P. R. and Lafortune, M. A. (1980) Ground reaction forces in distance running. *Journal of Biomechanics* **13**, 397–406.

Davenport, J., Munks, S. A. and Oxford, P. J. (1984) A comparison of the swimming of marine and freshwater turtles. *Proceedings of the Royal Society B* **220**, 447–475.

Davis, R. R. and Hull, M. L. (1981) Measurement of pedal loading in bicycles, II: Analysis and results. *Journal of Biomechanics* **14**, 857–872.

Dawson, T. J. and Taylor, C. R. (1973) Energetic cost of locomotion in kangaroos. *Nature* **246**, 313–314.

Demes, B., Jungers, W. L., Gross T. S. and Fleagle, J. G. (1995) Kinetics of leaping primates—influence of substrate orientation and compliance. *American Journal of Physical Anthropology* **96**, 419–429.

De Mont, M. E. (1990) Tuned oscillations in the swimming scallop *Pecten maximus*. *Canadian Journal of Zoology* **68**, 786–791.

De Mont, M. E. and Gosline, J. M. (1988) Mechanics of jet propulsion in the hydromedusan jellyfish, *Polyorchis penicillatus*, I–III (three papers). *Journal of Experimental Biology* **134**, 313–332.

Denny, M. W. (1980a) The role of gastropod pedal mucus in locomotion. *Nature* **285**, 160–161.

Denny, M. W. (1980b) Locomotion: The cost of gastropod crawling. *Science* **208**, 1288–1290.

Denny, M. W. (1984) Mechanical properties of pedal mucus and their consequences for gastropod structure and performance. *American Zoologist* **24**, 23–36.

Denny, M. W. (1993) *Air and Water*. Princeton University Press, Princeton, NJ.

Denton, E. J. (1974) On buoyancy and the lives of modern and fossil cephalopods. *Proceedings of the Royal Society B* **185**, 273–299.

Denton, E. J. and Gilpin-Brown J. B. (1973) Floatation mechanisms in modern and fossil cephalopods. *Advances in Marine Biology* **11**, 197–268.

Denton, E. J. and Marshall, N. B. (1958) The buoyancy of bathypelagic fishes without a gas-filled swimbladder. *Journal of the Marine Biological Association of the UK* **37**, 753–769.

Denton, E. J., Liddicoat, J. D. and Taylor, D. W. (1970) Impermeable "silvery" layers in fishes. *Journal of Physiology* **207**, 64P.

Dial, K. P., Biewener, A. A., Tobalske, B. W. and Warrick, D. R. (1997) Mechanical power output of bird flight. *Nature* **390**, 67–70.

Dial, K. P., Kaplan, S. R., Goslow, G. E. and Jenkins, F. A. (1988) A functional analysis of the primary upstroke and downstroke muscles in the domestic pigeon (*Columba livia*) during flight. *Journal of Experimental Biology* **134**, 1–16.

Dickinson, J. A., Cook, S. D. and Leinhardt, T. M. (1985) The measurement of shock waves following heel strike in running. *Journal of Biomechanics* **18**, 415–422.

Dickinson, M.H. and Lighton, J. R. B. (1995) Muscle efficiency and elastic storage in the flight motor of *Drosophila*. *Science* **268**, 87–89.

Dickinson, M. H. and Götz, K. G. (1996). The wake dynamics and flight forces of the fruit fly *Drosophila melanogaster*. *Journal of Experimental Biology* **199**, 2085–2104.

Dickinson, M. H., Lehmann, F.-O. and Sane, S. P. (1999) Wing rotation and the aerodynamic basis of insect flight. *Science* **284**, 1954–1960.

Chang, Y. H., Bertram, J. E. A. and Ruina, A. (1997) A dynamic force and moment analysis system for brachiation. *Journal of Experimental Biology* **200**, 3013–3020.

Chapman, G. (1950) Of the movement of worms. *Journal of Experimental Biology* **27**, 29–39.

Chapman, G. (1958) The hydrostatic skeleton in the invertebrates. *Biological Reviews* **33**, 338–371.

Cheer, A. Y. L. and Koehl, M. A. R. (1987) Paddles and rakes: fluid flow through bristled appendages of small organisms. *Journal of Theoretical Biology* **129**, 17–39.

Cheng, J.-Y. and DeMont, M. E. (1996) Jet-propelled swimming in scallops: Swimming mechanics and ontogenetic scaling. *Canadian Journal of Zoology* **74**, 1734–1748.

Cheng, J.-Y., Davison, I. G. and DeMont, M. E. (1996) Dynamics and energetics of scallop locomotion. *Journal of Experimental Biology* **199**, 1931–1946.

Cheng, J.-Y., Pedley, T. J. and Altringham, J. D. (1998) A continuous dynamic beam model for swimming fish. *Philosophical Transactions of the Royal Siciety B* **353**, 981–997.

Chorlton, F. (1967) *Textbook of Dynamics*. Van Nostrand, London.

Clark, L. J. and Alexander, R. McN. (1975) Mechanics of running by quail (*Coturnix*). *Journal of Zoology* **176**, 87–113.

Clark, R. B. (1964) *Dynamics in Metazoan Evolution*. Clarendon Press, Oxford, UK.

Clarke, B. D. and Bemis, W. (1979) Kinematics of swimming of penguins at the Detroit Zoo. *Journal of Zoology* **188**, 411–428.

Clarke, R. B. and Tritton, D. J. (1970) Swimming mechanisms in nereidiform polychaetes. *Journal of Zoology* **161**, 257–271.

Close, R.I. (1972) Dynamic properties of mammalian skeletal muscles. *Physiological Reviews* **52**, 129–197.

Corner, E. D. S., Denton, E. J. and Forster, G. R. (1969) On the buoyancy of some deep-sea sharks. *Proceedings of the Royal Society B* **171**, 415–429.

Crompton, R. H., Sellers, W. I. and Günther, M. M. (1993) Energetic efficiency and ecology as selective factors in the saltatory adaptation of prosimian primates. *Proceedings of the Royal Society B* **254**, 41–45.

Crowninshield, R. D., Johnston, R. C., Andrews, J. G. and Brand, R. A. (1978) A biomechanical investigation of the human hip. *Journal of Biomechanics* **11**, 75–85.

Dadswell, M. J. and Weihs, D. (1990) Size-related hydrodynamic characteristics of the giant scallop, *Pecten magellanicus* (Bivalvia, Pectinidae). *Canadian Journal of Zoology* **68**, 778–785.

Daniel, T. L. (1983) Mechanics and energetics of medusan jet propulsion. *Canadian Journal of Zoology* **61**, 1406–1420.

Daniel, T. L. (1985) Cost of locomotion: unsteady medusan swimming. *Journal of Experimental Biology* **119**, 149–164.

Daniel, T. L. (1988) Forward flapping flight from flexing fins. *Canadian Journal of Zoology* **66**, 630–638.

Chang, Y. H., Bertram, J. E. A. and Ruina, A. (1997) A dynamic force and moment analysis system for brachiation. *Journal of Experimental Biology* **200**, 3013–3020.

Chapman, G. (1950) Of the movement of worms. *Journal of Experimental Biology* **27**, 29–39.

Chapman, G. (1958) The hydrostatic skeleton in the invertebrates. *Biological Reviews* **33**, 338–371.

Cheer, A. Y. L. and Koehl, M. A. R. (1987) Paddles and rakes: fluid flow through bristled appendages of small organisms. *Journal of Theoretical Biology* **129**, 17–39.

Cheng, J.-Y. and DeMont, M. E. (1996) Jet-propelled swimming in scallops: Swimming mechanics and ontogenetic scaling. *Canadian Journal of Zoology* **74**, 1734–1748.

Cheng, J.-Y., Davison, I. G. and DeMont, M. E. (1996) Dynamics and energetics of scallop locomotion. *Journal of Experimental Biology* **199**, 1931–1946.

Cheng, J.-Y., Pedley, T. J. and Altringham, J. D. (1998) A continuous dynamic beam model for swimming fish. *Philosophical Transactions of the Royal Siciety B* **353**, 981–997.

Chorlton, F. (1967) *Textbook of Dynamics*. Van Nostrand, London.

Clark, L. J. and Alexander, R. McN. (1975) Mechanics of running by quail (*Coturnix*). *Journal of Zoology* **176**, 87–113.

Clark, R. B. (1964) *Dynamics in Metazoan Evolution*. Clarendon Press, Oxford, UK.

Clarke, B. D. and Bemis, W. (1979) Kinematics of swimming of penguins at the Detroit Zoo. *Journal of Zoology* **188**, 411–428.

Clarke, R. B. and Tritton, D. J. (1970) Swimming mechanisms in nereidiform polychaetes. *Journal of Zoology* **161**, 257–271.

Close, R.I. (1972) Dynamic properties of mammalian skeletal muscles. *Physiological Reviews* **52**, 129–197.

Corner, E. D. S., Denton, E. J. and Forster, G. R. (1969) On the buoyancy of some deep-sea sharks. *Proceedings of the Royal Society B* **171**, 415–429.

Crompton, R. H., Sellers, W. I. and Günther, M. M. (1993) Energetic efficiency and ecology as selective factors in the saltatory adaptation of prosimian primates. *Proceedings of the Royal Society B* **254**, 41–45.

Crowninshield, R. D., Johnston, R. C., Andrews, J. G. and Brand, R. A. (1978) A biomechanical investigation of the human hip. *Journal of Biomechanics* **11**, 75–85.

Dadswell, M. J. and Weihs, D. (1990) Size-related hydrodynamic characteristics of the giant scallop, *Pecten magellanicus* (Bivalvia, Pectinidae). *Canadian Journal of Zoology* **68**, 778–785.

Daniel, T. L. (1983) Mechanics and energetics of medusan jet propulsion. *Canadian Journal of Zoology* **61**, 1406–1420.

Daniel, T. L. (1985) Cost of locomotion: unsteady medusan swimming. *Journal of Experimental Biology* **119**, 149–164.

Daniel, T. L. (1988) Forward flapping flight from flexing fins. *Canadian Journal of Zoology* **66**, 630–638.

Davenport, J., Munks, S. A. and Oxford, P. J. (1984) A comparison of the swimming of marine and freshwater turtles. *Proceedings of the Royal Society B* **220**, 447–475.

Davis, R. R. and Hull, M. L. (1981) Measurement of pedal loading in bicycles, II: Analysis and results. *Journal of Biomechanics* **14**, 857–872.

Dawson, T. J. and Taylor, C. R. (1973) Energetic cost of locomotion in kangaroos. *Nature* **246**, 313–314.

Demes, B., Jungers, W. L., Gross T. S. and Fleagle, J. G. (1995) Kinetics of leaping primates—influence of substrate orientation and compliance. *American Journal of Physical Anthropology* **96**, 419–429.

De Mont, M. E. (1990) Tuned oscillations in the swimming scallop *Pecten maximus*. *Canadian Journal of Zoology* **68**, 786–791.

De Mont, M. E. and Gosline, J. M. (1988) Mechanics of jet propulsion in the hydromedusan jellyfish, *Polyorchis penicillatus*, I–III (three papers). *Journal of Experimental Biology* **134**, 313–332.

Denny, M. W. (1980a) The role of gastropod pedal mucus in locomotion. *Nature* **285**, 160–161.

Denny, M. W. (1980b) Locomotion: The cost of gastropod crawling. *Science* **208**, 1288–1290.

Denny, M. W. (1984) Mechanical properties of pedal mucus and their consequences for gastropod structure and performance. *American Zoologist* **24**, 23–36.

Denny, M. W. (1993) *Air and Water*. Princeton University Press, Princeton, NJ.

Denton, E. J. (1974) On buoyancy and the lives of modern and fossil cephalopods. *Proceedings of the Royal Society B* **185**, 273–299.

Denton, E. J. and Gilpin-Brown J. B. (1973) Floatation mechanisms in modern and fossil cephalopods. *Advances in Marine Biology* **11**, 197–268.

Denton, E. J. and Marshall, N. B. (1958) The buoyancy of bathypelagic fishes without a gas-filled swimbladder. *Journal of the Marine Biological Association of the UK* **37**, 753–769.

Denton, E. J., Liddicoat, J. D. and Taylor, D. W. (1970) Impermeable "silvery" layers in fishes. *Journal of Physiology* **207**, 64P.

Dial, K. P., Biewener, A. A., Tobalske, B. W. and Warrick, D. R. (1997) Mechanical power output of bird flight. *Nature* **390**, 67–70.

Dial, K. P., Kaplan, S. R., Goslow, G. E. and Jenkins, F. A. (1988) A functional analysis of the primary upstroke and downstroke muscles in the domestic pigeon (*Columba livia*) during flight. *Journal of Experimental Biology* **134**, 1–16.

Dickinson, J. A., Cook, S. D. and Leinhardt, T. M. (1985) The measurement of shock waves following heel strike in running. *Journal of Biomechanics* **18**, 415–422.

Dickinson, M.H. and Lighton, J. R. B. (1995) Muscle efficiency and elastic storage in the flight motor of *Drosophila*. *Science* **268**, 87–89.

Dickinson, M. H. and Götz, K. G. (1996). The wake dynamics and flight forces of the fruit fly *Drosophila melanogaster*. *Journal of Experimental Biology* **199**, 2085–2104.

Dickinson, M. H., Lehmann, F.-O. and Sane, S. P. (1999) Wing rotation and the aerodynamic basis of insect flight. *Science* **284**, 1954–1960.